A series of student texts in

Contemporary Biology

General Editors:
Professor Arthur J. Willis
Professor Michael A. Sleigh

The Diversity of Green Plants

Third Edition

Peter R. Bell, M.A.

Quain Professor of Botany,
Department of Botany and Microbiology, University College, London

Christopher L.F. Woodcock, Ph.D.

Professor, Department of Zoology,
University of Massachusetts, Amherst, U.S.A.

Edward Arnold

© Peter R. Bell and Christopher L.F. Woodcock 1983

First published in Great Britain 1968
by Edward Arnold (Publishers) Limited
41 Bedford Square
London WC1B 3DQ

Reprinted 1969
Second edition 1971
Reprinted (with corrections) 1972
Reprinted (with additions) 1975
Reprinted 1976, 1978, 1980
Third edition 1983
Reprinted 1986

Edward Arnold (Australia) Pty Ltd
80 Waverley Road
Caulfield East
Victoria 3145

Edward Arnold
3 East Read Street
Baltimore
Maryland 21202
U.S.A.

British Library Cataloguing in Publication Data

Bell, Peter R.
    The diversity of green plants.—3rd ed.—(A series of
student texts in contemporary biology)
    1. Botany—Anatomy
    1. Title   II. Woodcock, C.L.F.
    581.4   QK641

ISBN 0-7131-2866-6

Printed in Great Britain by
Thomson Litho Ltd, East Kilbride, Scotland

Preface to the third edition

In the almost twenty years since *Diversity of Green Plants* was first written striking advances in our knowledge of plants have put Botany well ahead in the Biological Sciences. At the fundamental level of membranes, the use of electron spin resonance spectroscopy is revealing the secrets of the extremely fast reactions which bring about the transfer of energy in photosynthesis. In plant cytology studies of the differentiation of the helical gametes of the lower land plants have shown the crucial role of microtubules in bringing about fundamental change in cell shape. It seems likely that these remarkable cells will also reveal the origin and composition of centrioles, a problem to which animal cytology has been able to give no firm answer. Genetical maps of chloroplasts are now as familiar as those of mitochondria, and the discovery that key nucleotide sequences are common to each has thrown new light on the possible evolutionary origins of these organelles. Even concerning the ancient plants exciting discoveries continue to be made. New knowledge of the curious forms at the end of the Palaeozoic which may have paved the way for the ultimate appearance of the flowering plants has come from Australian deposits. These examples from widely different aspects of the science are representative of the active research which is bringing outstanding liveliness and enthusiasm to current Botany.

As teachers we have been confronted with the problem of how this new knowledge is to be presented to the student. We remain convinced that a basic knowledge of plant morphology and evolution provides the structure essential for the orderly development of our science. Not only does an awareness of the inter-relationships of the organisms used in current research facilitate an understanding of the significance of the results, it also stimulates the recognition of new lines of profitable enquiry. Our purpose, therefore, is to promote our discipline in all its aspects by providing a concise account of the structure and reproduction of the varied groups of photoautotrophic plants, both living and extinct, proceeding from the simple to the complex.

Although we have adopted a form of classification as a framework for our account, we have thought of the photoautotrophic plants as a whole. Each is the morphogenetic expression of inherited information, much of it presumably common to all photoautotrophs. The fossil record does, however, give undeniable evidence of an evolutionary progression. We have therefore discussed the likely origin of each level of organization and its possible selective advantages, but unwarranted phylogenetic speculation has naturally been avoided. We have referred throughout to experimental work relevant to the problems of growth and form, and to the detection of evolutionary affinities, but detailed consideration of these topics has had to be omitted as beyond the scope of this book.

The Diversity of Green Plants is intended primarily for undergraduate students, but it will also be useful to all those seeking information about the relationships of the photoautotrophs familiar to the experimental botanist. Fuller accounts of the plants we describe can be found in the standard works to which our superscripts refer.

London and   P.R.B.
Amherst, Massachusetts   C.L.F.W.
1983

Acknowledgements

We are grateful to Professor Dr K. Mühlethaler for Fig. 1.1, to Professor G. Drews for Fig. 1.2, to Mr A. J. J. Rees for Fig. 3.7, to the Cambridge Instrument Company for Fig. 3.9, and to Dr H.G. Dickinson for Fig. 8.24.

In addition to the authors and publishers cited, we acknowledge with thanks permission to reproduce figures from the following: the Councils of the Linnean Society of London (Figs 4.7, 7.46, 8.14, 8.15); the Royal Society (Fig. 5.24); the Trustees of the British Museum (Natural History) (Figs 2.22, 2.25, 2.37, 2.40, 3.17, 3.18, 3.28, 8.5); and the University of Michigan Press (Figs 2.38, 2.39, 2.41).

We have also been fortunate in being able to call on the excellent draughtsmanship of J.G. Duckett (Figs 4.2, 4.10, 4.15, 4.17, 4.18, 4.19, 4.20, 4.21, 4.26 (in part), 4.29, 5.32b, 5.33, 6.22, 6.28 (in part), 7.11, 7.25b,c, 8.30), H.G. Dickinson (Figs 7.26, 8.22, 8.31), J. Fagg (Figs 1.3, 2.1, 2.2, 2.3, 2.4, 2.5, 2.6, 2.7, 2.8, 2.11, 2.12, 2.13, 2.14, 2.16, 2.17, 2.19, 2.32, 2.33, 2.42b, 3.1, 3.2b, 3.3a, 3.4, 3.5, 3.6, 3.11, 3.12, 3.13, 3.19, 3.24, 3.25, 3.26b, 3.27, 3.29, 5.8), and of Elizabeth Harrison (Figs 2.15, 2.43, 4.6, 4.8, 4.14, 5.3, 5.11, 5.13).

Our colleagues in various places gave us valuable criticism and advice, but we take full responsibility for any errors that may remain in the text. The preparation of the third edition would have been impossible without the skilled secretarial assistance of Lilian Jennings.

London and   P.R.B.
Amherst, Massachusetts   C.L.F.W.
1983

Contents

1

The principles governing the evolution of autotrophic plants

Significance of autotrophic nutrition

The living state is characterized by instability and change. A living cell consumes and releases energy by means of numerous chemical reactions, called collectively metabolism, taking place within it. Metabolism is synonymous with life. Even the apparently inert cells of seeds show some metabolism, but admittedly only a fraction of that which occurs during germination and subsequent growth. At normal temperatures, only in a dead cell, provided it remains sterile, does metabolism stop.

To maintain the dynamic state of a living cell it must be provided with sources of energy. The most usual sources are chemical and frequently consist of sugars. Together with these sources of energy, a cell also requires water, since much of metabolism is dependent on the maintenance of an aqueous phase in the cell, and those materials necessary for the repair and maintenance of its structure which it is unable to make for itself. These vary with the nature of the cell, but examples are certain metals which are essential components of important enzymes, the nitrogen of the proteins, and in some instances certain complex molecules called vitamins. All these nutritional requirements of the cell must be met; if not, the dynamic state we call life ultimately ceases.

It is a remarkable property of a large part of the Plant Kingdom that the cells, or in a multicellular plant at least some of the cells, are able to incorporate the energy of incident light into their metabolism. The energy, absorbed by the pigment chlorophyll, is used to generate ATP and reducing power in the form of NADPH + H⁺ (the light reactions). These two products then bring about the reduction and assimilation of carbon dioxide in the form of carbohydrate (the dark reactions). The ability to metabolize atmospheric carbon dioxide in this way (*photosynthesis*) releases the organisms concerned from the necessity of an external source of carbohydrate, and their nutritional demands are consequently relatively simple. Photosynthetic plants are therefore examples of *autotrophs*, the general term for organisms which require only simple molecules with single carbon atoms for their organic nutrition. Organisms which require complex carbon compounds (such as sugars) are termed *heterotrophs*.

So far as is known, photoautotrophs utilize for photosynthesis principally that radiant energy falling within the wave-band called 'visible light'. Consequently, photoautotrophs can exist only in an environment continuously or intermittently illuminated. Some small photoautotrophs can, it is true, also exist as heterotrophs, so we may distinguish between obligate photoautotrophs, such as higher plants, normally unable to utilize sugar supplied externally, and facultative heterotrophs which, although normally photoautotrophs, can adapt their metabolism to the nature of their environment.

Chlorophyll is a complex pigment, existing in a number of slightly different forms. The molecule is in part similar to that of the active group of the blood pigment haemoglobin, but chlorophyll contains magnesium instead of iron. The photosynthetic pigment is not free in the cells, but is always associated with lipoprotein membranes, which seem generally to contain arrays of particles of two distinct size classes (Fig. 1.1). Pure chlorophyll is green, and absorbs particularly in the red part of the spectrum. In cells its colour may be masked by accessory pigments. Although the function of these is not precisely known, in certain instances part of the energy which they absorb is undoubtedly transmitted to the chlorophyll and contributes to its excitation. Chlorophyll, however, is the pigment principally concerned in photosynthesis. It is present in all photoautotrophs, and absent from all obligate heterotrophs, including the whole of the Animal Kingdom.

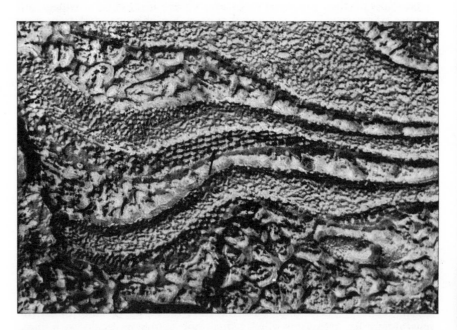

Fig. 1.1   Part of a granum in a chloroplast of spinach fractured obliquely. The replica has been prepared by the freeze-etch technique. The larger particles (indicated by arrow), sometimes in regular arrays, may be the sites at which light energy is transformed into chemical energy. (× 94 000)

The possession of chlorophyll and the consequent ability to utilize the sun's energy clearly bestow a great advantage on photoautotrophs. There is no necessity to forage for a supply of carbohydrates and Mereschkowsky's image[28] of the restless lion and the placid palm as contrasting the consequences of heterotrophic and autotrophic nutrition respectively, although poetic, is nevertheless apt. Photoautotrophs, of course, as all living organisms, remain dependent upon a supply of water and minerals, but such motility as occurs is in relation to light rather than mineral sustenance. Green unicellular organisms, for example, often show marked positive phototaxis.

The structure of the photoautotrophic cell

Two fairly distinct kinds of cellular organization occur amongst the photo-autotrophs as a whole. In the first, termed *prokaryotic,* [13, 44] the cell possesses no distinct nucleus, although a region irregular in outline and of differing density occurs at the centre of the cell. This is referred to as a nucleoid, and the genetic material lies therein. In the electron microscope this region appears fibrillar rather than granular, and the fibrils indicate the site of the deoxyribonucleic acid. The protoplast of such cells is bounded by a membrane and in photoautotrophic cells this membrane here and there invaginates into the cytoplasm (Fig. 1.2). These membranous invaginations are the sites of photosynthesis, and they disappear or become very reduced if the cells are grown in the dark. This simple kind of photoautotrophic cell is found in both the photosynthetic bacteria and the Cyanophyta, perhaps representative of the most primitive kind of photosynthesizing organisms. Remains very suggestive of photoautotrophic prokaryotes have been found in Canadian rocks believed to be about 2000×10^6 years old. Geochemical evidence of photosynthesis, together with what are possibly remains of bacterial cells, comes from even more ancient rocks in South Africa and Australia. The evidence for cellular life now extends as far back as 3500×10^6 years.

In the cells of all other photoautotrophic plants the nucleus, the photo-synthetic apparatus, and the membranes incorporating the electron transport chain of respiration are separated from the remainder of the cytoplasm by distinct envelopes. Such cells, termed *eukaryotic*, have evidently been capable of giving rise to much more complicated organisms than the prokaryotic. The photosynthetic apparatus, which consists of numerous lamellae running parallel to one another, is contained in one or more *plastids* (or chromatophores). The envelope of the plastid consists of two unit membranes, the inner of which invaginates into the central space and generates the internal lamellar system. The respiratory membranes and enzymes are contained within another distinct organelle, the *mitochondrion*. Both plastids and mitochondria contain nucleic acids, and have a prokaryote-like biochemistry. They possess genetic systems of their own, partly independent of that of the nucleus.

The simplest eukaryotic photoautotroph is thus a single cell containing a

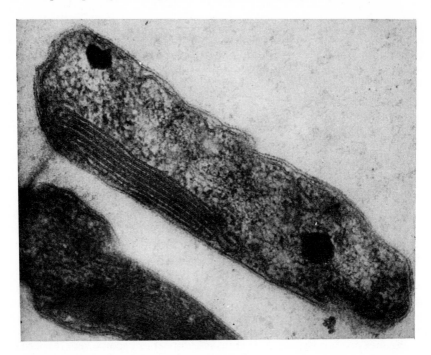

Fig. 1.2   Median section of a photosynthetic bacterium, *Rhodopseudomonas viridis*. The chromatophore is formed from an invagination of the cell membrane (plasmalemma). The folding of the membrane gives rise to a stack of lamellae, similar to the grana seen in the chloroplasts of higher plants. (× 95 000)

plastid and a mitochondrion. This condition is represented in *Micromonas pusilla* (Fig. 1.3), a minute alga ubiquitous in the sea. Electron micrographs of this organism undergoing fission show the plastid and mitochondrion dividing at the same time.

The evolutionary consequences of photosynthesis

It seems beyond doubt from the fossil record of life, and from the biological and geological inferences that can be drawn from it, that life began in water. Although the early forms of life were probably heterotrophic, photosynthesis undoubtedly originated in an aquatic environment, probably in organisms similar to the existing purple bacteria. The plants which later evolved in this aquatic environment, and still in the main exploit it, have many biochemical and structural features in common, and are collectively termed algae. Our treatment of the autotrophic plants consequently begins at this level of organization (Chapters 2 and 3).

At some stage, possibly in the Silurian period (Table 1.1) or even earlier, vegetation began to colonize the land. These early colonists, and conse-

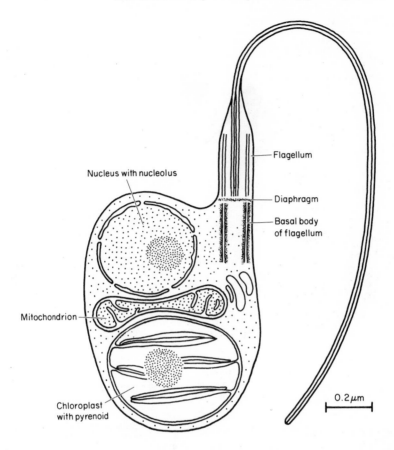

Fig. 1.3 *Micromonas pusilla.* Form and internal organization. Only the central microtubules run into the extension of the flagellum. *Micromonas* belongs to a small group of green algae of doubtful affinity. (From electron micrographs by Manton (1959). *Journal of the Marine Biological Association of the United Kingdom*, **38**, 319.)

quently the whole of our existing land flora, almost certainly emerged from that group of aquatic plants today represented by the green algae (Chlorophyta).[8] The Chlorophyta and the land plants (a term which means plants suited to life on land and not merely plants growing on land) have the same photosynthetic pigments, and basically the same photosynthetic apparatus. Moreover, at least one green alga (*Cladophorella*; see p. 39) which grows on damp mud is covered on its upper surface by a material which, judging from its resistance to acids and oxidizing agents, closely resembles cutin. This perhaps indicates the way in which the cuticle, ubiquitous in land vegetation, was derived.

Any consideration of the evolution of a photosynthesizing land flora must thus necessarily take into account the physiological features of the green algae, and how these may have been modified in the transition to terrestrial

Table 1.1   The Geological Table

ERA	PERIOD		AGE (in 10^6 years)	First authentic appearance
QUATERNARY	Pleistocene and Recent		0–2.5	
TERTIARY	Pliocene ⎫	Upper		
(or Cenozoic)	Miocene ⎬	Tertiary	2.5–28	
	Oligocene ⎫	Lower		
	Eocene ⎬	Tertiary	28–65	
	Paleocene ⎭			
MESOZOIC	Cretaceous	Upper	65–135	Angiosperms*
		Lower		
	Jurassic	Upper	135–190	
		Middle		
		Lower (Lias)		
	Triassic		190–225	
PALAEOZOIC	Permian	Upper	225–280	Cycad- and
		Lower		*Ginkgo*-like plants, *Glossopteris*
	Carboniferous	Upper (Pennsylvanian)	280–345	Conifers
		Lower (Mississippian)		Pteridosperms
	Devonian	Upper	345–395	Ferns, seeds
		Middle		Sphenopsids Heterospory
		Lower		Vascular plants: Lycopods and Psilopsids
	Silurian		395–440	Triradiate spores
	Ordovician		440–500	
	Cambrian		500–600	
PRE-	Proterozoic		600–1500	Calcareous algae
CAMBRIAN	Archaeozoic		1500–3000+	Fungi, algae, bacteria (see p. 3)

*See pp. 326–30 for a discussion concerning the first appearance of the angiosperms

life. Recent research into algal environments is yielding much information relevant to this problem. It is commonly found, for example, that from 5 to 35% of the light striking the surface of a lake or sea is reflected, the actual amount lost depending upon the angle of incidence. The light penetrating the water is then gradually absorbed as it advances, so that up to 53% of the radiation passing the surface may be dissipated as heat in the first metre.[33] Consequently, in warm and temperate regions, the rate of photosynthesis of submerged plants is normally controlled by the amount of light reaching them, and not by the amount of carbon dioxide in the water. We can see at once that the first colonists of land, emerging on to bare mineral surfaces, would almost certainly have had to contend with irradiances strikingly higher than those experienced by their aquatic ancestors. This would have provided opportunities for greatly increased photosynthesis.

Another discovery of recent research, also very relevant to the problem of the colonization of the land, is the surprising extent to which algae release photosynthesized materials into the surrounding water. In Windermere, for example, up to 35% of the total carbon fixed is continuously lost in this way. Losses of this order are clearly possible only from aquatic plants. As vegetation advanced from marshes, or from littoral belts subject to periodic inundation, on to relatively dry substrata, much more of the fixed carbon must have been conserved within the plant body.

The emergence of green plants from seas or lakes must therefore have involved striking changes in carbohydrate metabolism. The increased assimilation of carbon, resulting from the increased irradiances on land, and a diminishing loss of photosynthesized carbohydrates by outward diffusion as plants reached relatively dry areas, could have resulted in embarrassingly large, and possibly toxic, quantities of carbohydrates in the cells. The physiology must therefore have changed simultaneously to meet this new situation.

We have no evidence that adaptation to terrestrial life was accompanied by any reduction in the amount of chlorophyll in the chloroplasts (the term used for plastids that are unambiguously green), or in the efficiency of the photosynthetic apparatus. It is therefore not surprising to find that in addition to the condensation of sugar to starch, already encountered in the algae, the migration of plants on to land was accompanied by other ways of removing fixed carbon from the general metabolism. Rigid cell walls, making possible multicellular plant bodies penetrated by air spaces, are composed of cellulose and hemicellulose, both sugar products segregated more or less permanently in non-metabolizable form. Resin, phlobaphene and lignin are also derived from sugars. In addition to their structural and sealing properties these substances, which are often prominent in the more primitive land plants, are antiseptic and may have been important protectants to the early colonists. Massive plant bodies, which seem to have appeared relatively soon in the evolution of the land flora, also made possible the confinement of photosynthesis to specialized regions, such as leaves, so that the amount of assimilation per unit mass was reduced. Natural selection would, of course, have ensured that those forms survived in which the cells containing the condensed carbohydrates, and the materials derived from them, assisted the functioning of the plant as a whole.

We can envisage how this led to the evolution of xylem, the principal lignified tissue of land plants. Although largely dead, xylem is of paramount importance in the structure of plants since it provides both a skeleton supporting the plant in space, and an effective system for the transport of water and solutes. Cutin and sporopollenin, condensation products of a fatty nature, are also given essential roles in land plants. They serve to lessen the loss of water by evaporation from the living cells and, with the stomata, conserve a saturated atmosphere within the intercellular spaces. Plants were thus able to move into areas of lower humidity.

In the course of evolution many complex and bizarre forms of growth have appeared in land plants, but the material from which they are fashioned has remained predominantly carbon, extracted from the atmosphere. This diversity can be related to the tetravalent nature of carbon, and the great range of compounds that can be formed from it. Had not the photosynthetic fixation of this versatile element arisen on the Earth's surface, plant life, and that of animals which depend upon it, would have been impossible. Indeed, it is difficult to conceive of any alternative form of life appearing in its absence.

The mobility of plants

Although the earliest plants probably soon acquired motility of the kind seen today in *Chlamydomonas* and *Euglena* (Chapters 2 and 3), this appears to have been rapidly lost in the evolution of higher forms. Photosynthesis, together with the ability to manufacture a skeleton and a vascular system from the assimilated carbon, eventually made the large and firmly anchored terrestrial plant a practical possibility. Such an organism is, of course, immobile, and thus suffers a serious disadvantage, not shared by the higher animal, at times of natural catastrophe, such as volcanic eruption or fire. Plants, however, very frequently possess a remarkable mobility, or at least a ready transportability by agencies such as wind and water, in their reproductive bodies. Fern spores, for example, have been caught in aeroplane traps in quantity at 1500 m and even higher, and the hairy spikelets of the grasses *Paspalum urvillei* and *Andropogon bicornis* have been encountered at 1200 m above Panama. Lakes, seas and the coats and feet of animals also play their part in distributing plants. The immobility of the individual is thus frequently compensated for by the mobility of the species, and devastated areas and new land surfaces become colonized with amazing rapidity and effectiveness.

Life cycles

A life cycle, involving segregation and recombination of the genetic material, is as basic to the evolution of plants as to that of animals. In one part of the cycle the nucleus contains a single set of chromosomes (and is consequently termed *haploid*), and in the other two sets of chromosomes (and is termed *diploid*). The cycle is seen at its simplest in the unicellular algae of aquatic environments (Chapters 2 and 3), where haploid individuals in certain circumstances behave as gametes and fuse, so forming a *zygote*. The zygote, which contains a diploid nucleus, either undergoes meiosis at once, or only after some delay, in which case the diploid condition can be thought of as having an independent existence. Either the haploid or the diploid phase, or both, may be multicellular. The multicellular plant is called a *gametophyte* if it produces gametes directly, and a *sporophyte* if it produces, following meiosis, individual cells which either behave as gametes immediately or develop into gametophytes. Each phase may also multiply itself asexually. These various possibilities are summarized in Fig. 1.4.

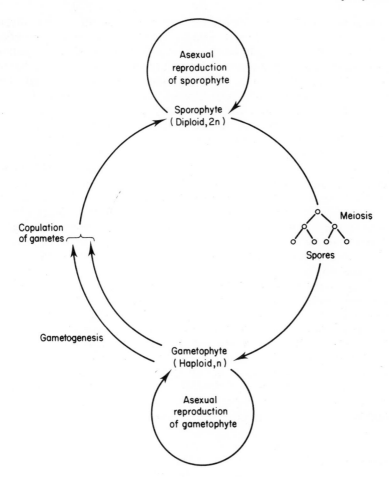

Fig. 1.4 The life cycle of autotrophic plants generalized. The large circle represents sexual reproduction. Only relatively few species display all the reproductive potentialities shown.

A life cycle is thus basically a nuclear cycle, and it is not necessarily accompanied by any morphological change. In the alga *Dictyota* (p. 92), for example, the gametophyte and sporophyte are identical, and it is necessary to observe the manner of reproduction in order to identify the place in the cycle which any individual occupies. Such a life cycle is termed *isomorphic* (or homologous). Frequently, however, the two phases of the cycle have different morphologies, one often being less conspicuous than the other, and sometimes parasitic upon it. These cycles are termed *heteromorphic* (or antithetic). Although the algae show both isomorphic and heteromorphic life cycles, those of land plants are exclusively heteromorphic. Occasionally there may be a morphological cycle without a corresponding nuclear cycle, as in the apogamous ferns (see p. 220), but this is regarded as a derived condition.

Gametes are always uninucleate, and, when motile, usually naked cells. In the simplest form of sexual reproduction, termed *isogamy*, the two gametes involved in fusion are free cells and morphologically identical. Nevertheless, detailed investigations continue to show that gametes from the same parent rarely fuse. Some measure of self-incompatibility, and hence physiological differentiation between the parents, appears to be the general rule.

Isogamy was probably the most ancient condition, and this appears to have been succeeded by *anisogamy*. Here the gametes, although still free cells, are morphologically dissimilar, but usually differ in little more than size. The larger, which may also be less mobile, is called the female. The extreme form of anisogamy is *oogamy*, in which the female gamete, now called an egg cell or ovum, is large, non-motile, and filled with food materials. The egg cell may either float freely in water, as in the alga *Fucus*, or be retained in a chamber, as in some algae and all land plants. The chamber bears various names according to the group of plants being considered. Since the progression from isogamy is accompanied in many algal groups by an increase in somatic complexity, it seems very probable that this morphological progression is also a phylogenetic one.

In several instances of sexual reproduction it has been shown that one or both gametes produce traces of chemical substances, termed pheromones (or gamones), which cause the appropriate gametes to approach each other. The chemistry of these pheromones varies widely. In some algae they are peptides. In the ferns the male gametes are attracted to the opened egg chambers by a pheromone which may be malic acid. This substance is known to have a striking chemotactic effect *in vitro*.

The life cycles of the transmigrant forms

The transition to a terrestrial environment clearly presented a number of problems in relation to sexual reproduction. Although all land plants are oogamous, and are presumably derived from oogamous algae, fluid was still necessary in the initial land plants to allow the motile male gametes to reach the stationary female. This problem appears to have been met first by the egg becoming enclosed in a flask-shaped chamber, the *archegonium*, in the neck of which the male gametes accumulate, and second by the male gamete becoming a highly motile cell. The male gametes of the lower archegoniate plants (Chapters 4, 5 and 6), termed *spermatozoids* (or antherozoids), are remarkable cytological objects. Each is furnished with two or more highly active flagella, and both the cell and nucleus have an elongated snake-like form, well suited for penetration of the archegonial neck. Dependence upon water is thus reduced to the necessity for a thin film in the region of the sex organs at the time of maturity of the gametes.

The archegonium is common to all the lower land plants, but its origin remains tantalizingly obscure. It may have appeared immediately before the colonization of the land, possibly as a consequence of morphogenetic tendencies seen today in association with the eggs of some Chaetophorales

(p. 43), the Charales (p. 60), and certain red algae (p. 97). If so, these antecedents of the transmigrants are no longer represented amongst living algae. Whatever the exact time of the evolution of the archegonium, however, there are no compelling reasons for regarding it as having been evolved more than once. Since the fossil record indicates that the most primitive forms of land plants were probably all archegoniate, it follows that the migration of plants on to land was also very likely an unique event.

If the transmigrant forms were archegoniate, what was the nature of their life cycles? This is largely a matter for conjecture. However, as will be seen in later chapters, except for one approach to isomorphy (in the living Psilopsida), the lower archegoniate plants possess markedly heteromorphic life cycles in which the conspicuous generation is either the gametophyte (Bryophyta), or the sporophyte (Lycopsida, Sphenopsida, ferns). The transmigrants possibly had an intermediate position, with more or less isomorphic cycles. Nevertheless, heteromorphic cycles were probably very soon developed as terrestrial vegetation diversified and exploited particular features of the new environment. The cycle in which the sporophyte was the predominant phase clearly had the greater evolutionary potential, since, with the exception of the Bryophyta, it is characteristic of all existing terrestrial vegetation.

Sexual reproduction in later terrestrial vegetation

An important step in the evolution of sexual reproduction on land was undoubtedly the emergence in the archegoniate plants of heterospory. This involves the production of spores of two sizes, the larger giving rise to a wholly female gametophyte and the smaller to a male (see pp. 132, 155, 215). In homosporous archegoniate plants, quite considerable growth of the gametophyte is often required before it acquires the ability to produce egg cells. In the primitive heterosporous plants, however, the small female gametophyte formed on germination of the megaspore produces archegonia in a very short time. The microspore also develops rapidly, and spermatozoids are soon liberated from the diminutive male gametophyte. The time involved in the gametophytic phase is thus reduced to a minimum. In plants in which the sporophytic phase is dominant, speedy reproduction of this kind has the obvious advantage of accelerating the establishment of new forms and hence the rate of evolution.

In higher archegoniate plants (Chapter 6) we see how sexual reproduction becomes increasingly independent of water. These archegoniates are exclusively heterosporous, but the megaspore is retained and germinates within a specialized sporangium called an ovule. In some forms (*Cycas*, p. 237; *Ginkgo*, p. 263) fertilization is still effected by flagellate male gametes, but the only fluid necessary is a small drop, immediately above the archegonia, into which the gametes are released. Other higher archegoniate plants escape even from this requirement. The male gametophyte is filamentous, and, as a consequence of its growing towards the female gametophyte, it liberates the male gametes (which now lack any specialized means of locomotion) directly into an archegonium. In a few allied plants (e.g. *Gnetum*, p. 270) modifications of the female gametophyte result in the disappearance of the archegonium, and ultimately we arrive at the embryo sac and finely ordered cytology that is characteristic of the sexual reproduction of the flowering plants. Comparative morphology and the fossil record indicate that the morphological sequence we have considered here also represents the evolutionary development of sexual reproduction in land plants. Compared with the cytological elegance of fertilization in an angiosperm, the clumsy spermatozoid of *Cycas* is thus not only barbarous, but also primitive.

The classification of autotrophic plants

Having outlined the general principles which have governed the evolution of autotrophic plants we can now proceed to consider them according to their level of organization and in their natural alliances. We require a general classification to provide a framework for this information. Of the several general classifications available, all of which have some virtues, we have chosen that shown in Table 1.2. This is convenient for our purposes, and we do not wish to imply that the relative ranking given to the higher categories has any special significance. The classification depends firstly upon the presence or absence of a differentiated vascular system (which separates the Tracheophyta from the remainder), and subsequently upon what appear to be the natural affinities of the plants concerned. In the arrangement of the algae we have followed Round.[37]

Table 1.2

Division	Sub-Division	Class	Order
Cyanophyta		Cyanophyceae	Chroococcales
			Hormogonales
Prochlorophyta			
Chlorophyta			Volvocales
			Chlorococcales
			Ulotrichales
			Cladophorales
			Chaetophorales
			Oedogoniales
			Mesotaeniales
			Desmidiales
			Zygnemales
			Siphonales
			Charales
Chrysophyta			
Xanthophyta			
Haptophyta			

Table 1.2 *continued*

Division	Sub-Division	Class	Order
Dinophyta (Pyrrophyta)		Desmophyceae Dinophyceae	
Bacillariophyta			
Cryptophyta			
Euglenophyta			
Phaeophyta			Ectocarpales Laminariales Fucales Dictyotales Cutleriales
Rhodophyta		Rhodophyceae Bangioideae Florideae (Sub-Classes)	
Bryophyta ·		Hepaticae Anthocerotae Musci	
Tracheophyta	Psilopsida		Psilotales Psilophytales
	Lycopsida		Lycopodiales Selaginellales Isoetales Lepidodendrales Pleuromeiales
	Sphenopsida		Equisetales Calamitales Sphenophyllales Pseudoborniales
	Pteropsida	Filicinae	Cladoxylales Coenopteridales Ophioglossales Marattiales Filicales
		Gymnospermae	Pteridospermales Cycadales Bennettitales Caytoniales Cordaitales Coniferales Ginkgoales Gnetales
		Angiospermae Dicotyledonae Monocotyledonae (Sub-Classes)	

2
The Algae, I

The simplest photoautotroph imaginable is a single cell floating in a liquid medium, synthesizing its own sugar, and reproducing at intervals by binary fission. Such organisms do in fact exist in both fresh and salt waters. Examples are provided by *Chroococcus* (Fig. 2.3), the minute marine *Micromonas* (Fig. 1.3), and by *Euglena* (Fig. 3.15).

These organisms are examples of algae, [17, 25, 38] a group of plants showing the greatest diversity of any major division of the Plant Kingdom. They range from minute, free-floating, unicellular forms to large plants, exclusively marine, several metres in length. Nevertheless, despite this enormous variation in size, the algae remain predominantly aquatic in habit and comparatively simple in structure. In the smaller multicellular species (e.g. *Pediastrum*, Fig. 2.16) the cells resemble each other in appearance and function, and they can be regarded as forming little more than an aggregate of independent units. In the larger, however, there is morphological and cellular differentiation, although usually less extensive than in most land plants. The few heterotrophic forms, mostly small, are regarded as derived.

The larger algae are found only in the sea, commonly anchored to a firm substratum. This is because the holdfast, by which they are attached, is able to make a tight bond with a hard surface, but not to penetrate soft material as is a root. The restriction of these larger forms to a marine environment is perhaps accounted for by the relative impermanence of inland waters in geological time, and the consequent limiting of the opportunity for the evolution of complex freshwater forms. Although marine algae are sometimes able to withstand inundation in fresh water (e.g. *Fucus*, p. 88), and occasionally may even become adapted to permanently low salinity (e.g. *Ulva* and *Enteromorpha*, p. 37), they do not normally survive indefinitely or grow in these conditions. Presumably fresh waters are unable to supply minerals at a rate adequate for their metabolism. A large alga in European seas is *Laminaria* (Fig. 3.20a), some species of which may reach 4 m in length. Off the west coast of North America are found the gigantic *Nereocystis* and *Macrocystis*, with thalli commonly exceeding 50 m. Maintaining the integrity of a thallus of this size raises substantial mechanical problems. Although the sea provides considerable supporting upthrust, currents and turbulence cause more sustained tensions and pressures than similar movements in a gaseous medium. The toughness, and hard rubbery resistance to any kind of distortion found in the larger algae, are thus necessary qualities for survival in the Oceans. These attributes arise principally from the general properties of the cell walls and of the surface, and not from any specialized strengthening elements.

As would be expected of a group exploiting the aquatic habitat, algae have a number of distinctive biochemical characteristics.[45] Many, for example, accumulate fats and oils rather than starch and others polyhydric alcohols (see p. 82). The microfibrils of algal cell walls contain in several instances the polysaccharides mannan and xylan in addition to cellulose. The nitrogenous polysaccharide chitin is found as an outer layer of the wall in *Cladophora prolifera* and possibly in *Oedogonium*. Pectin, a polymer based on galacturonic acid, is a common component of algal cell walls, sometimes forming a distinct outer sheath (e.g. *Scenedesmus*, Fig. 2.15). Colloids such as fucin and fucoidin, unknown outside the algae, occur in the amorphous matrices of the walls of brown algae. Alginic acid, which occurs in quantity in the middle lamellae and primary walls of several brown algae, is extracted commercially and finds a wide range of uses as an emulsifier in industry. Complex mucilaginous polysaccharides rich in galactan sulphates are characteristic of the red algae.

There is no evidence that the major groups of algae have any close relationship with each other. Nevertheless, there are sufficient morphological, physiological and ecological similarities between these plants to make the term 'alga' a useful one. Earlier the algae were grouped with the fungi and bacteria in the Thallophyta, but current opinion attributes to each algal group the status of a Division. We have followed this practice in Table 1.2 (pp. 12–13).

Study of the structure and reproduction of the algae reveals a number of ways in which these simple autotrophs have increased their morphological and reproductive complexity. We shall in the main be concerned with the illustration and discussion of these trends, and we shall not attempt a complete taxonomic or morphological survey of any group. Nevertheless, in the first of these two chapters devoted to the algae the treatment of the Cyanophyta and Chlorophyta is fuller than that of the remainder of the algae in the second. The Cyanophyta justify attention because of their relevance to the possible origin of the chloroplast and their importance as nitrogen-fixers in many tropical soils. The Chlorophyta yield clues to the origin of the land flora.

The prokaryotic algae

The Cyanophyta (blue-green algae)[6] are representative of the simplest photo-autotrophs. Although the cells may occur in aggregates, sometimes even arranged linearly in a filament, there is little coordination of activity. The Cyanophyta are thus considered to be capable at most of forming colonies of individuals rather than true multicellular organisms.

Cyanophyta

Habitat	Water, swamps, soil, occasionally endolithic.
Pigments	Chlorophyll *a*, *β*-carotene; myxoxanthin, zeaxanthin; biliproteins (phycoerythrin and phycocyanin).
Food reserves	Cyanophycean starch (similar to glycogen), polyphosphate granules (volutin), cyanophycin (a protein).
Cell wall components	Murein, hemicelluloses.
Reproduction	Asexual. Genetic recombination observed, but mechanism unknown.
Growth forms	Unicellular, cellular aggregates, filamentous.

CYTOLOGY   The cells, rarely exceeding 10 μm in diameter, are less differentiated than those of any other algal Division. The nuclear material is not enclosed in an envelope. The Cyanophyta are thus prokaryotic, and in many features resemble the photosynthetic bacteria (see p. 3). They are indeed often classified with the bacteria rather than with the algae, and the resemblance is enhanced by many being attacked by viruses similar to the bacteriophages, and the bearing by some species of filamentous appendages (*fimbriae*) otherwise found only in Gram-negative bacteria. The Cyanophyta differ from the photosynthetic bacteria however in possessing chlorophyll *a* and photosystem II (the light reaction responsible for the deoxygenation of water). Nevertheless the metabolic primitiveness of the Cyanophyta is shown by their ability to use pyrophosphate in place of ATP as a source of energy.

Under the light microscope, partly because of the small size of the cells, the photosynthetic pigments appear dispersed in the cytoplasm, but electron microscopy revealed that this is erroneous. The cells contain membranes which form flattened vesicles, directly comparable with the thylakoids of eukaryote chloroplasts. The amount of chlorophyll in the cells is related to the extent of this lamellar system, and direct chemical analysis has confirmed that the photosynthetic pigments are associated with the lipoproteins of the membranes. The thylakoids, which probably arise as, and may remain, invaginations of the plasmalemma (as in *Rhodopseudomonas*, Fig. 1.2), form stacks parallel to the longitudinal axis of the cell, or a number of concentric whorls, or less commonly, a three-dimensional reticulum. In aged cells the thylakoids may aggregate sporadically forming crystalline regions about 0.3 μm in diameter resembling the prolamellar bodies of etioplasts of higher plants. In some forms small particles (phycobilisomes) about 30 nm in diameter can be seen lying between the thylakoids. These are aggregates of the biliprotein pigment phycobilin. Since the cells do not contain mitochondria, the electron transport component of respiration may also be located upon the internal membrane system. Clusters of minute cylinders

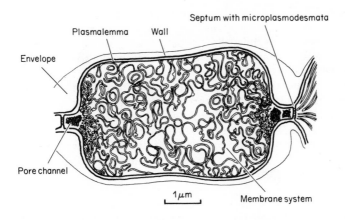

Fig. 2.1 *Anabaena*. Longitudinal section of a heterocyst. (After Fay, in Carr and Whitton (1973). *The Biology of Blue-Green Algae*. Blackwell Scientific Publications, Oxford.)

containing gas (gas vesicles) are present in the cells of some species, and regulate buoyancy. These vesicles are bounded by a sheet consisting solely of protein, and collapse if the cells are subjected to sudden mechanical shock. Vacuoles similar to those in cells of eukaryotes appear to be absent.

In many filamentous species the chains of cells are interrupted by occasional conspicuously larger cells called *heterocysts* (Fig. 2.1). There are often indications of regularity in their spacing, and the relative simplicity of the system may facilitate the identification of the factors leading to their initiation. The heterocysts contain chlorophyll *a* and some other pigments but are altogether paler, and by contrast may appear empty. Electron microscopy reveals that they contain an elaborate and often reticulate membrane system. The wall of a heterocyst is conspicuously thickened except for a pore at one or both poles. Its protoplast is there separated from those of its neighbours solely by a thin septum, in which fine channels (microplasmodesmata) can sometimes be discerned.

Heterocysts have attracted considerable attention since they are the site of the fixation of atmospheric nitrogen. The enzyme responsible, nitrogenase, is irreversibly inactivated by oxygen, and it is significant that the heterocysts lack photosystem II and therefore generate no intracellular oxygen. Indeed their frequently elaborate membrane systems may indicate considerable respiratory activity. This would in turn promote an anaerobic environment within the cells, which the conspicuously thickened walls (Fig. 2.1) may help to maintain. Of particular note is that those species without heterocysts which are able to fix nitrogen can do so only at very low oxygen tensions. In a few instances heterocysts have been observed to regain pigmentation and then to germinate. On these grounds some have considered them to be vestigial reproductive cells.

The cell walls of the Cyanophyta contain a layer of the peptidoglucan murein adjacent to the plasmalemma. The walls are thus chemically and structurally similar to those of Gram-negative bacteria, and their formation is similarly disorganized by penicillin. The presence of murein also accounts for the dissolution of the wall, as those of bacteria, by lysozyme. The outer part of the wall is commonly a layered distinct sheath, sometimes pigmented, and frequently consisting largely of gelatinous or slimy hemicelluloses. The creeping movements of some species (reaching up to 4 μm sec^{-1}) are caused by localized excretion of mucilage through fine pores, about 4 nm in diameter, in the wall. The gelatinous material of the outer wall frequently holds cell colonies together (Figs 2.3 and 2.4). The morphological integrity of the more strikingly filamentous forms often depends upon the cells being held in linear sequence by the toughness of the sheath. 'Branching' occurs at lesions in the sheath (cf. *Tolypothrix*, Fig. 2.7).

DISTRIBUTION   The Cyanophyta are widely distributed, occurring in both soil and water, though rarely at a pH below 4.0. Some are marine (e.g. *Trichodesmium*, responsible for the colour of the Red Sea). Others are found under such extreme conditions as snowfields and hot springs (where they can survive temperatures of up to 85°C), and beneath the eroded surfaces of rocks in Antarctic deserts. Many occur as slimes on rocks, damp soils and tree-trunks, and as scums on stagnant water. Some live as endosymbionts in the protoplasts of colourless flagellates and amoebae (*Glaucocystis* (see p. 32) is often regarded as a composite organism of this kind), and others as colonies in the tissues of higher plants (e.g. the presence of *Anabaena* in the fern *Azolla* and the gymnosperm *Cycas*). Cyanophyta are also components of some lichens. These symbiotic associations are probably related to the ability of many species to both photosynthesize and fix atmospheric nitrogen, and the subsequent release of metabolites to the host. The fertility of many swampy tropical soils (e.g. rice padi) is dependent upon nitrogen fixation by Cyanophyta. Conversely in temperate regions they often contaminate fresh water, particularly where these become enriched by mineral nutrients from agricultural drainage (eutrophication). Some cyanophytes secrete calcium carbonate or silica, and so build up laminated rocks (stromatolites). Living examples are found in Western Australia and North America, and fossil stromatolites occur as far back as the Precambrian.

REPRODUCTION   Although genetic recombination has been reported, the observed reproduction of the Cyanophyta is entirely asexual, in its simplest form involving nothing more than cell division. This, correlated with the absence of distinct nuclei, is not associated with the formation of any recognizable mitotic figure or chromosomes. In some species, mainly those that are unicellular, the cell enlarges, and the protoplast meanwhile divides to form many naked daughter cells termed *endospores*. When these are eventually released they develop a cell wall and become new individuals. In unfavourable conditions, thick-walled cysts (*akinetes*) may be formed. Experiments have shown that these resting spores are very resistant to desiccation and extremes of temperature. In filamentous forms the akinetes

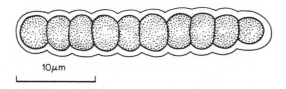

10μm

Fig. 2.2 *Nostoc.* Hormogonium.

frequently germinate to form a short thread of rounded cells, termed a *hormogonium* (Fig. 2.2). This has some capacity for movement which, since there are no flagella or cilia, seems attributable solely to the streaming of mucilage. The hormogonia eventually settle and give rise to filaments. Also in filamentous forms whole short side branches may become resting organs (*hormocysts*). As already mentioned, heterocysts occasionally serve as reproductive organs.

The Cyanophyta contain two principal Orders, the Chroococcales, mainly single or aggregated spherical cells, and the filamentous Hormogonales.

Chroococcales

Although containing the simplest blue-green algae, some of the Chroococcales are nevertheless colonial forms with a regular and conspicuous symmetry. In *Chroococcus* (Fig. 2.3) single cells are occasionally seen, but more usually, and always in the similar *Gloeocapsa*, the cells remain held together after division in a mucilaginous matrix. These aggregates of indefinite size and shape are referred to as *palmelloid* forms. In

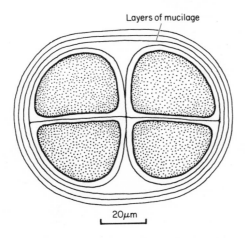

Layers of mucilage

20μm

Fig. 2.3 *Chroococcus.* Colony held together by successive sheets of mucilage.

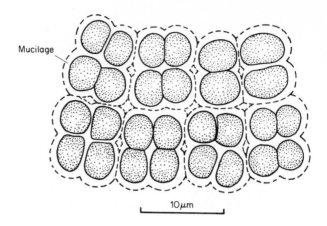

Mucilage

10 μm

Fig. 2.4   *Merismopedia.*

Merismopedia (Fig. 2.4) the cells are arranged in regular rows to form a plate, and in *Coelosphaerium* a hollow sphere. Other geometric arrangements are characteristic of further genera in this Order. Although the cells in these colonial forms appear to be all of similar status and function, there is evidence of polarity, since in vegetative reproduction divisions appear to take place more readily in certain directions than others.

Hormogonales

In the filamentous forms the chain of cells, as distinct from the sheath, is termed the trichome. The width of the trichome ranges from 20 μm to 1 μm,

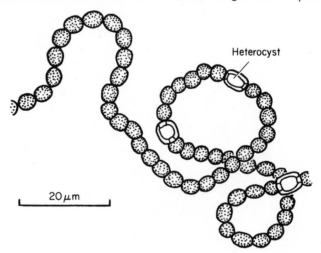

Heterocyst

20 μm

Fig. 2.5   *Nostoc.*

(a)

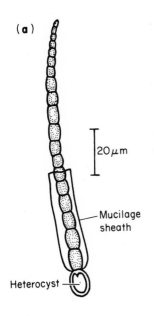

20 μm

Mucilage
sheath

Heterocyst

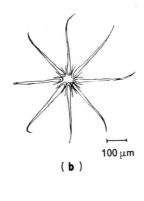

100 μm

(b)

Fig. 2.6 *Gloeotrichia.* **(a)** Single trichome with basal heterocyst. **(b)** Star-shaped cluster of trichomes.

sometimes in the same plant so that the filaments taper to fine points or hairs (as in *Gloeotrichia*, Fig. 2.6a). The simplest filamentous form is *Nostoc* (Fig. 2.5). Here all the cells are of equivalent status and divisions occur sporadically along the filament, but in others, especially those with tapering trichomes, divisions may be largely confined to the basal regions. The extent to which filaments associate varies. In *Anabaena* the filaments at most are clustered irregularly, whereas in *Gloeotrichia* the trichomes radiate from a central plate formed by a coalescence of the larger basal cells (Fig. 2.6b). Many genera, for example *Tolypothrix*, have branched filaments, but 'branching' is usually caused by growth of one part of the cell chain being deflected by the other part of the mucilage sheath (Fig. 2.7). True lateral branching, initiated by an oblique division of a cell in the filament, occurs only rarely. *Spirulina* is a form with a characteristic spiral filament. It forms mats in some freshwater lakes in Africa, and compressed into cakes is used as a human food.

Slow mobility is a feature of many Hormogonales. Although this is particularly noticeable in respect of the hormogonia, it is also shown by normal filamentous cultures. *Anabaena*, for example, grown in an agitated medium is uniformly dispersed, but if the shaking is stopped the filaments soon group themselves into a few tight bunches. This is probably a consequence of the chance cohesion of filaments, followed by the gliding of the filaments along each other within the coalescent mucilage. The oscillatory movements characteristic of *Oscillatoria* appear to arise from changes in curvature propagated along the filament, but are otherwise unexplained.

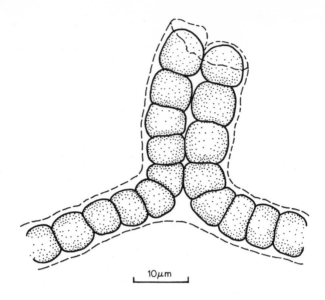

Fig. 2.7   *Tolypothrix* showing 'branching'.

The relationship of the Cyanophyta with other algae

Except in respect of the photosynthetic pigments and the presence of photosystems I and II, the Cyanophyta have little in common with any other Division of the algae. The closest resemblance is with the chloroplasts of the Rhodophyta (p. 93). These also contain biliprotein pigments, and in some species phycobilisomes are present between the thylakoids. If the theory that the chloroplast of the eukaryotes evolved from an invading photosynthetic prokaryote is correct, it seems likely that, the Rhodophyta excepted, the prokaryotic ancestor did not closely resemble the present day Cyanophyta.

Prochlorophyta

A further prokaryotic alga recently discovered is *Prochloron*.[23] It is unicellular and occurs associated, probably symbiotically, with ascidians in warmer seas. The number of species is not yet known.

The Prochlorophyta differ from the Cyanophyta in the absence of the biliproteins and the presence of chlorophylls *a* and *b*, both of which are also found in the chloroplasts of the green algae and higher plants. The thylakoids are more or less concentric within the almost spherical cell, and show local stacking. The central region of the cell is clearer, but there is yet no firm evidence that it is the site of the genetic material. The ribosomes are similar in size to those of the cytoplasm of a eukaryotic cell.

The eukaryotic algae

The cells of all algae other than the Cyanophyta and Prochlorophyta are eukaryotic, and in respect of organization have much in common with those of higher plants.

A striking feature of many of the unicellular eukaryotic algae, and of the zoospores and gametes of many multicellular forms, is the presence of flagella. An unexpected and remarkable discovery of electron microscopy is that flagella from all eukaryotic organisms have a common basic structure, providing a characteristic picture in transverse section (Fig. 2.8). Nine pairs of fibrils, each pair orientated tangentially, are arranged concentrically near the periphery of the flagellum. Usually two more fibrils lie symmetrically at the centre.

It is now known that these fibrils are identical with microtubules, and consist of the protein tubulin. The two microtubules of the peripheral pairs differ in profile. Viewed from the base of the flagellum outwards, the microtubule on the right (the 'A' tubule) usually appears circular in outline, whereas that of the 'B' tubule on the left is not completely so. The portion of the periphery shared with the 'A' tubule commonly follows the curvature of the latter. Usually, but not always in plant flagella, two short arms can be made out on the 'A' tubule. These consist of a special protein, dynein, which is an ATPase. The 'A' and 'B' tubules are 18–25 nm in diameter, but the free tubules at the centre are often a little wider.

The movement of flagella is probably caused by the paired microtubules sliding over one another. The mechanism is not however entirely understood, and more detailed information will probably come from the study of mutants in which the structure of flagella is in some way defective. The microtubules of flagella appear to be quite similar to others in the cell, but they are not so sensitive to colchicine. In some instances flagellogenesis may even continue in the presence of this anti-microtubular drug.

Formerly two classes of flagella were recognized, 'whip-lash' considered to be smooth, and 'Flimmer' furnished with rows of minute hairs

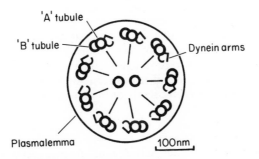

Fig. 2.8 Diagram of transverse section of a eukaryotic flagellum viewed from the base. The 'spokes' radiating from the centre to the peripheral doublets can usually be made out, and sometimes an ill-defined sheath is present around the central pair.

(*mastigonemes*). It now seems doubtful whether algal flagella are ever entirely smooth, but appendages are certainly much more conspicuous in some groups than others. Appendages other than hairs are also known. The single flagellum of *Micromonas*, for example, is covered with minute scales, and that of the spermatozoid of *Dictyota* (p. 92) has both Flimmer hairs and, in the plane of symmetry of the flagellum, a row of spines. Electron microscope studies show that mastigonemes begin to be formed in the perinuclear space. Their assembly and manner of transference to the surface of the flagellum have been followed in detail in *Ochromonas* (p. 64).

The nature of the surface, and other features of the flagella, such as their number, arrangement, and method and kind of insertion, provide considerable assistance in identifying the relationships of the algae.

Chlorophyta

Habitat	Aquatic (mainly freshwater), terrestrial in moist situations, a few epiphytic.
Pigments	Chlorophylls *a* and *b*, β-carotene (α-carotene less prominent); lutein, violaxanthin, neoxanthin, zeaxanthin (siphonaxanthin in siphonaceous forms).
Food reserves	Starch, rarely inulin, oils and fats.
Cell wall components	Cellulose, crystalline glycoproteins, various hemicelluloses.
Reproduction	Asexual and sexual (isogamy, anisogamy and oogamy).
Growth forms	Flagellate, coccoid, filamentous, rarely foliaceous or siphonaceous.
Flagella	2 or 4, occasionally numerous, hairs or scales inconspicuous.

The Chlorophyta (green algae) in respect of metabolism, photosynthetic pigments and ultrastructure show much in common with the vascular plants and the bryophytes. We shall here consider representatives of the commoner Orders and other species of interest. The classification of the Chlorophyta depends upon the degree of development of the thallus, and, where present, the nature of the sexual reproduction. The shape of the chloroplast (chromatophore) is also a useful feature. In some genera they are characteristically large and are present singly or in very small numbers in each cell.

Volvocales

This Order contains unicellular and colonial forms, and also multicellular individuals. They are widely distributed, and common in fresh water. In those species which are free-swimming, two or four single flagella of equal length are attached at the anterior end of the cells.

Unicellular forms

Because of its common occurrence, the genus frequently used to illustrate the unicellular state is *Chlamydomonas* (Fig. 2.9), of which about 400 species have been described. The cells, which rarely exceed 30 μm in major diameter, contain a single chloroplast, usually basin shaped, and one or more mitochondria. Towards one side of the chloroplast, in which the thylakoids show loose stacking forming irregular grana, lies a conspicuous pyrenoid. This proteinaceous body, a common feature of algal chloroplasts, acts as the centre of starch formation. A granular stigma ('eye-spot'), associated with carotenoid pigments, also lies within the chloroplast. Although possibly photosensitive, this property is not confined to the stigma, since phototaxis persists in mutants in which it is lacking. Each cell also contains two pulsating vacuoles which discharge their contents at short intervals, and so play an important role in the osmoregulation of the cell.

The cell wall of *Chlamydomonas* appears to lack cellulose and to consist of a number of layers, some of which form a crystalline lattice made up of glycoprotein sub-units. The sugars involved are mannose, arabinose and galactose. Forms related to *Chlamydomonas* have a similar wall, but with differences in detail.

Cultures of *Chlamydomonas* can be raised so that the cells divide synchronously, facilitating the study of metabolic changes during the cell cycle. *Chlamydomonas* has also figured prominently in research into chloroplast genetics, and an insight into the control of wall formation has been gained from wall-less mutants. Knowledge of the nature and regulation of flagellar movement has also come from *Chlamydomonas*. The whole flagellar apparatus, its function unimpaired, can be isolated and studied *in vitro*. The

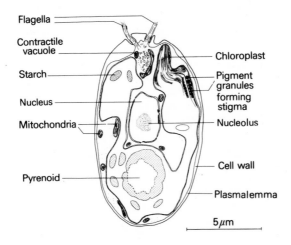

Fig. 2.9 *Chlamydomonas reinhardtii.* Longitudinal section of cell. (From electron micrographs by Ursula W. Goodenough.)

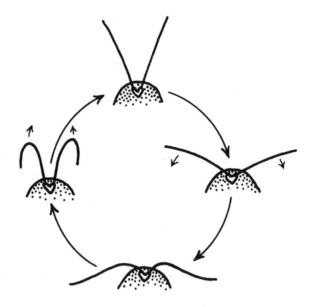

Fig. 2.10　The beating cycle of the paired flagella of *Chlamydomonas*.

breast stroke motion of the flagella (Fig. 2.10), which is uniplanar, can be reversed by increasing the concentration of calcium ions in the medium.

REPRODUCTION　*Chlamydomonas* multiplies asexually by mitosis: the parent flagella are resorbed and the protoplast divides in a manner resembling the cleavage division of animal cells to form 2–8 daughter cells (*aplanospores*). These secrete cell walls and acquire flagella before being liberated, but on an agar surface they may continue to lack flagella and form palmelloid colonies. Sexual reproduction involves the fusion of gametes. These are structurally similar to vegetative cells, but smaller. Gametogenesis can be induced by culturing in nutrient-(particularly nitrogen-)deficient medium. Fusion of morphologically similar gametes (isogamy), the simplest method of combining nuclear information, is found in some species of *Chlamydomonas*, but other species show anisogamy, and a few oogamy. Aggregation of gametes is promoted by agglutins, probably complexes of membrane and glycoprotein shed from the flagella of activated cells. A medium in which the + strain has been grown will cause aggregation (iso-agglutination) of the − strain, and *vice versa*. When the cultures of the + and − strains are mixed there is first animated cell movement, and then pairing of opposed strains (Fig. 2.11). The first contact is between the tips of the flagella. These then extend laterally and come to lie side by side, the cell walls in some species being simultaneously shed. As the cells become accurately aligned a protoplasmic bridge forms between the apposed apical papillae and the flagella simultaneously cease to pair. Fusion is completed in

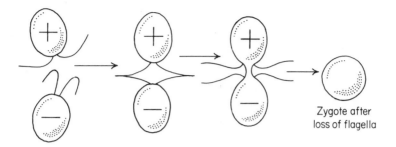

Zygote after
loss of flagella

Fig. 2.11 Mating and zygote formation in isogamous *Chlamydomonas*.

15–20 minutes, leading to a zygote with 4 flagella. Zygotes continue to swim, but show negative phototaxis and eventually settle, although in one species (*C. variabilis*) the swimming phase lasts long enough for the zygotes to have been formerly described as a separate organism. Settling of the zygotes is accompanied by loss of flagella and the secretion of a thickened wall.

Reproduction is anisogamous in *C. braunii*, the – strain producing 4 macrogametes and the + strain 8 microgametes. The oogamous state is approached in *C. coccifera* in which the cells of one strain produce a single macrogamete and those of the other 16 or 32 microgametes. It is reasonable to envisage anisogamy having evolved from isogamy, and experiments with isogamous strains have shown how this may have come about. A large number of mitoses before gametogenesis results in smaller gametes. Controlling the number of mitoses in cultures of different mating types can thus lead to gametes differing in size. Anisogamy (and ultimately oogamy) could have resulted from differences of this kind having become genetically fixed and linked with mating type.

Meiosis occurs on germination of the zygote and four new individuals are produced. There is 2:2 segregation of mating type,[22] indicating Mendelian inheritance of this feature. The chloroplast DNA of one parent however appears to be absent from the zygote and is probably depolymerized following syngamy.

Although representative of the simplest eukaryotic green plants, *Chlamydomonas* clearly possesses features of structural and functional complexity far greater than those found in the most advanced blue-green algae. It would appear that the attainment of the nucleate state represents an advance in organization without which others, such as flagellate motility and sexual reproduction, are impossible.

An alga similar to *Chlamydomonas*, but lacking a cell wall, is *Dunaliella*. In conditions of extreme salinity (e.g. in salt pans) *Dunaliella* accumulates glycerol to such a higher concentration that the cells are able to resist desiccation by exosmosis. In these conditions the cells often become pigmented bright orange. On the basis of its structure, *Polytoma*, a heterotrophic form, is also probably related to *Chlamydomonas*. As well as revealing similarities, electron microscopy also reveals profound differences between flagellate

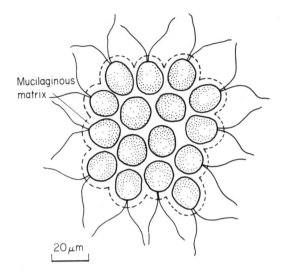

Mucilaginous
matrix

20 μm

Fig. 2.12 *Gonium.*

organisms, and their assignment to one class, the Flagellatae, as proposed by some taxonomists, is clearly unwarranted.

Colonial forms

The Volvocales also contain motile colonies, composed of identical cells, each similar in morphology to *Chlamydomonas*. *Gonium* (Fig. 2.12), for example, has a flat plate of 4 or 16 cells (depending on the species) which are regularly arranged, and held together in a tough mucilaginous matrix. The 16 cone-shaped cells of *Pandorina* (Fig. 2.13), however, are arranged in a

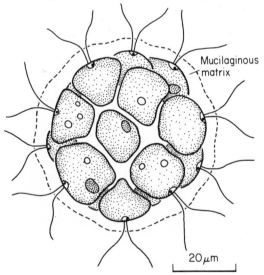

Mucilaginous
matrix

20 μm

Fig. 2.13 *Pandorina.*

sphere. In both asexual reproduction (in which a single cell gives rise to a new spherical colony) and gametogenesis, all the cells of these simple colonies become involved simultaneously.

Volvox: a multicellular individual

Evidence for both coordination and division of labour between cells is seen in *Volvox* (Fig. 2.14). The number of cells in this spherical organism is always a power of 2, indicating synchronized division during growth. The cells, which individually resemble *Chlamydomonas*, are held together by mucilage. The refractive boundaries between the sheaths of individual cells often give a hexagonal pattern to the surface. The centre of the sphere is filled with less viscous mucilage. The cells remain connected by cytoplasmic strands, a feature of significance in relation to the coordination of their activity. The flagella, for example, beat in unison producing a steady rolling motion. Despite the radial symmetry the organism has an anterior and posterior, and in the anterior region the stigmata are more conspicuous than elsewhere. Some species may reach diameters of almost 1 mm.

REPRODUCTION   Asexual reproduction takes place by the enlargement and repeated longitudinal division of a vegetative cell so that eventually a new sphere of cells is formed bulging into the hollow centre. Initially the cells of this daughter individual are oriented inversely in relation to those of the mother, but at the completion of cell division the new sphere invaginates and the orientation of the cells consequently reverses. The daughter is simultaneously liberated into the cavity, and may even itself generate a daughter before it is released by the death of the original mother. On release the daughter organism expands to its mature volume.

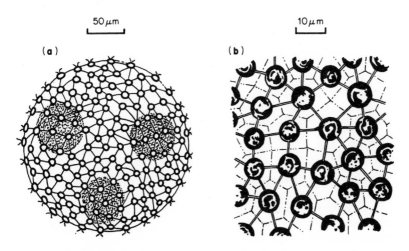

50 μm 10 μm

(a) (b)

Fig. 2.14   *Volvox.* **(a)** Asexual reproduction, three daughters lying within the cavity of the parent. **(b)** Detail of the connections between individual cells.

Sexual reproduction in *Volvox*, in which both monoecious and dioecious species occur, is truly oogamous. The gametangia may develop from any cell or, when the distinction of anterior and posterior is well marked, be confined to the posterior region. The mature oogonia are large flask-shaped cells without flagella. Each is developed from a single vegetative cell, and remains in position up to and during fertilization. The male organ, the antheridium, similarly develops to form a vegetative cell. Successive divisions lead to a bowl-shaped mass of biflagellate spermatozoids (antherozoids). After release, the aggregate migrates to an oogonium and there breaks up into individual gametes. The details of fertilization are little known. Subsequently the zygote forms an oospore which in some species acquires a thickened wall and is able to survive unfavourable conditions. Meiosis occurs at germination, but since only one flagellate swarmer is released three of the meiotic products evidently fail to survive. Successive divisions of the swarmer lead to a small spherical individual, but after repeated asexual reproduction the mature size characteristic of the species is ultimately regained.

In some dioecious species individuals entering the sexual phase and becoming female can induce others still vegetative to become male, a consequence of a 'male inducing substance', protein in nature, secreted by the female. This is only one example of such correlative systems in *Volvox*, and the situation in the genus as a whole is evidently complex.

Volvox differs from simpler Volvocales in possessing somatic cells which play no part in reproduction. These cells consequently perish when an individual ruptures to release a daughter, or when its integrity is destroyed by the liberation of numbers of oospores. Because of the evident organization in the thallus, and the presence of somatic cells, the term colony is inappropriately applied to *Volvox*. This organism has reached the level of a multicellular individual. In the Volvocales as a whole, the transition from colonial forms to *Volvox* is a gradual one, and although the status of *Volvox* is unambiguous that of many of the intermediates is less clearly defined. The term **coenobium** is often used for a colony of cells in which there is some degree of coordination.

Possibly allied to the Volvocales is a small group of unicellular green algae which commonly bear four flagella inserted in an apical groove. The body and flagella are covered with minute scales. They are sometimes placed in a separate division (Prasinophyta).

Chlorococcales

This heterogeneous Order contains lowly green algae in which motility is confined to zoospores and gametes. Although there are common features in reproduction, the Chlorococcales show much diversity, and interrelationships appear distant. Most members of the Order occur in fresh waters, a few live in the oceans and moist places on land, and others are endophytic in the intercellular spaces of higher plants, symbionts with lower animals, or constituents of lichens. *Chlorococcum* and *Chlorella* are unicellular, the latter

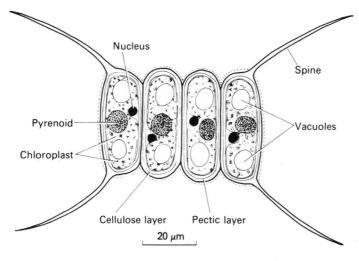

Fig. 2.15 *Scenedesmus quadricauda*. Section of coenobium in the plane of the spines, based on photo- and electron micrographs. The middle layer of the three-layered wall is micellar in structure, but its chemical nature is not certain. The spines consist of bundles of micelles emerging from the middle layer.

occurring both free and as the endosymbiont in the coelenterate *Hydra*. *Chlorella*, one of the plants with which Priestley and Ingenhousz first demonstrated photosynthesis in 1779, has been used extensively in studies of algal metabolism. Another unicellular form, *Botryococcus*, is commonly planktonic, and sometimes forms water blooms. Up to 75% of its dry weight may consist of hydrocarbons, and the dried remains of blooms resemble sheets of crude rubber. *Oocystis*, similar to *Chlorella*, has figured prominently in studies of cell wall formation. *Prototheca*, also unicellular, is an heterotrophic parasite of fish.

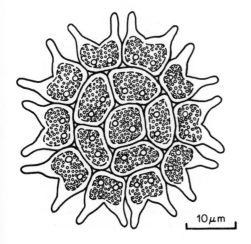

Fig. 2.16 *Pediastrum*

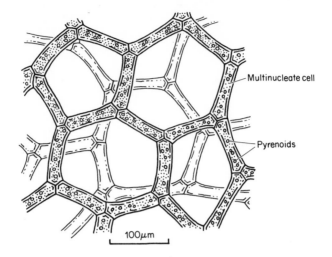

Fig. 2.17　*Hydrodictyon.* Portion of the cylindrical reticulum of a growing individual showing the multinucleate cells with conspicuous pyrenoids. (After Esser, (1976). *Kryptogamen.* Berlin.)

Scenedesmus (Fig. 2.15) and *Pediastrum* (Fig. 2.16) are representative of the simpler coenobial forms. Lignin has been detected in the spines of *Scenedesmus*, and may also be present in the cell walls of other members of the Order. *Hydrodictyon* is altogether remarkable. The multinucleate cells, each with a reticulate chloroplast containing numerous pyrenoids, form a hollow cylindrical reticulum with closed ends (Fig. 2.17). Individual cells may reach 5–10 mm in length, and the whole coenobium extends to 20 cm or more. At the other extreme is the diminutive *Protosiphon*. The thallus consists of a single sphere, about 0.3 mm in diameter, anchored to the substratum by a colourless rhizoid which may reach a length of 1 mm (Fig. 2.18). The sphere contains several nuclei, and a reticulate chloroplast with several pyrenoids. *Protosiphon* is frequent on damp mud and walls, and some forms can withstand extremes of heat and salinity in desert soils. Pure cultures have been shown to produce bacteriostatic substances.

The curious unicellular *Glaucocystis*, found in acid pools, is also believed to be a member of this Order. It contains no chloroplast of its own, but has a blue-green alga as an endosymbiont. It may alternatively be a curious flagellate allied to the red algae (see p. 99).

REPRODUCTION　Asexual reproduction by simple fission does not occur in the Chlorococcales. In *Chlorella* the nucleus undergoes two mitoses, and the cytoplasm is divided between the four daughter nuclei. A cell wall forms around each of these non-motile **aplanospores**. They are then liberated from the mother cell and grow to the mature size.

In the coenobial forms each cell produces aplanospores or zoospores, depending upon species. In *Scenedesmus* the aplanospores aggregate before or immediately after release to form a new coenobium. In *Hydrodictyon* the

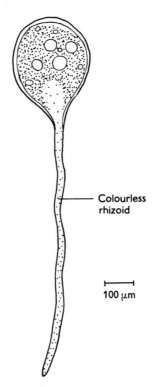

Colourless
rhizoid

100 μm

Fig. 2.18 *Protosiphon botryoides.* (After Klebs, from
Fritsch (1935). *The Structure and Reproduction of the
Algae, I.* Cambridge University Press.)

multinucleate cells become transformed into a mass of biflagellate
zoospores. These are retained, shed their flagella, and then form a new
coenobium within the mother cell. This eventually ruptures and the young
coenobium is free to expand to its mature size.

Sexual reproduction is isogamous or anisogamous, the biflagellate
gametes conjugating laterally and not at their flagellate poles. Meiosis takes
place as the zygote germinates, yielding four swarmers. These lose their
flagella, become multinucleate, and divide to form zoospores. As in sexual
reproduction, these are retained and their walls fuse to form a juvenile
coenobium, the cells of which are initially uninucleate. This 'polyeder' stage
in the life cycle is particularly conspicuous in *Hydrodictyon*.

Reproduction in *Protosiphon* is little specialized. Simple asexual reproduction takes place by budding, but biflagellate swarmers are often produced
after flooding. These may behave as either zoospores or gametes. There is no
evidence for the existence of mating types since gametes from the same plant
may fuse. Germination of the zygote, probably the occasion of meiosis, may
be immediate or delayed.

Ulotrichales

The Ulotrichales show considerable advances in organization over the algae described so far, both in vegetative complexity and in the elaboration of the life cycle. All the various forms, including the macroscopic, such as *Ulva*, can be regarded as being derived from a filamentous system. In the simpler forms sexual reproduction is isogamous, and the zygote undergoes meiosis on germination. However, in many of those genera where the thallus shows greater morphological complexity, the zygote germinates without a reduction division, and consequently yields a diploid organism. These forms show a life cycle in which the two phases, gametophyte and sporophyte, are morphologically identical (see p. 9).

Ulothrix

This genus is representative of the simple filamentous forms which lack branching, and in which all the cells are of equal status. Vegetative division is intercalary and all the cells are involved in reproduction, except the basal attachment cell usually present in young filaments. The vegetative cells, which are often wider than long, are uninucleate and have a single peripheral chloroplast (Fig. 2.19a). Fragmentation of the filament is a common method of vegetative propagation, but this is caused principally by accidental breakage; simultaneous dissociation of the filament into segments has rarely been observed.

REPRODUCTION   In asexual reproduction 1 to 32 zoospores (the number depending upon the species) are produced in each cell by division of the protoplast. The mature zoospores are pear-shaped and quadriflagellate (macrozoospores) (Fig. 2.19b). In some species smaller biflagellate zoospores (microzoospores) are also produced, intermediate in size between macrozoospores and gametes. The zoospores are liberated through a pore in the wall of the parent cell, each zoospore surrounded by a mucilaginous sheath. The free zoospore closely resembles a unicellular member of the Volvocales, devoid of its cell wall. After settling, the zoospore attaches itself by the posterior end (to which the stigma has now shifted), and grows out laterally, producing a holdfast cell on one side, and new vegetative cells on the other. Sometimes, after the initial division in the parent cell in a filament, aplanospores with resistant walls are produced instead of zoospores.

In sexual reproduction, gametogenesis resembles the production of zoospores in asexual reproduction. The gametes, however, are uniformly biflagellate, and 8, 16, 32 or 64 (the number again depending upon the species) are produced in each gametangium (Fig. 2.19b). *Ulothrix* is physiologically heterothallic, gametes from the same filament being unable to unite. The zygote is mobile for a short while, but then secretes a resistant wall and enters a resting period during which it may become attached and form a small unicellular plant. Germination commences with a reduction division, followed by one or two mitoses and the formation of zoospores or aplanospores, the mating types again segregating. The diploid state is thus represented only by the zygotic cell and has no prolonged existence. According to some accounts the biflagellate swarmers are exclusively gametic.

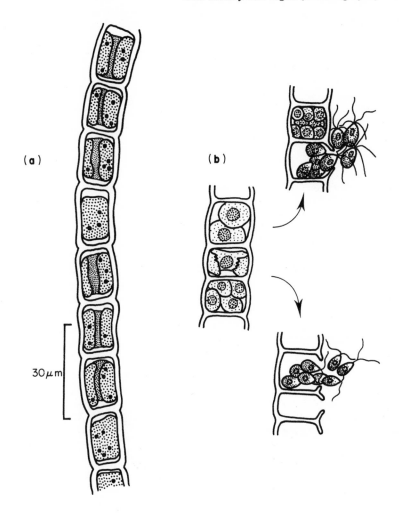

Fig. 2.19 *Ulothrix zonata*. **(a)** Vegetative filament showing the girdle-shaped chloroplast. **(b)** *Above,* production of quadriflagellate zoospores; *below,* production of gametes (diagrammatic).

Forms related to *Ulothrix* are *Microspora* (Fig. 2.20), in which the cell walls disjoin into characteristic H-shaped pieces when the filament fragments, and *Cylindrocapsa*, where the cells have conspicuously thick walls and a dense chloroplast. *Cylindrocapsa* is also outstanding in showing well-developed oogamy.

Ulva

At first sight, there is little to relate the macroscopic foliaceous thallus of *Ulva* to the microscopic filaments characteristic of so many Ulotrichales. The

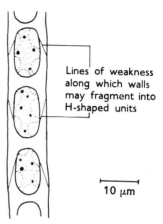

Lines of weakness
along which walls
may fragment into
H-shaped units

10 μm

Fig. 2.20 *Microspora pachyderma*. Portion of filament. (After West and Fritsch (1927). *A Treatise on the British Freshwater Algae*. Cambridge University Press.)

general structure of the cells, however, is the same, and each contains a single chloroplast. At cell division each chloroplast divides at the same time as the nucleus. *Ulva* was the first multicellular plant in which it was possible to follow unambiguous chloroplast division with the electron microscope.

Young plants of *Ulva* always begin their development as simple filaments. Division in principally two dimensions results in a flattened expanse of tissue, two cells thick, expanding from a narrow stalk and holdfast (Fig. 2.21a).

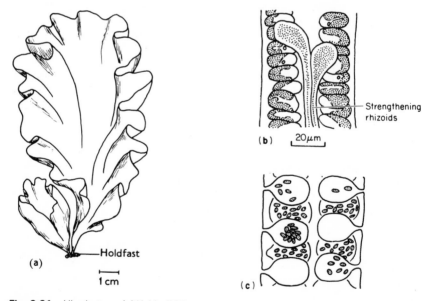

Strengthening rhizoids

(b)   20μm

Holdfast

(a)

1 cm

(c)

Fig. 2.21 *Ulva lactuca*. **(a)** Habit. **(b)** Transverse section of vegetative thallus. **(c)** Transverse setion of thallus producing zoospores. **((a), (c)** After Newton (1973). *A Handbook of the British Seaweeds*. British Museum; **(b)** after Thuret, from Fritsch (1935). *The Structure and Reproduction of the Algae*, I. Cambridge University Press.)

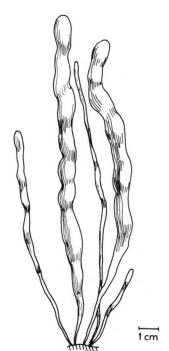

Fig. 2.22 *Enteromorpha intestinalis.* The thallus consists of hollow thin-walled cylinders. (After Newton (1937). *A Handbook of British Seaweeds.* British Museum.)

1 cm

Some of the vegetative cells produce multinucleate rhizoids which grow between the two cell layers, strengthening the thallus (Fig. 2.21b). This occurs particularly in the stalk. Mutants are known which remain filamentous, the cells evidently no longer acquiring the ability to divide in more than one direction.

The marine intertidal zone is the characteristic habitat of both *Ulva*, and of the closely related *Enteromorpha* which has a peculiar tubular thallus (Fig. 2.22). Both species can, however, tolerate wide variations in salinity, and may be found far up tidal estuaries.

REPRODUCTION   Asexual reproduction occurs as in *Ulothrix*, most of the vegetative cells taking part (Fig. 2.21c). After the zoospores have been discharged, the parent thallus remains as a bleached framework of empty cells.

Sexual reproduction is again similar to that of *Ulothrix*, and *Ulva* is also physiologically heterothallic. The zygote, however, does not undergo reduction division on germination, but grows instead into a diploid thallus identical with that of the haploid plant. The production of zoospores is accompanied by meiosis (although there are some irregularities), equal numbers of both mating strains being produced. *Enteromorpha* differs from *Ulva* in being anisogamous.

Prasiola, found on soil and shore lines, is perhaps related to the Ulvales. The life cycle, although not fully known, appears to be isomorphic. The chloroplasts are stellate.

The origin of the isomorphic life cycle

There are no intermediate forms in the Ulotrichales indicating how an isomorphic life cycle, such as that of *Ulva*, might have originated. It is possible that a relatively simple mutation prevented meiosis at the zygotic stage, and that the inhibition remained effective until after many cell generations. Since there is no evidence that chromosome number itself determines the form of growth in any group of plants, this delaying of mitosis would have allowed the development of a diploid thallus closely resembling that of the haploid. In the evolution of *Ulva* the emergence of a distinct sporophyte was probably accompanied by the progressive elaboration of the vegetative structure of both phases of growth.

Cladophorales

This Order, consisting predominantly of branched filamentous forms, has affinities with both the Ulotrichales and, in respect of the multinucleate cells, with the siphonaceous algae. Nevertheless the Cladophorales have sufficient distinctive characters of their own to merit treatment as a separate Order.

Cladophora

Of the 160 or so species of this genus, some are marine and others freshwater. The filaments show true branching, buds developing towards the anterior end of the elongated, cylindrical vegetative cells (Fig. 2.23). The bundle of filaments is usually attached below by a rhizoid-like cell to a firm surface in the substratum. The internal structure of the cells is complex, each cell being

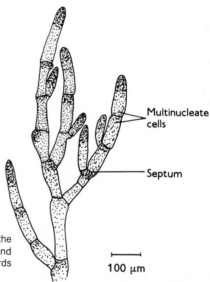

Multinucleate cells

Septum

100 μm

Fig. 2.23 *Cladophora* sp. Upper part of the thallus showing characteristic branching and the greater density of the cytoplasm towards the apical ends of the cells.

multinucleate and having a single reticulate chloroplast with many pyrenoids. The cell wall comprises three layers, the inner of cellulose, a central layer of hemicellulose, and finally an outer coating, possibly chitinous, which gives the alga its characteristic crisp feeling, and provides a surface which is rapidly colonized by small epiphytes, notably diatoms. Division of the protoplast, which is not regularly related to nuclear division, is accompanied by the formation of transverse septa. These develop from the margin towards the centre, and gradually acquire a complicated lamellate structure at the periphery, which may impart some flexibility to the older filaments. Growth of the filaments is apical in *Cladophora*, but intercalary growth is general elsewhere.

A very curious feature exhibited by some species of *Cladophora* and certain other genera is the tendency, when subjected to a gentle rolling motion in water, to aggregate into cushions or spheres. These are termed 'aegagropilous' species because of the fancied resemblance of these aggregates to the balls of wool found in goats' stomachs. *Cladophorella*, a plant of damp places, is of interest because its upper cells secrete what appears to be a true cuticle, otherwise unknown in the algae.

REPRODUCTION Both the asexual and sexual reproduction of Cladophorales resemble that of *Ulothrix*. The zoospores and gametes are produced in non-specialized cells, and copulation is usually isogamous, rarely anisogamous. Fusion of the gametes is lateral. An isomorphic life cycle has been demonstrated in some species, but in others the cycle is heteromorphic. Before these were understood the gametophytic and sporophytic phases, being unrecognized, were assigned to separate genera.

Vegetative reproduction by fragmentation of branches also occurs, and seems to be the sole means of reproduction of the aegagropilous species. In some species survival through unfavourable periods is afforded by the formation of thick-walled resting spores packed with food reserves.

Chaetophorales

A distinct advance in the organization of a filamentous thallus is found in the Chaetophorales where the thallus is composed of both prostrate and erect components, and is consequently termed **heterotrichous**. The prostrate system is typically a flat plate attached to the substratum, from which arises the erect branching system bearing the reproductive organs and often characteristic hairs. Although elements of an aerial and a prostrate system are always detectable, many genera show greater development of one component than of the other, sometimes almost to its exclusion. The maintenance of the dominance of one system has been shown in some species to depend upon environmental factors. In *Stigeoclonium*, for example, magnesium depresses the growth of the erect system, while an excess of nitrogen inhibits that of the prostrate component.

Environmental factors, as well as affecting the quantitative relationships of the two systems, also influence the extent to which the erect system produces hairs and branches. Correlated with this morphological plasticity, the Order shows a remarkable range of habitats extending from the littoral zones of seas and lakes to damp soil (e.g. *Fritschiella*), tree trunks (e.g. *Pleurococcus*), and particularly in the Tropics and sub-Tropics the surfaces of leaves. Sexual reproduction is correspondingly diverse, both isogamy and well-developed oogamy being represented. The life cycle shows little development of the zygotic phase.

Stigeoclonium

This genus is representative of those members of the Chaetophorales in which both the prostrate and erect components of the thallus are easily recognizable, and each is more or less well developed. The species differ widely in external morphology. In general the erect system ends in long, thin, hyaline hairs, and is less branched than the prostrate system (Fig. 2.24). The latter often forms a pseudo-parenchymatous sheet as a consequence of the close packing of the branches, but detailed investigation of its structure in natural habitats is rendered difficult by the tenacity with which it adheres to the substratum. The vegetative cells contain a peripheral girdle-shaped chloroplast with one or more pyrenoids.

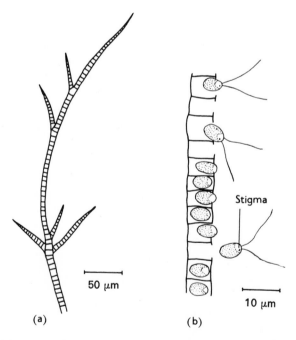

Fig. 2.24   *Stigeoclonium tenue*. **(a)** Terminal portion of erect system. **(b)** Release of gametes. (After West and Fritsch (1927). *A Treatise on the British Freshwater Algae*. Cambridge University Press.)

REPRODUCTION   In asexual and sexual reproduction *Stigeoclonium* resembles *Ulothrix*. Production of zoospores or gametes is confined to the erect system and occurs simultaneously in a large number of cells. The gametes are noticeably smaller than the zoospores, and copulation is isogamous. Sometimes, however, flagellate cells identical in all but behaviour with gametes have been observed to act as zoospores and yield thick-walled resting spores. The zygote is commonly bright orange. Meiosis occurs on germination, and biflagellate swarmers are released. The various reproductive bodies usually begin to develop first the prostrate system and then the erect.

Allied to *Stigeoclonium* is the terrestrial alga *Fritschiella* in which the prostrate system, buried in damp mud, produces nodules of cells which serve as perennating organs.

Draparnaldia and its close relation *Draparnaldiopsis* are two aquatic members of the Order in which the erect system is dominant, and in the latter genus of great complexity.

The upright axes of *Draparnaldia* consist of large barrel-shaped cells, from which arise the highly branched whorls of laterals with much smaller cells (Fig. 2.25). Frequently these laterals are so profuse that the axis is quite obscured. The chloroplasts of the lateral branches are notably better

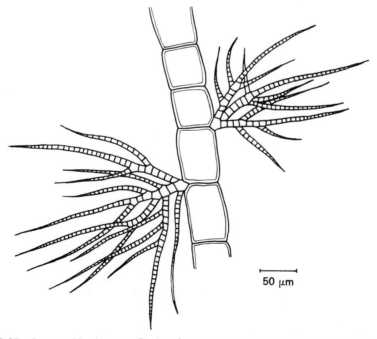

50 μm

Fig. 2.25   *Draparnaldia glomerata.* Portion of erect system. (After West and Fritsch (1927). *A Treatise on the British Freshwater Algae.* Cambridge University Press.)

developed than those of the axes, indicating some localization of function within the upright system. *Draparnaldiopsis* is similar, but the axis consists of long and short cells, and the whorls of fine laterals, usually originating in four tufts, arise only from the short cells. The tufts consist of radiating long and short branches. Other branches, arising particularly from the bases of the longer laterals, turn down, branch freely, and invest the internode in a cortical sheath, often thicker than the axis itself. In both genera the prostrate system is vestigial, and it is represented by a holdfast, the function of which is assisted by the outgrowth of rhizoids from adjacent cells of the main axis.

Reproduction of the two genera, similar to that of *Stigeoclonium*, is confined to the lateral branches.

Aphanochaete (Fig. 2.26)

This genus contains a number of common epiphytes, and is representative of those members of the Chaetophorales in which the prostrate component is dominant. In vegetative structure it resembles the prostrate system of *Stigeoclonium*, long unicellular hairs being the only traces of the erect system. Sexual reproduction is anisogamous.

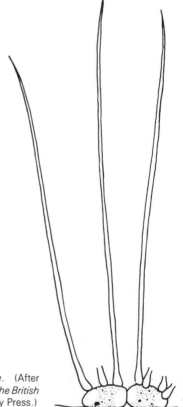

20 μm

Fig. 2.26 *Aphanochaete polychaete.* (After West and Fritsch (1927). *A Treatise on the British Freshwater Algae*. Cambridge University Press.)

Coleochaete

Many freshwater epiphytes are referable to *Coleochaete* (Fig. 2.27). Here the prostrate system is again generally dominant, although in a few species the habit is more typically heterotrichous. In *C. pulvinata*, for example, the erect system is well developed and forms a hemispherical cushion. Sheathed bristles are a unique and characteristic feature of the thallus of this genus.

REPRODUCTION   *Coleochaete* is also outstanding in its sexual reproduction. It is alone amongst the Chaetophorales in being oogamous, and the structures associated with the egg are highly specialized. The cells which differentiate into oogonia terminate short branches, although they may ultimately appear lateral because of continued growth from the penultimate vegetative cell. The oogonium develops a long neck, termed a *trichogyne* (Fig. 2.27b), which opens at the tip when the egg is ready for fertilization. The antheridia arise in the filaments, often just above the oogonium, and each produces a

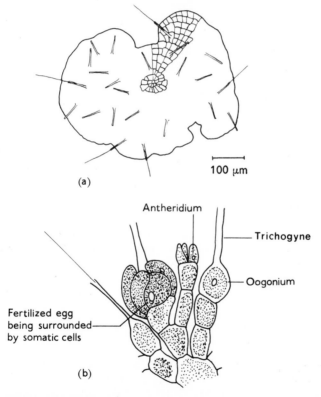

Fig. 2.27 *Coleochaete.* **(a)** *C. scutata.* Surface view. **(b)** *C. pulvinata.* Sexual reproductive structures. (**(a)** After West and Fritsch (1927). *A Treatise on the British Freshwater Algae.* Cambridge University Press; **(b)** after Pringsheim, from Fritsch (1935). *The Structure and Reproduction of the Algae, I.* Cambridge University Press.)

single colourless spermatozoid. On liberation, they are attracted, presumably chemotactically, to the receptive spot at the tip of the trichogyne, and fertilization follows.

As a consequence of fertilization, but in a manner still unknown, the vegetative cells adjacent to the oogonium are stimulated into growth and they envelop the zygote in a continuous parenchymatous sheath. This sheath eventually dies, but its cells contribute to a thick wall, partly labyrinthine, which forms around the resting zygote (see p. 117). After a resting period, the nucleus enters meiosis and the zygote germinates to give rise to a plate of 16 or 32 wedge-shaped cells, each of which liberates a zoospore.

Trentepohlia

Adaptation to its terrestrial habitat is the most noteworthy feature of *Trentepohlia*. The thallus, which is differentiated into prostrate and erect components, grows attached to rocks and bark, as well as on soil and humus. It is common in warm moist situations, and is conspicuous for its orange, rather than green, appearance. The cell walls are thick and often layered, and this feature probably accounts for the ability of the alga to resist desiccation. If growth is inhibited, as in conditions of drought, there is a marked increase in the pigmentation of the cells, and concurrently an accumulation of fat, in which the pigment (haematochrome) is dissolved.

REPRODUCTION   Asexual reproduction is by zoospores, and takes place only in damp conditions. Several morphologically distinct types of zoosporangia have been described, and *Trentepohlia* appears to be advanced in respect of this feature. In dry conditions intact sporangia may be distributed by wind, and even purely vegetative reproduction occurs by the dispersal of single cells or groups of cells from the prostrate system. Isogamous sexual reproduction has been observed in some species.

Pleurococcus (*Desmococcus*)

This genus contains what is probably the commonest green alga *P. vulgaris*, familiar as a friable incrustation on the windward side of walls and tree-trunks. *Pleurococcus*, which consists of small complanate aggregations of cells, is regarded on the basis of its cytology as a very reduced member of the Chaetophorales. It has no known method of reproduction other than simple cell division.

The relationships of the Chaetophorales

The morphological diversity of the Chaetophorales, and the restricted knowledge about their life cycles, make it difficult to assess their relationships with other green algae. Isolated specialized forms, such as *Draparnaldia* and *Coleochaete*, are a conspicuous feature of the Order. Although the cells of the Chaetophorales usually contain only a single basin-shaped chloroplast there are indications that they represent the diverse relics of a kind of algal organization from which the earliest land plants arose. The method of reproduction of *Coleochaete* bears a resemblance to that of the archegoniates. Further a structure consisting of an array of minute plates and microtubules at the root of the flagella in zoospores and gametes is very similar to the multilayered structure found in archegoniate spermatozoids, and it is not seen elsewhere in the algae. Also in contrast to other algae the Chaetophorales contain a glycolate dehydrogenase similar to that of vascular plants. The heteromorphic life cycle of the primitive land plants may have been derived from two components of the heterotrichous system. The erect system, for example, possibly became associated with the diploid phase of the nucleus and hence the sporophyte, and the prostrate with the haploid phase and the gametophyte.

Oedogoniales

The Oedogoniales are a well-defined Order, possessing several unique features. Some have attached so much weight to these as to consider that the Order should be removed from the Chlorophyta. The Oedogoniales are not, however, anomalous in such fundamentals as wall structure and pigmentation, and they are probably better regarded as a small group that has diverged from the main line of evolution, the intermediate stages being no longer represented. The Oedogoniales comprise only three genera, *Oedogonium*, *Oedocladium*, and *Bulbochaete*. Of these, *Oedogonium* is by far the commonest and best known.

Oedogonium

THE ORGANIZATION OF THE FILAMENT   The thallus of *Oedogonium* is an unbranched filament (Fig. 2.28). When young, the filaments are attached by

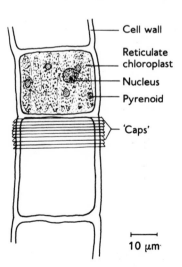

- Cell wall
- Reticulate chloroplast
- Nucleus
- Pyrenoid
- 'Caps'

10 μm

Fig. 2.28   *Oedogonium* sp. Portion of filament.

a basal holdfast, but unless the water they inhabit is flowing, the mature condition is free-floating. The individual cells have a single, large nucleus, and a reticulate chloroplast. Pyrenoids are frequent at the interstices of the reticulum. Cell division is normally intercalary, but the way in which the new cell walls are produced is highly peculiar (Fig. 2.29). The first indication that a cell is about to divide is the formation of a ring of wall material towards the upper end of the cell, just below the septum. The nucleus then divides and a septum forms between the daughter nuclei but this septum remains free at the periphery. During the division of the nucleus the ring in the upper part of the cell becomes larger and crescent-shaped in vertical section. Eventually the cell wall breaks transversely at this level, and the ring is drawn out longitudinally to form a cylinder of new wall material. Meanwhile the septum between the nuclei moves up the cell, reaches the bottom of the newly formed cylinder, and then fuses peripherally with the longitudinal walls. The wall of the lower cell is thus largely that of the mother cell, while that of the upper, except for a conspicuous cap of original wall at the anterior end, is wholly new. The presence of caps, which if the daughter cell goes on dividing may be several in number, at the anterior ends of the cell is a feature diagnostic of the Oedogoniales.

REPRODUCTION Both   asexual   and   sexual   reproduction   occur   in *Oedogonium*. Asexual reproduction is by means of zoospores which, like others of the Oedogoniales, are unique in possessing an apical ring of many short flagella. They arise singly in cells at various sites along the filament. While in the sporangium the zoospore is surrounded by a mucilage sheath giving the wall a characteristic two-layered appearance. The zoospore is

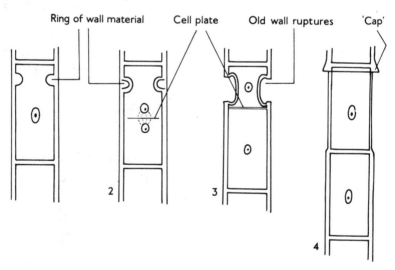

Fig. 2.29 *Oedogonium*. Stages in cell division. Diagrammatic.

released by rupture of the sporangium, thereby causing a break in the filament. It emerges still surrounded by mucilage, but this is soon shed and the zoospore becomes almost spherical (Fig. 2.30a). After a brief motile phase the zoospore settles and irregular rhizoid-like outgrowths from the flagellate end attach it to the substratum. A new filament then develops.

Both monoecious and dioecious species of *Oedogonium* are known; some of the dioecious species exhibit a difference of morphology between the male and female filaments (dimorphism). In the monoecious, and isomorphic dioecious species, large, almost spherical oogonia are produced from the upper cell of a vegetative division. The lower cell may divide again, and another oogonium develop, but more usually it remains as the supporting cell. A single, dense, oosphere is formed in each oogonium (Fig. 2.30b). The oosphere shrinks away from the cell wall, and develops a colourless receptive spot as it matures. When it is fully mature, a pore appears at the anterior end of the oogonium adjacent to the receptive spot of the oosphere. Meanwhile, antheridia are produced by division of a vegetative cell into several disc-shaped portions, each of which normally produces two male gametes. These are similar, both in morphology and method of liberation, to the zoospores. After fertilization the zygote forms an oospore, often reddish in colour and regularly surrounded by a thickened wall.

Although oospores are sometimes able to germinate immediately, a long resting period seems generally to be necessary. There is evidence that chilling hastens germination. When growth is eventually resumed, the normal course is for four haploid protoplasts to be extruded, each of which develops a crown of flagella, swims away, and settles to produce a new plant.

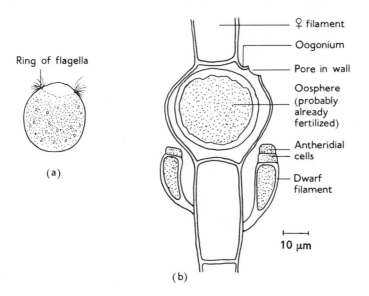

Fig. 2.30 *Oedogonium* sp. **(a)** Zoospore. **(b)** Oogonium with adjacent dwarf males.

In dimorphic (or nannandrous) species, the female filaments are of the usual size, but the males are much smaller. These dwarf males arise from the female plants by way of a special propagule called an androspore. This is produced in a manner similar to that of the male gametes, except that only one androspore is produced per cell. Resembling zoospores, except for their smaller size, and yellowish colour, the androspores come to settle either on the oogonium or its supporting cell, and give rise to small filaments (Fig. 2.30b). After a few cells have been produced, antheridia are cut off, and reproduction proceeds as in the isomorphic (macrandrous) species.

In laboratory cultures the sexual phase can be induced by increasing the carbon dioxide concentration of the medium. With dioecious species it is also necessary to intermix male filaments with female to stimulate the production of oogonia. The secretion of hormone-like substances into the medium seems indicated (cf. *Volvox*, p. 30).

Bulbochaete and *Oedocladium*

Bulbochaete resembles *Oedogonium* in essentials, but the filaments are branched and terminate in hairs, similar to those of the Chaetophorales. *Oedocladium* differs from *Oedogonium* in having both branched filaments and a heterotrichous system.

The relationships of the Oedogoniales

The hairs of *Bulbochaete* and the heterotrichous system of *Oedocladium* recall features of the Chaetophorales, and perhaps indicate a distant affinity. Another feature in common is the peripheral girdle-shaped chloroplast in each cell, and the small gamete-like zoospores of some Chaetophorales may indicate the origin of nannandry in some Oedogoniales.

Mesotaeniales

This small Order encompasses the saccoderm desmids. These are mostly unicellular organisms containing two plate-like chloroplasts symmetrically placed on each side of the nucleus (Fig. 2.31). The cell walls are without ornamentation and lack pores (cf. Desmidiales, p. 49). The cell sap is some-

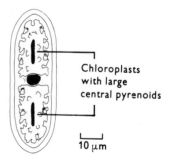

Chloroplasts
with large
central pyrenoids

10 μm

Fig. 2.31 *Cylindrocystis brebissonii.* The two prominent stellate chloroplasts are symmetrically placed in the cell and the nucleus lies between them.

times pigmented, and these desmids are in part responsible for the 'red snow' seen on Alpine glaciers and in the Arctic. They also occur in bog pools.

Asexual reproduction is by a simple division, several cells often being held together within a mucilage sheath. Sexual reproduction involves conjugation of two cells. Meiosis and division of the zygote into four daughter cells take place after a resting period. Mendelian segregation of mating type has been demonstrated in pure cultures.

Desmidiales

Most of the desmids[4] (the placoderm desmids) are referable to this Order. They are typically unicellular (although sometimes the cells may be aggregated in chains) and are conspicuous in the phytoplankton of oligotrophic meres. The cells have complex shapes and a precise and striking symmetry. In many forms the cell is divided into two halves, one being the mirror image of the other, connected by a narrow central portion, the isthmus. The nucleus usually lies in the isthmus, and one or more chloroplasts are present in each half cell. Even in those forms (e.g. *Cosmarium*, Fig. 2.32) in which the central constriction is less conspicuous the cells still display the exact bilateral symmetry.

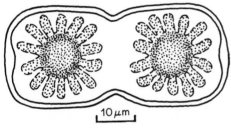

10 μm

Fig. 2.32 *Cosmarium diplosporum.* (After West and Fritsch (1927). *A Treatise on the British Freshwater Algae* Cambridge University Press.)

The cell wall is made up of an inner cellulose layer and an outer layer of variable composition, frequently containing iron compounds or silica. It is particularly this outer layer, often patterned with spines and other protuberances, which provides the specific characteristics, and which puts the desmids amongst the most beautiful of microscopic objects. The cells are frequently surrounded by an investment of mucilage secreted by minute pores in the wall. It is probably the localized secretion of such mucilage which enables the cells to perform slow movements.

Asexual reproduction is by fission. The details of the process are not yet fully understood, but in outline it is as follows. After nuclear division a ring of wall material develops at the centre of the isthmus and grows inwards to form a septum. When complete, the daughter cells separate. Each then regenerates the missing half and the symmetry is restored (Fig. 2.33). This results in two adult organisms in each of which half of the thallus is inherited from the previous generation and the other half newly synthesized. This is an example of semi-conservative replication of an organism recalling that of the molecule of double-stranded DNA.

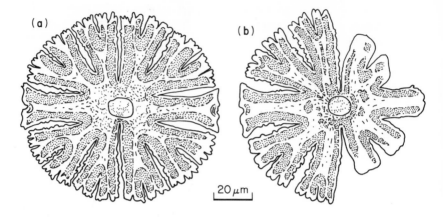

Fig. 2.33 *Micrasterias.* **(a)** Mature individual preparatory to division. **(b)** Following division, the beginning of regeneration of new half cell.

Sexual reproduction involves conjugation of cells. A protuberance may be formed from one cell and directed towards the other, but in many species the walls break down at the isthmus and the protoplasts emerge and fuse. Following fusion a thick-walled zygospore is formed. Meiosis and germination occur after a resting period. As in the saccoderm desmids, segregation of mating types has been demonstrated.

Zygnemales

The thalli of this Order consist of unbranched and free-floating filaments, although rhizoid-like outgrowths may attach the basal cell to the substratum in the young state. Almost all are found in fresh water.

The cells are often markedly elongate, having an axile ratio of 5:1 or more. The nucleus lies near the centre suspended by cytoplasmic strands, and the conspicuous chloroplast is in the form of a flat plate (e.g. *Mougeotia*), a helical band at the periphery of the cell (e.g. *Spirogyra*, Fig. 2.34), or it consists of two stellate portions (e.g. *Zygnema*). Numerous pyrenoids occur in each form of chloroplast. In strong light the chloroplast of *Mougeotia* turns so that its narrow edge is presented to the source. Experiments with polarized light have shown that the stimulus to this movement comes from an array of photoreceptive molecules at the periphery of the cytoplasm. The proteinaceous pigment phytochrome is also involved.

Growth of the filament is intercalary, all the cells except the basal cell being capable of division. The dividing wall grows in from the margin, but the central region of the septum is completed by the formation of a phragmoplast.

REPRODUCTION   Asexual reproduction is solely by fragmentation of the filament, and is preceded by changes in the transverse walls. In some species

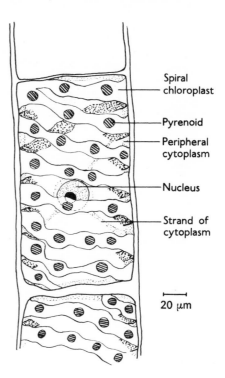

Spiral
chloroplast

Pyrenoid

Peripheral
cytoplasm

Nucleus

Strand of
cytoplasm

20 μm

Fig. 2.34 *Spirogyra* sp. Cytology of
vegetative cell.

the middle lamella of the transverse wall becomes gelatinous and a turgor
difference arises between the cells. This causes one cell to bulge into its neigh-
bour and the junction between the cells becomes strained to breaking point.
In other species a ring or collar of wall material forms on each side of the
septum and in a manner not yet understood leads to separation. The
comparative ease with which the cells separate has led some to regard the
Zygnemales as basically unicellular.

Sexual reproduction involves conjugation, occurring normally towards the
end of the growing season, but factors such as nitrogen deprivation, pH, and
altitude (presumably because of its relationship with mean temperature) have
all been found to influence it. Before conjugation the two participating filaments must lie parallel and come close together (Fig. 2.35). The necessary
adjusting movements are probably brought about by the localized secretion
of mucilage. Since solitary filaments show no directed movement the first
approach of one filament to another, presumably of opposed mating type,
must involve some form of chemotaxis. Once alignment has taken place,
further mucilage secretion holds the filaments in position and opposing cells
begin to form protuberances directed towards each other. The production of
protuberances is not necessarily simultaneous, and a cell may be touched by
the protuberance of its partner before it has itself become active. The production of a similar structure is then immediately induced. The growth of the
protuberances pushes the filaments apart at the site of conjugation. If an

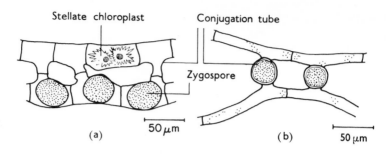

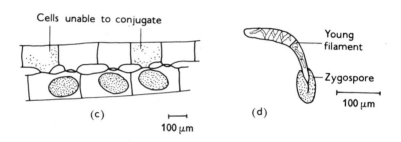

Fig. 2.35   Sexual reproduction in the Conjugales. **(a)** *Zygnema stellinium*. **(b)** *Mougeotia parvula*. **(c)** *Spirogyra setiformis*. **(d)** *Spirogyra velata*. Germination of zygospore. (All after West and Fritsch (1927). *A Treatise on the British Freshwater Algae*. Cambridge, University Press.)

unmated protuberance touches a cell which has already formed a connection there is no further response. Eventually growth of the protuberances ceases, the opposed distal walls dissolve, and a conjugation tube is formed between the two cells. With the formation of the tube complex physiological changes take place within the protoplasts. There is an increase in starch, and a decrease in permeability and the osmotic pressure of the cell sap. Before completion of the conjugation tube no sexual differentiation of the two protoplasts is apparent, but as soon as communication is established one protoplast withdraws from its wall, and passes slowly through the tube to fuse with its partner. Since all the protoplasts in one filament behave in the same way it is legitimate to regard one filament as male and one as female. After fusion the chloroplast of the male gamete is resorbed, and the zygote contracts.

The process of conjugation is subject to a number of variations. In *Spirogyra*, for example, adjacent cells of the same filament may sometimes conjugate. In *Zygnema* and *Mougeotia* the zygote is often formed in the conjugation tube. Sexual differentiation of a morphological nature is seen in *Sirogonium*. Here the male gametangium, regularly associated with a sterile cell, is notably smaller than the female. In all instances the zygote becomes a thick-walled zygospore and enters a period of dormancy. The germinating spore gives rise to a single filament. In those species investigated, meiosis has been shown to occur before germination, but three nuclei degenerate. The survival of only one meiotic product prevents a direct analysis of Mendelian segregation.

There are evident affinities in relation to sexual reproduction between the Mesotaeniales, Desmidiales and Zygnemales. These three Orders have often been grouped together as the Conjugales.

The siphonaceous algae (Siphonales)

Although in current classifications the siphonaceous algae are distributed amongst several Orders, they can be conveniently considered together. The thalli normally contain multinucleate protoplasts and are consequently termed *coenocytic*. Dividing walls are absent until the formation of the reproductive organs. The range in form of the thallus is considerable; it may consist of a simple unbranched tube or a complex mass of interwoven filaments. Those species with complex thalli are invariably marine, and the thalli are often mechanically strengthened by superficial deposits of calcium carbonate. There are numerous chloroplasts, and in some species they have particularly tough envelopes. They can remain functional following ingestion by marine invertebrates. The siphonaceous algae also differ from the rest of the Chlorophyta in containing both α- and β-carotene, instead of β-carotene alone, and at least one additional xanthophyll. Almost all are confined to tropical or warm seas.

Valonia, the simplest of the group, consists of a bladder-like cell reaching a few centimetres in diameter. It often bears clusters of daughter vesicles, and short basal branches from a rhizoid-like holdfast. *Valonia* has been extensively used in experiments on wall structure, permeability and absorption of electrolytes.

Acetabularia (Fig. 2.36) has a mushroom-shaped thallus when mature, the cap reaching a diameter of about 1 cm. The whole delicate plant is stabilized by a shell of calcium carbonate. The central axis produces whorls of deciduous branches during growth. Unlike most of the siphonaceous algae, *Acetabularia* is uninucleate throughout its vegetative development, the nucleus, large and conspicuous, remaining at the base of the stalk. It has been found that if the stalk is removed from the nucleate portion the stalk remains capable of some growth and the apex may even begin to form a cap. Nevertheless the nucleus is clearly essential for continued growth and morphogenesis. Decapitation and grafting experiments have shown that the nucleus produces a sequence of 'morphogenetic substances' (probably messenger ribonucleic acids) which ascend into the stalk and determine the kind of growth which occurs at the top. Grafts made between the stalk of one species and the nucleate portion of another have revealed the extent to which the cytoplasm of the first species can affect the expression of the genetic information contained in the nucleus of the second. The fructosan inulin has been

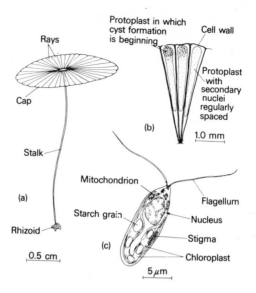

Fig. 2.36   *Acetabularia mediterranea*. **(a)** Cell with mature cap. **(b)** Detail of cap during cyst formation. **(c)** Longitudinal section of gamete.

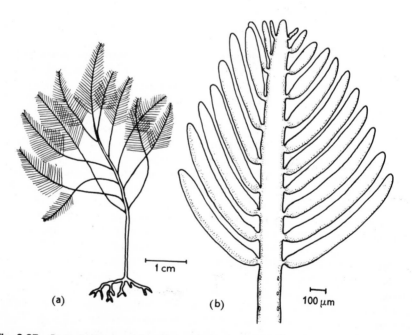

Fig. 2.37   *Bryopsis plumosa*. **(a)** Habit. **(b)** Terminal portion of branch. (After Newton (1937). *A Handbook of the British Seaweeds*. British Museum.)

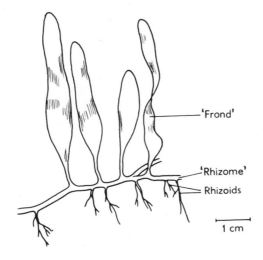

Fig. 2.38 *Caulerpa prolifera.* Portion of plant. (After Taylor (1960). *Marine Algae of the Eastern Tropical and Subtropical Coasts of the Americas.* Michigan.)

reported as a reserve product in *Acetabularia*. The chloroplasts can be isolated comparatively easily. They were the first with which it was possible to demonstrate normal photosynthesis *in vitro*.

Bryopsis (Fig. 2.37), locally common, shows a considerable advance in vegetative organization. The thallus possesses a main axis, from which branches arise pinnately. These also branch, leading to a complanate bipinnate, and in some species tripinnate, condition. At the insertion of each branch the cell walls are constricted and conspicuously thickened. This feature undoubtedly has mechanical significance, since a simple branched tube lacking septa and with walls of uniform thickness, would become structurally unstable beyond a certain point. In the vegetative condition the cytoplasm is spread as an even layer containing numerous nuclei and chloroplasts. The frond-like thallus of *Bryopsis* rises more or less vertically and is anchored to the substratum by rhizoids produced from a small prostrate filament. Fragments of the thallus will regenerate as new plants, but rhizoidal outgrowths are formed only from the morphologically lower end. This is an example of polarity, widespread in the Plant Kingdom, but particularly open to investigation in *Bryopsis* because of its occurrence in a relatively simple unicellular system.

Caulerpa has a large creeping 'rhizome' from which both rhizoids and upright 'fronds' arise (Fig. 2.38). The 'fronds' show a great variation in shape when mature, different species having been named from their distinctive forms (Fig. 2.39). Internally, all species of *Caulerpa* show ingrowths of the cell wall forming, throughout the thicker portion of the thallus, a web of interconnecting bars. This adds to the mechanical stability of the thallus, and the increased surface area of the protoplast may facilitate the passage of

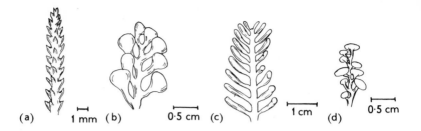

Fig. 2.39   *Caulerpa.* Types of 'fronds'. **(a)** *C. cupressoides.* **(b)** *C. racemosa* var *macrophysa.* **(c)** *C. ashmeadii.* **(d)** *C. peltata.* (All after Taylor (1960). *Marine Algae of the Eastern Tropical and Subtropical Coasts of the Americas.* Michigan.)

minerals. The fibrillar polysaccharide of the wall is not cellulose but a β-1:3 xylan, a polymer of a pentose sugar. The walls also contain callose, a β-1:3 glucan, also found in the sieve tubes and reproductive structures of land plants.

   Codium (Fig. 2.40), which extends into cooler seas, is representative of the most elaborate vegetative organization found in the siphonaceous algae. The thallus is made up of closely packed interwoven filaments (hyphae), although the outward morphology varies widely with species. In *C. tomentosum* the thallus is a system of dichotomously branched axes, each about 0.5 cm in diameter, anchored at the base. Other species are flattened, forming a cushion or plate, or spherical. In all species a weft of branched filaments gives rise to a

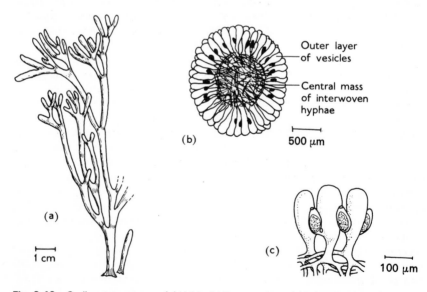

Fig. 2.40   *Codium tomentosum.* **(a)** Habit. **(b)** Transverse section of thallus. **(c)** Gametangia. (All after Newton (1937). *A Handbook of the British Seaweeds.* British Museum.)

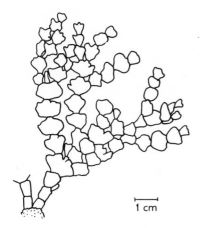

Fig. 2.41   *Halimeda   simulans.*   Portion   of thallus (After Taylor (1960). *Marine Algae of the Eastern Tropical and Subtropical Coats of the Americas.* Michigan.)

1 cm

continuous covering of elongated vesicles at the exterior. *Halimeda* (Fig. 2.41) is a related tropical alga with a complex thallus of variable morphology. Calcium carbonate is deposited in the side walls of the outer vesicles. These deposits survive after the death of the plant and yield one of the several forms of coral. The fossil record of lime-secreting siphonaceous algae goes back as far as the Upper Silurian (see Table 1.1).

REPRODUCTION   Specialized methods of asexual reproduction are absent from the siphonaceous algae. In *Valonia*, for example, vegetative multiplication takes place by the sporadic detachment of the daughter vesicles. In the larger forms segments, branches or parts of the 'rhizome' become detached and establish themselves as new plants. In *Codium* and *Caulerpa* this seems to be the principal method of reproduction.

Sexual reproduction is isogamous or anisogamous; oogamy is unknown. In *Bryopsis* terminal pinnae, normally the smallest, become transformed into gametangia and are cut off from the rest of the thallus by a septum. The gametangia are conspicuously opaque as a consequence of the considerable multiplication of plastids. The gametes are liberated by dissolution of the apex of the gametangia. They are biflagellate, but there is striking anisogamy and, since each plant usually produces gametes of only one size, a clear trend to dioecy. The larger gametes also contain a distinctly green plastid, whereas the smaller gametes are yellowish. After fusion the zygote develops immediately into a new plant. Sexual reproduction in *Codium* and *Caulerpa* appears to be basically similar to that in *Bryopsis*. Meiosis has been detected in gametogenesis, so the thallus in these forms is probably normally diploid.

Sexual reproduction in *Acetabularia* has many curious features. During vegetative growth the chromosomes in the single nucleus undergo continued endoreplication. When growth is completed this basal nucleus then divides into several thousand secondary nuclei. These ascend into the cap which eventually becomes cleft into uninucleate compartments. Each compartment then becomes a cyst from which biflagellate gametes are ultimately liberated.

Although the gametes are morphologically identical, their pairing behaviour indicates the presence of + and − mating strains, a situation recalling that already encountered in *Chlamydomonas*. Despite much research it is still not absolutely certain when meiosis occurs, but it is either preceding the formation of the cysts or during gametogenesis. The zygote therefore is diploid. There is no resting period and the formation of a new plant begins immediately.

The relationships of the siphonaceous algae

The relationships of the siphonaceous algae are obscure, but the origin may have been in a form resembling the present Chlorococcales, an Order in which tendencies towards the coenocytic habit are evident. An ancestor may have been a simple coenocyte resembling *Protosiphon* (p. 32).

Charales

Although many authors regard the Charales as having Divisional rank, they may be included in the Chlorophyta because they are basically similar in pigmentation, metabolism and the limited nature of their anatomy to the algae. Nevertheless, they have many striking features unrepresented else-where in the Plant Kingdom. There are only six living genera, the remainder being fossil. The living *Chara* and *Nitella* are common in base-rich waters. Many species develop an exoskeleton of calcium carbonate. This is clearly an ancient feature since Charalean exoskeletons, which are readily preserved as fossils, can be traced back as far as the Devonian.

THE VEGETATIVE ORGANIZATION   The thallus of the Charales always possesses a clear main axis, growing from a dome-shaped apical cell (Fig. 2.42). Segments are cut off at regular intervals and in a single column from the flat base of this cell, and each segment immediately divides again by a transverse wall. The upper cell of this division becomes a nodal initial, and the lower, which does not divide again, an internode cell. The nodal initials remain meristematic and give rise to whorls of branches, mostly of limited growth. The mature plant thus has a morphology reminiscent of *Draparnaldiopsis*. The internode cells often achieve remarkable lengths (up to 10 cm or more in *Nitella*), and the cytoplasm commonly shows vigorous streaming, rates of up to 100 μm per second being common. The chloro-plasts, which are discoid and lack pyrenoids, are however little affected since they lie in an outer less active layer (ectoplasm). The streaming is associated with actin microfilaments, about 50 μm of actin, accompanied by myosin, being present in a single cell. The nucleus often fragments in the mature cells.

The large size of the internode cells of *Nitella* makes them ideal for studying permeability. It is even possible to cut off the ends and wash out the contents, leaving the plasmalemma intact. The biophysical properties of this membrane can then be investigated without interference from the cytoplasm.

In *Nitella* the internodes consist only of the single internode cell, but in

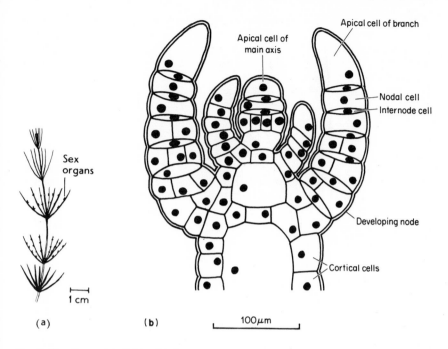

Fig. 2.42 *Chara.* **(a)** Habit. **(b)** Detail of apex of main axis (**(b)** After Esser (1976). *Kryptogamen.* Berlin.)

Chara this becomes surrounded by a cortex formed of rows of cells growing down from the lower cells of the node.

Unlike most benthic algae the Charales are usually anchored to a soft substrate. Attachment is by colourless branched rhizoids which issue from the lower subterranean nodes. The rhizoids are negatively geotropic and contain aggregated crystals of barium sulphate. These may act as statoliths, and by deflecting the flow of Golgi vesicles containing wall materials to one side of the cell cause differential growth, and hence curvature.

REPRODUCTION   Vegetative propagation, the only method of asexual reproduction, is by the formation of tuberous outgrowths on subterranean nodes or rhizoids. These become filled with starch and can withstand unfavourable conditions. The nodal outgrowths often have a stellate symmetry ('starch stars').

Sexual reproduction is oogamous, and highly specialized. The male and female reproductive structures, which consist of antheridia and oogonia surrounded by envelopes of sterile tissue, are often brightly pigmented and sufficiently large to be seen with the naked eye. In both monoecious and dioecious species, the oogonia develop just above the insertions of the lateral appendages, and the antheridia just below (Fig. 2.43). A complicated sequence of divisions leads to the mature antheridium. In the centre a tangled

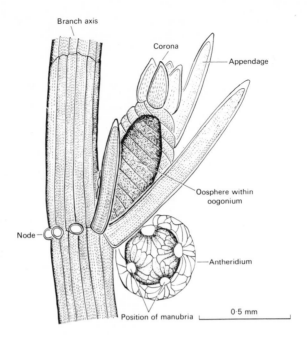

Branch axis

Corona

Appendage

Oosphere within
oogonium

Node

Antheridium

Position of manubria

0·5 mm

Fig. 2.43   *Chara* sp. Male and female reproductive organs.

mass of threads arises from lateral receptacles (manubria) (Fig. 2.44). Each
cell of each thread differentiates into a spermatozoid, which in structure and
shape resembles those produced by some archegoniate plants (Fig. 2.44c).
Oogonia, when mature, are surrounded by a spiral of elongated vegetative
cells, at the tip of which short columns of rounded cells come together as a
corona (Fig. 2.43). Within this sheath is a single egg cell which accumulates
food reserves and becomes largely opaque. Spermatozoids penetrate slits
which appear in the coronal region, and fertilization occurs at a small clear
area near the apex of the egg, the so-called 'receptive spot'. After fertilization
a cellulose membrane is secreted by the zygote. This, together with the wall of
the oogonium and the inner walls of the spiral cells, which become thickened
and indurated, serves to enclose the zygote in an almost impervious jacket,
often reinforced externally with calcium carbonate. Red light promotes
germination, indicating that it is controlled by a phytochrome system.

Germination, which occurs after a resting period, is almost certainly
accompanied by reduction division, but only one of the tetrad of nuclei so
produced remains intact; the remainder degenerate. Division of the intact
nucleus leads to the production of two cells, which, as the membranes of the
zygote break open at the apex, yield from the one a rhizoid and from the other
an erect green filament. This filament is the protonema, a stage of development peculiar, in the algae, to the Charales. Cell divisions occur in the protonema, and it differentiates into nodes and internodes. The first node gives

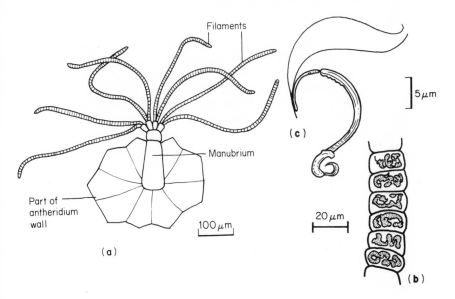

Fig. 2.44 *Chara*. **(a)** Antheridial filaments. **(b)** Differentiating spermatozoids in cells of filament. **(c)** Biflagellate spermatozoid.

rise to rhizoids and additional protonema, and the second to a whorl of laterals, one of them developing into a normal main axis, and the others into yet more protonema. The mature plants thus arise at lateral branches of protonemal filaments. At least one species of *Chara* has been shown to be parthenogenetic.

The relationships of the Charales

The relationships of the Charales are largely speculative. The fossils show that they are an extremely ancient type of algal organization, even fossilized zygotes (*Gyrogonites*) being recognizable in very early rocks. The small discoid chloroplasts are more like those of the higher plants than of Chlorophyta generally, but there is no evidence that terrestrial plants came from *Chara*-like ancestors. The Charales share with the desmids and *Spirogyra* the presence of barium sulphate crystals in the cells, and whorled branching and an archegonium-like oogonium are represented in the Chaetophorales (*Draparnaldia* and *Coleochaete* respectively). Although there may be distant relationships the Charales are best regarded as a highly specialized, and in certain ecological situations highly successful, Order of aquatic plants, long separated from the main trends of algal evolution. They are possibly now represented by fewer forms than earlier in geological time.

Evolution within the Chlorophyta

It is reasonable to regard the unicellular forms as relics of the most primitive green algae. Diversification can be envisaged as being accompanied by a progression towards colonial, coenobial and filamentous organization leading to the present major Orders, each with its distinctive features. In Table 2.1 these Orders have been arranged according to their morphological complexity. Without implying firm phylogenetic pathways, the possible relationships between Orders indicated by cytology and cytochemistry are indicated by broken lines. The central vertical line represents the trends in vegetative and reproductive organizations as seen in the living green algae which may ultimately have led to the land plants. The grounds for regarding the Chaetophorales as closer to the lower archegoniate plants than other green algae are discussed in relation to the origin of the bryophytes (Chapter 4, see also pp. 44–5).

Table 2.1   Possible relationships between the living green algae, and the trends in morphological and reproductive organization leading to the land plants.

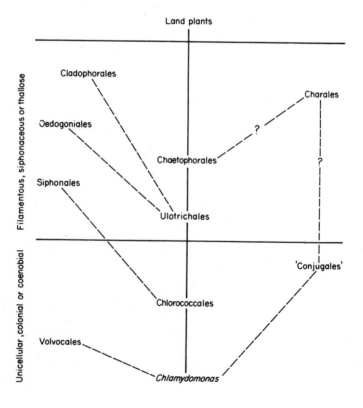

3
The Algae, II

In this chapter we shall be concerned with those eukaryotic algae which lie outside the Chlorophyta. They show a number of organizational trends resembling those already seen in the Chlorophyta, together with some features not represented in living Chlorophyta, but possibly present at some stages in their evolutionary history.

Chrysophyta

Habitat	Aquatic (mainly freshwater).
Pigments	Chlorophylls a, c; β-carotene; fucoxanthin.
Food reserves	Fat, chrysolaminarin (leucosin).
Cell wall components	Cellulose, hemicelluloses, often with siliceous scales.
Reproduction	Asexual, occasionally sexual.
Growth forms	Flagellate, coccoid, colonial, rarely filamentous; amoeboid stages in some forms.
Flagella	2 unequal (Flimmer), or 1, anterior; in some uniflagellate forms the second a stump detectable only with the electron microscope.

Although the Chrysophyta are strikingly different in colour (the cells usually contain two golden-brown chloroplasts (chromatophores)) the members of this Division show many growth forms resembling those found in the Chlorophyta. The highest level of organization attained however is only the simple filament, suggesting that evolution, although parallel, has also been very much slower than in the green algae.

The Chrysophyta are of considerable interest from the physiological and biochemical points of view, and they are often important components of plankton.

The flagellate forms

The flagellate habit is represented by a number of species, often referred to as

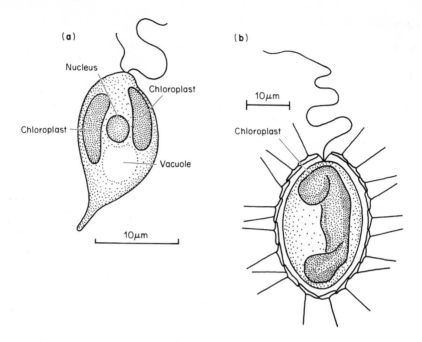

Fig. 3.1 **(a)** *Ochromonas.* **(b)** *Mallomonas.* (After Scagel, Bandoni, Rouse, Schofield, Stein and Taylor (1966). *An Evolutionary Survey of the Plant Kingdom.* Belmont.)

the chrysomonads, some of which are able to alter their shapes and method of locomotion. *Ochromonas*, for example, is commonly pear-shaped, with two unequal flagella (one up to six times the length of the other) emerging from a depression at the anterior end (Fig. 3.1a). Sometimes however it may lose this shape and produce narrow outgrowths (rhizopodia), the flagella being lost. This amoeboid phase may also pass over temporarily into a thick-walled cyst. *Ochromonas* is also nutritionally versatile and at least one species is capable of feeding on a unicellular blue-green alga.

In the common planktonic *Mallomonas* (Fig. 3.1b) the pectinaceous wall is covered with imbricating siliceous scales, some of which bear delicate hinged needles, often as long as the cell itself, possibly assisting flotation. In marine plankton are also found the curious Silicoflagellineae (Dictyochales). These are uniflagellate and possess an intracellular siliceous skeleton. The cell has little or no wall, and produces rhizopodia. The skeletons are known as fossils from the Tertiary onwards.

The flagella in *Ochromonas* are furnished with mastigonemes which themselves bear minute filaments. Electron microscopy has shown that, after beginning their assembly in the perinuclear space, the mastigonemes are transferred to Golgi vesicles, where the filaments are added. They then pass, still in vesicles, to the site of emergence of the flagella; and appear in two rows on the flagellar surface.

Palmelloid and coccoid forms

A number of these are commonly encountered in the field. *Synura* forms a free-floating colony in the plankton of drinking-water reservoirs. The individual cells are similar to *Mallomonas*, but are held together in a gelatinous mucilage. *Hydrurus*, which forms brownish layers on rocks in alpine streams, is similar but the matrix branches irregularly. In *Dinobryon* (Fig. 3.2a), frequent in both freshwater and the sea, the cells are elongate, and each is enclosed in a cellulose sheath. The sheaths are commonly connected head to tail.

Filamentous forms

The only filamentous form of note is *Phaeothamnion* (Fig. 3.2b), found in fresh waters. The club-shaped cells form branching filaments which readily disarticulate when disturbed.

Reproduction of the Chrysophyta is predominantly asexual, longitudinal fission being characteristic of the unicellular flagellates. In *Dinobryon* one product of the division moves out of the sheath and forms a new sheath

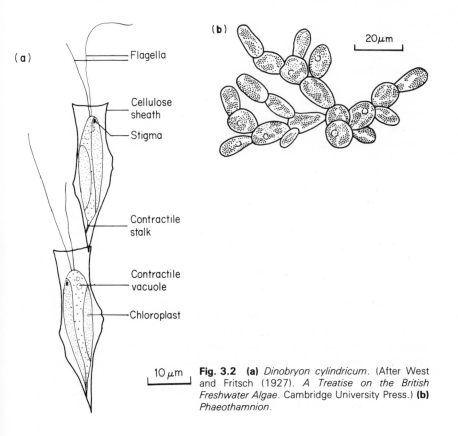

Fig. 3.2 (a) *Dinobryon cylindricum.* (After West and Fritsch (1927). *A Treatise on the British Freshwater Algae.* Cambridge University Press.) (b) *Phaeothamnion.*

attached to the rim of the old, accounting for the characteristic form of the colony. Copulation of flagellated cells has been observed in some instances, but little is known of life cycles.

Xanthophyta

Habitat	Aquatic (freshwater), damp soil.
Pigments	Chlorophylls *a*, *c*, *e*; β-carotene; heteroxanthin, diadinoxanthin, diatoxanthin, vaucheriaxanthin.
Food reserves	Fats, chrysolaminarin.
Cell wall components	Cellulose, hemicellulose.
Reproduction	Asexual and sexual, in some forms oogamous.
Growth forms	Flagellate, coccoid, colonial, filamentous, siphonaceous.
Flagella	Typically 2, strikingly unequal, the larger usually Flimmer.

The Xanthophyta, although a relatively small Division, again show a series of growth forms parallelling those of the Chlorophyta. In the Xanthophyta the organization attained exceeds that in the Chrysophyta. *Vaucheria*, for example, has a well established siphonaceous habit and complex oogamy.

The inequality of the flagella, and their insertion to one side of the apex are distinguishing features of this Division, accounting for the earlier name Heterokontae. The cell walls in many forms show a tendency to fall into two halves. The chloroplasts are green, but in contrast to those of the Chlorophyta they will often turn blue in dilute hydrochloric acid.

Flagellate and coccoid forms

Representative of the simplest flagellate forms is *Heterochloris*, and of the coccoid *Chlorobotrys* (Fig. 3.3a), common in bog and fen pools. *Chlorobotrys* has a spherical cell resembling *Chlorella*. Several cells are usually held together in mucilage forming a free-floating colony.

Filamentous forms

Of the filamentous forms *Tribonema* (Fig. 3.3b) is the most notable. The short unbranched threads are encountered in freshwater and on damp earth, usually in base-rich situations. The cell wall is clearly made up of two overlapping halves. Consequently the filament, when disrupted, breaks up into a number of pieces H-shaped in optical section, closely resembling the situation in the green alga *Microspora* (p. 35). It is the H-shaped piece, rather than the complete cell, which is the basic unit of the filament so far as the wall is concerned, and at cell division, instead of the usual septum being formed, a

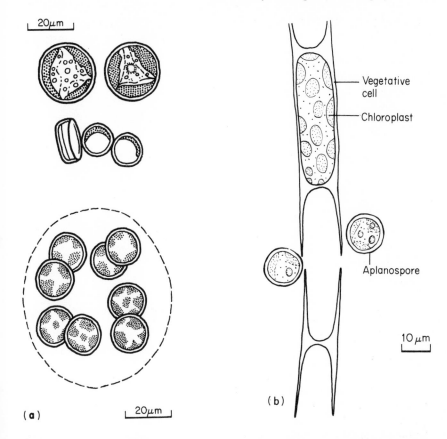

Fig. 3.3 (a) *Chlorobotrys* sp. *Above* Individual cells and germinating cyst. *Below* Aggregate of cells held together in mucilage. **(b)** *Tribonema bombicinum*. (After West and Fritsch (1927). *A Treatise on the British Freshwater Algae*. Cambridge University Press.)

new H-shape is produced at the centre of the parent cell. Internally, the cell organelles are disposed in much the same way as they are in *Microspora*, except that several lenticular plastids occupy the peripheral cytoplasm, instead of a single band-like plastid. Although the cells normally have a single nucleus there is evidence in some species of multinucleate cells, perhaps indicating a transition to a coenocytic condition.

Siphonaceous forms

The simplest of these is *Botrydium* (Fig. 3.4), superficially similar to *Protosiphon* (p. 32), but differing in its forked rhizoid, and its ability to divide vegetatively. The multinucleate cells, reaching up to 2 mm in diameter and containing numerous disc-shaped chloroplasts, are frequently encountered on damp mud. *Vaucheria* (Fig. 3.5) is filamentous and

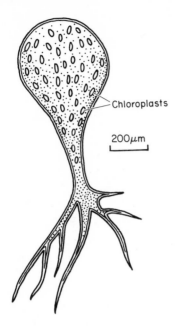

Chloroplasts

200μm

Fig. 3.4 *Botrydium granulatum.*

branched, the single cell containing numerous nuclei and chloroplasts. It is found in freshwater or on damp soil, often becoming a glasshouse pest. In terrestrial situations colourless rhizoids attach the thallus to the substratum. Some species become encrusted with calcium carbonate and contribute to the formation of tufa.

REPRODUCTION   Asexual reproduction in *Botrydium* is by heterokont zoospores generated and released when the parent plant is flooded. In dry conditions aplanospores may be produced, or even cysts, from the rhizoidal regions. *Tribonema* reproduces vegetatively by fragmentation of the filament and asexually by the production of zoospores. These are produced singly or in pairs in the parent cells and released by separation of the two halves. In place of zoospores the cells of the filament may sometimes give rise to thick-walled aplanospores. As these germinate the protoplast enlarges and causes the wall to fall into two pieces revealing its structural similarity to the walls of the filament. Asexual reproduction in *Vaucheria* occurs frequently, and involves the formation of a peculiar zoospore (Fig. 3.5a). The apical region of a filament, rich in oil droplets and plastids, is cut off by a transverse septum. The many nuclei within this segment migrate to the periphery of the protoplast, and opposite each nucleus emerges a pair of flagella. The tip of the filament becomes gelatinous, and a multiflagellate zoospore, which can be regarded as a mass of unseparated uninucleate zoospores, escapes. On coming to rest the flagella disappear, a wall is secreted, and the zoospore begins to germinate. Two or three filaments emerge from the spore, one of which usually acts as a holdfast. In some species, particularly the terrestrial,

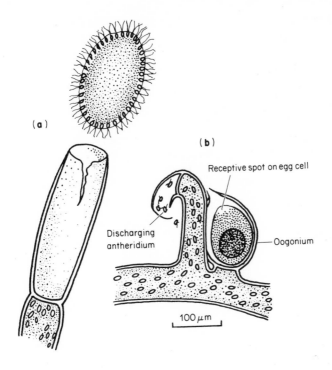

Fig. 3.5 *Vaucheria*. Reproductive organs. **(a)** Release of zoospore. **(b)** Gametangia.

aplanospores lacking flagella are produced in a similar manner. The spore is liberated by breakdown of the sporangial wall, and germination may begin before liberation is complete.

Sexual reproduction, involving isogamy, has been reported in *Tribonema* and *Botrydium*, but reaches its fullest expression in the oogamous *Vaucheria* (Fig. 3.5b). The onset of the reproductive phase is indicated by the formation of antheridia. The tip of a lateral branch, in which many nuclei, but few plastids, accumulate, is cut off by a transverse septum. The cytoplasm becomes apportioned between the nuclei, and each uninucleate protoplast then differentiates into a biflagellate gamete. Meanwhile the female organ develops, in many species next to the male on the same filament. Although the oogonium is initially multinucleate at maturity it contains a single uninucleate egg cell. When the egg is mature a short beak develops asymmetrically at the apex and the egg becomes accessible to the male gametes. Only one penetrates the egg cell, and its nucleus comes to lie by that of the egg, not fusing with it until it has swollen to an approximately equal volume. The zygote becomes surrounded by a highly impervious wall and may remain dormant for several months. Germination, which results in the formation of a new filament, is preceded by meiosis.

Since in most species fusion takes place between gametes from the same filament, sexual reproduction in *Vaucheria* provides for little more than the interpolation of a resting stage in the life cycle. A few species, however, are dioecious, allowing the possibility of genetic recombination.

The limited evolution of the Xanthophyta

If, as seems probable, the Xanthophyta are the result of a line of evolution from some ancestral motile form, parallel to that of the Chlorophyta, the problem immediately arises of why the rate of evolution has been so much slower. It is possible that relatively infrequent sexual reproduction has been a continuous feature of the Xanthophyta. This would have limited the opportunity for genetic recombination and the appearance of new forms. Biochemical features associated with the different pigmentation may also have restrained variation and evolutionary success.

Haptophyta

Habitat	Mostly marine.
Pigments	Chlorophylls *a* and *c*; β-carotene; fucoxanthin.
Food reserves	Fat, chrysolaminarin.
Cell wall components	Cellulose, hemicellulose, calcium carbonate.
Reproduction	Asexual, but life cycles little known.
Growth forms	Unicellular.
Flagella	Typically 2, mostly smooth, usually with a third (haptonema) lying between them.

The Haptophyta are distinguished by the haptonema, a flagellum-like organ of variable length. In some forms it is longer than the flagella and, although often coiled up like a proboscis (Fig. 3.6), it may also serve to attach the cells to the substratum. In transverse section the haptonema shows three concentric membranes enclosing a number of microtubules. The outer membrane is continuous with the plasmalemma and the inner pair represents a tubular extension of the endoplasmic reticulum.

The remarkable coccolithophorids belong to the Haptophyta. The walls of these organisms are closely beset with delicately sculptured plates of calcium carbonate. These are first laid down in Golgi vesicles and then excreted to the exterior. Deposits of these scales from dead organisms (coccoliths) form the principal component of calcareous rocks such as the chalk of the Cretaceous period (Fig. 3.7). It has been estimated that chalk contains up to 800×10^6 coccoliths in 1 cm^3.

Reproduction is probably largely asexual, motile stages often alternating with non-motile. Similar non-motile forms, particularly of the

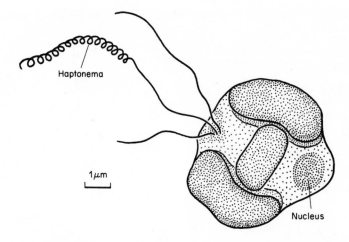

Haptonema

1 μm

Nucleus

Fig. 3.6 *Chrysochromulina* showing the haptonema between the flagella. The two parietal chloroplasts, central chrysolaminarin (leucosin) granule, and nucleus are also indicated. (After Scagel, Bandoni, Rouse, Schofield, Stein and Taylor (1966). *An Evolutionary Survey of the Plant Kingdom*. Belmont.)

coccolithophorids, often have dissimilar motile stages, a feature which makes a knowledge of the life cycle essential for classification.

Dinophyta (Pyrrophyta)

Habitat	Aquatic.
Pigments	Chlorophylls *a* and *c*; β-carotene; peridinin, other xanthophylls less prominent.
Food reserves	Starch, fat.
Cell wall components	Cellulose, hemicellulose.
Reproduction	Asexual, rarely sexual (probably isogamous).
Growth forms	Mostly unicellular, a few coccoid or filamentous.
Flagella	2, apical or lateral, in some forms absent.

The Dinophyta are predominantly unicellular planktonic[30] organisms, with walls characteristically furnished with longitudinal and transverse furrows. Of the two Classes, the Desmophyceae are notable for having cell walls consisting of two watchglass-like halves. The edges are sometimes extended as elaborate borders, possibly assisting flotation. Some species (principally tropical) emit light and are among the causes of marine luminescence.

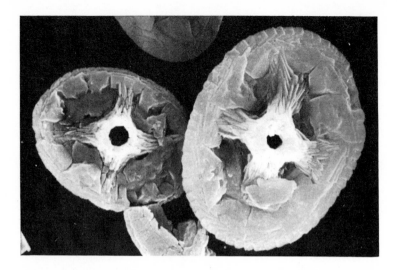

Fig. 3.7 *Eiffellithus eximus*, a coccolith from the Cretaceous. (× 6950)

In the Dinophyceae, which include the dinoflagellates, the cell walls are often reinforced with hexagonal polysaccharide plates (Fig. 3.8). The arrangement of the flagella in the dinoflagellates is unique. The attachment is lateral. One flagellum, usually smooth, is directed towards the posterior and its undulations push the cell forward. The other flagellum, usually Flimmer, lies in the transverse groove. This flagellum may serve to stabilize the cell, and improve the effectiveness of the extended flagellum. It has been observed that when the transverse flagellum is inactive the cell rotates in one direction, and when it is active in the converse.

The cells of the Dinophyta contain one or more chloroplasts, often dark brown in colour as a consequence of a large proportion of the pigment peridinin. The chloroplast envelope consists of three membranes, a feature found elsewhere only in the Euglenophyta. A stigma is often present in the motile species and in zoospores. Many species lack pigmented plastids and live heterotrophically, while others are fully pigmented and form symbiotic associations with invertebrates (e.g. *Amphidium klebsii* (which is also free-living) in the flatworm *Amphiscolops*). Some parasitic dinoflagellates are amoeboid. The nuclei of the dinoflagellates are peculiar in containing chromosomes which remain spiralized throughout the nuclear cycle and which, like those of prokaryotes, are deficient in histone proteins. Another prokaryote feature of some dinoflagellates is that during population growth DNA synthesis is continuous and shows no well defined pauses in relation to nuclear division. Many dinoflagellates on being irritated emit trichocysts. These are narrow proteinaceous needles, partly crystalline, which first appear in Golgi bodies and emerge to the exterior through fine pores. This protozoan feature is encountered sporadically amongst the unicellular flagellate algae.

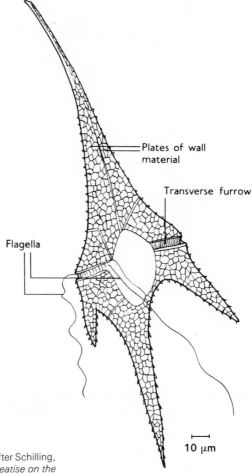

Plates of wall material

Transverse furrow

Flagella

10 μm

Fig. 3.8 *Ceratium hirundinella*. (After Schilling, from West and Fritsch (1927). *A Treatise on the British Freshwater Algae*. Cambridge University Press.)

Although motile unicellular forms predominate in the Dinophyta, a few planktonic representatives are coccoid, and in the genera *Dinothrix* (known only from marine aquaria) and *Dinoclonium* the cells, which lack flagella, are aggregated in short sparsely branched filaments.

Reproduction in the motile species is only by simple vegetative division, but zoospores are produced by the coccoid and filamentous forms. Isogamous reproduction may occur in a few species.

Bacillariophyta

The Bacillariophyta,[50] commonly known as the diatoms, are a large Division of microscopic algae with intricately sculptured, siliceous walls. The bilateral

Bacillariophyta

Habitat	Aquatic and terrestrial.
Pigments	Chlorophylls *a* and *c*; β- (and possibly ϵ-) carotene; diatoxanthin, diadinoxanthin, fucoxanthin.
Food reserves	Fat, chrysolaminarin.
Cell wall components	Hemicelluloses, silica.
Reproduction	Asexual and sexual (anisogamous and oogamous).
Growth forms	Unicellular, colonial.
Flagella	1 Flimmer (in some forms lacking the central pair of fibrils); anterior. (Present only in the male gametes of some members.)

or radial symmetry of the cells, and the regularity of the delicate markings on their walls (Fig. 3.9), make the diatoms very beautiful microscopic objects, rivalling even the desmids. The diatoms are frequent in freshwater and marine phytoplankton and are therefore of economic importance in the management of fisheries. The siliceous walls resist dissolution and decay after the death of the organism, and accumulate as fossils on beds of lakes and seas. Huge deposits of these 'diatomaceous earths' (Kieselguhr) are known from the Tertiary period, and some are mined for use as abrasives, filters and the refractory linings of furnaces.

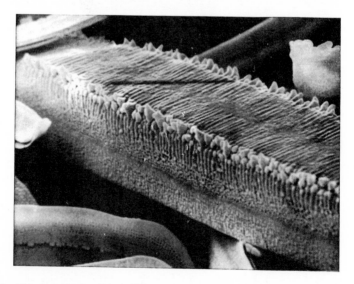

Fig. 3.9 The surface of a diatom (*Diatoma hiemale*) showing the sculpturing of the wall. ($\times$ 3335)

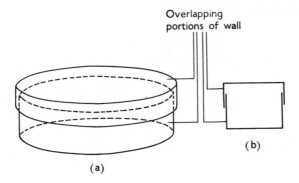

Fig. 3.10 Diagrammatic representation of the arrangement of the two halves of the wall in a diatom. **(a)** 3-dimensional view. **(b)** Transverse section.

Much study has been given to the cell wall of diatoms, and it is principally upon its characteristics that the classification rests. Although the several thousand species of diatoms occur in many different shapes, the walls of all consist basically of two parts which overlap like the halves of a Petri dish (Fig. 3.10). Consequently, the appearance of the cells from the side (girdle

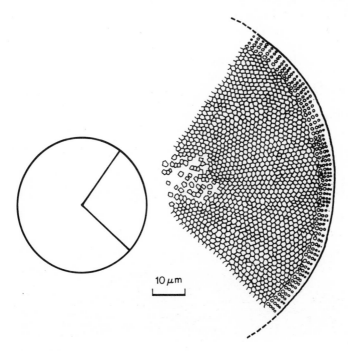

Fig. 3.11 *Coscinodiscus*. Portion of the valve (right) of a mature-sized individual showing the pattern of decoration.

view) is different from that from above (valve view). Two principal Orders, the Centrales (centric diatoms) and Pennales (pennate diatoms) are distinguished by the difference in valve view. In the Centrales the valve is circular, triangular or polygonal, and in the Pennales elongate. The decoration of the valve of the centric diatoms follows a radial pattern, while in the pennate diatoms the lines of ornamentation lie parallel and transverse to the plane of symmetry of the valve. The intersection of the plane of symmetry and the surface of the valve is usually marked by a longitudinal fissure (raphe) at which there may be contact between protoplast and medium. Elsewhere the sculpturing of the wall depends upon inequalities in the thickness of the outer siliceous layer. In some species pores may be present in the thinner regions.

Representative of the free-floating centric diatoms is *Coscinodiscus* (Fig. 3.11), widely distributed in the oceans. *Melosira*, found in both freshwater and marine plankton, is colonial. The cylindrical cells remain attached valve to valve, forming many-celled filaments (Fig. 3.12a). In *Chaetoceras*, also filamentous, the valves have spine-like outgrowths which project from the sides of the filament and possibly assist flotation. The common free-floating pennate diatom of fresh water is *Pinnularia* (Fig. 3.12b). A similar marine organism is *Pleurosigma*, whose walls are so delicately and regularly sculptured that the valves are used as specimens to test the resolution of light microscopes. Colonial forms are *Asterionella* (Fig. 3.13), frequent in Windermere, in which the cells are grouped in star-like aggregates, and the marine *Licmophora*, where the cells form tree-like colonies. Many pennate diatoms are able to perform small jerky movements, often returning irregularly to their starting point. Although the mechanism of this movement is not entirely clear, it is believed to result from jets of mucilage which issue from the raphe.

New valves forming

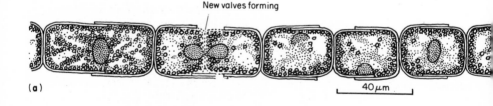

(a)   40 μm

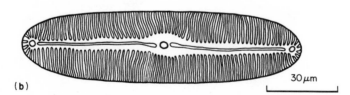

(b)   30 μm

Fig. 3.12   **(a)** *Melosira.* **(b)** *Pinnularia.* **((b)** After Esser (1976). *Kryptogamen.* Berlin.)

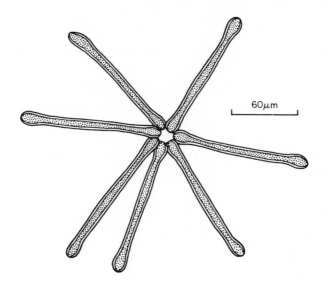

Fig. 3.13 *Asterionella.*

Within the cell the cytoplasm forms a thin layer containing one or more chloroplasts, usually brownish in colour because of accessory pigments, surrounding a large central vacuole. In the centric diatoms the nucleus is also held in this peripheral layer, but in the pennate diatoms the nucleus is suspended in the centre of the cell by a band of cytoplasm which traverses the vacuole.

REPRODUCTION   Simple asexual reproduction takes place by cell division. This involves the separation of the two halves of the wall, one half going to each daughter cell. Each daughter then immediately secretes another half wall, thus completing its envelope. Since the mature walls are quite inflexible, and each new half fits into a pre-existing one in the manner of the bottom half of a petri dish into the top, the size of the cells inevitably decreases with successive vegetative divisions. Eventually a regeneration process occurs which is normally sexual. Conditions of depleted nutrition may lead to the formation of resting spores. After two divisions, the second of which is unequal, the larger cell becomes the spore. The wall surrounding the spore is of the normal kind, but thicker.

Sexual reproduction in the centric diatoms is oogamous and meiosis occurs in the production of gametes. The antheridial cell produces four spermatozoids, each of which has a single Flimmer flagellum. In the cell giving rise to the oogonium, one daughter nucleus of the first division degenerates, and also one of the second. The remaining nucleus, together with the whole of the cytoplasm, becomes the egg. After fertilization, which may be within or without the oogonium, the zygote surrounds itself with an extensible polysaccharide wall. It then grows to as much as four times its original size, forming an *auxospore*. Normal siliceous walls are then laid down, and a renewed sequence of vegetative divisions, with progressively diminishing cell size, begins.

Reproduction of the pennate diatoms is isogamous. Two vegetative cells become held together in mucilage. The two nuclei divide meiotically, but only two products survive as gametes in each instance. The valves then open and the gametes (which lack flagella) fuse. As in the centric diatoms the zygotes grow into auxospores before the beginning of vegetative divisions. Sometimes two gametes of the same parent may fuse (autogamy), and formation of auxospores without sexual fusion (apogamy) has also been observed.

The relationships of the Bacillariophyta

Although the Bacillariophyta share basic biochemical and metabolic features with the Xanthophyta and Chrysophyta (particularly with some chrysomonads, see p. 64) they are clearly highly specialized. If a common ancestor existed, the Bacillariophyta probably diverged at an early stage. On the other hand, there are no well-established records of Palaeozoic diatoms, and they do not appear in quantity until the later Mesozoic and Tertiary eras. The earliest forms are centric, suggesting that the flagellate male gamete is a primitive feature.

Cryptophyta

Habitat	Aquatic.
Pigments	Chlorophylls *a* and *c*; α- (and rarely ϵ-) carotene; alloxanthin, other xanthophylls less prominent; biliproteins (phycoerythrin, phycocyanin).
Food reserves	Starch.
Cell wall components	Naked, but a pellicle may be present of unknown composition.
Reproduction	Asexual.
Growth forms	Flagellate, palmelloid stages occur.
Flagella	2, anterior, Flimmer, slightly unequal.

This Division includes a few species, referred to collectively as cryptomonads, occasional in freshwater and marine phytoplankton. They are unicellular and biflagellate, in some forms one of the flagella being directed towards the posterior. The flagella are inserted at one side of a depression ('gullet') lined with refractive granules (Fig. 3.14). The colour is variable depending upon the proportion of accessory pigments in the chloro-

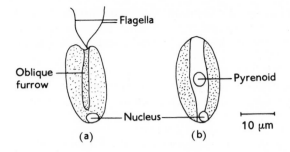

Fig. 3.14 *Cryptomonas anomala.* **(a)** Side view. **(b)** View from above. (After West and Fritsch (1927). *A Treatise on the British Freshwater Algae.* Cambridge University Press.)

plasts. The outer layer of the nuclear envelope extends as a sheet around the chloroplast ('chloroplast endoplasmic reticulum') so that each is contained in a membranous compartment. Starch is found within this compartment, but outside the chloroplast. Reproduction is solely by fission.

The cryptomonads are distinctive in respect of ultrastructure and biochemistry. In the presence of biliproteins they resemble both the Cyanophyta and Rhodophyta, and for this reason have attracted attention in relation to the evolution of the eukaryotic photosynthetic system.

Euglenophyta

Habitat	Freshwater, a few marine.
Pigments	Chlorophylls *a* and *b*; β-carotene; diadinoxanthin, other xanthophylls less prominent.
Food reserve	Paramylum.
Cell wall components	No cell wall, but a pellicle containing a helically arranged structural protein present in many.
Reproduction	Asexual, sexual doubtful, but if present isogamous.
Growth form	Predominantly unicellular, flagellate.
Flagella	2, but only 1 (rarely 2) emerging from the gullet, Flimmer, anterior.

The Euglenophyta[5] have been assigned to both the Plant and Animal Kingdoms, since the Division contains both green autotrophic and colourless heterotrophic forms. Many of the autotrophic Euglenophyta are also able to thrive in the dark if supplied with suitable metabolites, so they can be regarded as facultative heterotrophs. Races free of chloroplasts can be raised by growing cultures at 35°C, when cell division proceeds faster than chloroplast fission so that some cells eventually lack them altogether. Similar colourless races can also be produced by low concentrations of streptomycin, which inhibits chloroplast replication. It seems very likely that naturally heterotrophic species have evolved by spontaneous loss of plastids from autotrophic forms.

Although heterotrophic nutrition, a feature of animals, is common in the Euglenophyta, the evidence points to the ability to ingest food materials being a secondary feature. Heterotrophic nutrition, both facultative and obligate, is in fact known in many other algae. Examples are provided by the Cyanophyta, and by *Polytoma* (Volvocales) and *Prototheca* (Chlorococcales) in the Chlorophyta. The Euglenophyta are nevertheless outstanding in the extent to which they have developed this facility. Similarly a cell wall, absent in the Euglenophyta and animal cells generally, is by no means always present in other flagellate unicellular algae. It is lacking for example in *Dunaliella* (p. 27), otherwise similar to *Chlamydomonas*, and in *Micromonas* (p. 5). Flagellate zoospores and gametes are also commonly naked.

The mode of nutrition and presence or absence of a cell wall thus appear to be relatively plastic features at the flagellate level of organization. It should be noted however that although there is evidence that photosynthetic activity has been lost in the course of evolution, resulting in heterotrophic forms, there is no indication that it has been spontaneously acquired. Indeed it is difficult to envisage how a heterotrophic form lacking chloroplasts could generate a photosynthetic apparatus *ab initio*.

THE CYTOLOGY OF *EUGLENA*   *Euglena* is normally autotrophic, the cells containing several discoid or band-like chloroplasts (Fig. 3.15). Each has a compound envelope of three membranes, the outer of which is continuous with the endoplasmic reticulum. Some species abound in fresh water rich in

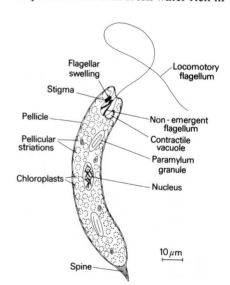

Fig. 3.15 *Euglena spirogyra.* Reconstruction of cell based on photo- and electron micrographs. (From Leedale, Meeuse and Pringsheim (1965). *Archiv fuer Mikrobiologie,* **50**, 68.)

Flagellar swelling
Locomotory flagellum
Stigma
Pellicle
Non-emergent flagellum
Pellicular striations
Contractile vacuole
Paramylum granule
Chloroplasts
Nucleus
Spine
10 μm

organic material, such as seepage from dunghills and farmyards, while others occur on damp mud by rivers, salt-marshes and similar places.

The cell is retained in a characteristic spindle-like shape by a rigid outer layer of cytoplasm termed a pellicle. Electron microscopy has shown that the pellicle contains a band-like proteinaceous component wound in several helices around the cell. This structural protein lies within the plasmalemma and is associated with microtubules. In those species in which the pellicle is flexible a flowing peristaltic movement is commonly observed, particularly if flagellar movement is constrained or the flagella are shed. The causes of this 'metabolic' movement are obscure. The chloroplasts lie towards the periphery of the less viscous cytoplasm, and the lamellae characteristically consist of three appressed thylakoids. A pyrenoid is present, but the storage product paramylum (a β-1, 3-linked glucan) is formed outside the chloroplast. The nucleus lies in or near the posterior half of the cell and chromosomes remain contracted throughout interphase, recalling the situation in the dinoflagellates. At the anterior end of the cell is a small invagination, the gullet, which in heterotrophic forms may serve for the ingestion of food. A system of vacuoles discharges at intervals into the gullet and provides for osmoregulation. There are two flagella, only one of which, normally Flimmer (Fig. 3.16), emerges through the mouth of the gullet. The base of this flagellum bears a thickening, believed to be a photoreceptor, close to the stigma in the adjacent cytoplasm.

Possibly belonging to the heterotrophic Euglenophyta is *Scytomonas*, a common intestinal parasite. Its cytology closely resembles that of *Euglena*, and paramylum and fat occur as food reserves.

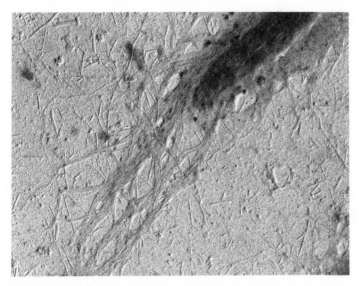

Fig. 3.16   The tip of the flagellum of *Euglena gracilis* showing the 'Flimmer' hairs. Shadowed preparation. (× 40 000)

REPRODUCTION   Binary fission is the common method of reproduction of the Euglenophyta. By the end of nuclear division the locomotor apparatus is replicated and cleavage of the whole cell proceeds from the anterior to the posterior end. Nuclear division resembling meiosis has been reported, but well-established instances of sexual reproduction in the Euglenophyta are lacking.

Other forms of Euglenophyta

Although predominantly flagellate, a few Euglenophyta are encapsulated and form dendroid colonies. These can be regarded as the result of developmental trends from the flagellate state which parallel those in other algal groups, but which have reached only a rudimentary level of morphological complexity.

The relationships of the Euglenophyta with other algae are obscure. In the presence of chlorophylls *a* and *b* they resemble the Chlorophyta, and this perhaps indicates a common origin in the remote past.

Phaeophyta

Habitat	Predominantly marine.
Pigments	Chlorophylls *a* and *c*; β-carotene; violaxanthin, fucoxanthin.
Food reserves	Polyols (e.g. mannitol), laminarin.
Cell wall components	Cellulose, hemicellulose, sulphated polysaccharides.
Reproduction	Asexual and sexual (oogamous).
Growth forms	Filamentous, parenchymatous.
Flagella	2 unequal, 1 smooth and 1 Flimmer (sometimes spiny); anterior or lateral.

Of the many genera of the Phaeophyta (brown algae), only three rare ones are freshwater, the remainder being seaweeds whose macroscopic flattened thalli are familiar inhabitants of the intertidal regions of rocky coastlines. Other genera inhabit the region just beneath the low-tide mark, while a few are found solely in mid-ocean.

The vegetative organization of the Phaeophyta surpasses that of any of the algae so far considered. The simplest thallus, consisting of branching filaments, is heterotrichous, and thus resembles the most complex found in the Chlorophyta. This morphologically advanced state is also reflected in the manner of reproduction. In the Phaeophyta oogamy is the general rule, and the alternation of phases in the life cycle is developed to the point where gametophyte and sporophyte begin to diverge morphologically.

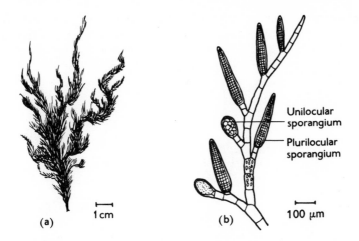

Unilocular
sporangium

Plurilocular
sporangium

1 cm

100 μm

(a)

(b)

Fig. 3.17 *Ectocarpus confervoides.* **(a)** Habit. **(b)** Sporangia. (After Newton (1937). *A Handbook of the British Seaweeds.* British Museum Publications.)

The Phaeophyta are divided into nine Orders, but representatives of only the five commoner ones will be considered here. They will, nevertheless, fully illustrate the probable evolutionary trends in the development of the morphology and of the reproductive cycles of the brown algae.

Ectocarpales

Species of *Ectocarpus* (Fig. 3.17), mostly small heterotrichous plants resembling in habit the green alga *Cladophora* (p. 38), are common along the Atlantic coast of America and in the colder seas of the northern hemisphere. Around ports and docks some forms have become tolerant of metal pollution, particularly of copper. Cell division in the erect filaments is limited to well-defined intercalary regions. This method of growth, frequent in the Ectocarpales, is termed **trichothallic**. The common *Pylaiella* is very similar to *Ectocarpus* but tidal action often causes the axes to roll together, forming a characteristic cable-like growth.

Other members of the Order are more complex. *Ascocyclus*, for example, has an elaborate prostrate system, and in this respect resembles the green alga *Coleochaete*. The gelatinous cushion-like thallus of *Leathesia*, although appearing parenchymatous, is formed by the adhesion of filaments. A truly parenchymatous condition is reached in *Punctaria* (Fig. 3.18). Here cell division takes place in various directions and results in a leaf-like thallus closely resembling *Ulva*.

The cells of *Ectocarpus* contain only a single ribbon-like chloroplast with many pyrenoids, but in other genera several chloroplasts may be present.

The life cycle of Ectocarpus

Ectocarpus exists in two forms, haploid and diploid. In favourable environ-

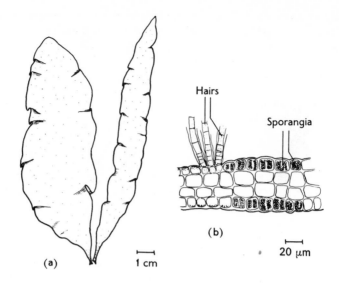

Fig. 3.18 *Punctaria latifolia.* **(a)** Habit. **(b)** Transverse section of thallus with sporangia. (After Newton (1937). *A Handbook of the British Seaweeds.* British Museum Publications.)

mental conditions the two phases, except in respect of the reproductive organs, are morphologically identical, but in colder regions the haploid plant may develop little, or even not at all. Reproduction of the diploid plant takes place solely by zoospores, but these are produced in two kinds of sporangia, termed respectively plurilocular and unilocular (Fig. 3.17b). The *plurilocular* sporangia arise from cells which undergo repeated transverse and subsequently longitudinal division, giving rise to spindle-shaped groups of small, more or less cubical compartments, each of which at maturity contains a single biflagellate zoospore. The plurilocular sporangia are commonly either lateral, or terminate in short lateral branches. Dehiscence is by an apical pore. Dissolution of the partitions allows the zoospores to escape. After a motile phase, the zoospores settle and yield diploid plants identical with their parent.

The cytology of the developing *unilocular* sporangium is more complex. The nucleus of the initial cell is conspicuously large, and its first divisions are probably meiotic. Although the cytoplasm becomes multinucleate, no walls are formed, and it is only at maturity that the protoplast becomes divided into uninucleate portions. Each portion differentiates into a biflagellate zoospore, typically bean-shaped. All are eventually released at an apical pore. Although these zoospores behave as those produced in plurilocular sporangia, they yield only haploid plants. Where the haploid plant is absent, as in the Irish Sea, the products of unilocular sporangia function as gametes.

The haploid plants bear only plurilocular sporangia. These produce motile cells which behave either as zoospores, reproducing the haploid phase asexually, or as gametes. The zygote germinates without meiosis or production of zoospores, and grows directly into a diploid plant.

Isogamy predominates in *Ectocarpus*, but anisogamy occurs in more complex forms. A pheromone has been demonstrated in *Ectocarpus* and has been identified chemically as an unsaturated hydrocarbon (ectocarpen). In some epiphytic species of *Ectocarpus* and *Pylaiella* the gametophytes and sporophytes, although similar, grow on different plants. The sporophytes of *Pylaiella littoralis*, for example, grow on *Fucus* and the gametophytes on *Ascophyllum*.

Representative of the small Order Sphacelariales, close to the Ectocarpales, is *Sphacelaria*, a heterotrichous alga of southern oceans. The upright filaments terminate in an apical cell conspicuous for its large size and brown pigmentation.

Cutleriales

Cutleria (Fig. 3.19), representative of this small Order, is an heterotrichous form of the Mediterranean region. There is a well developed life cycle, but only one component of the heterotrichous system is fully represented in each phase, so the cycle is markedly heteromorphic.

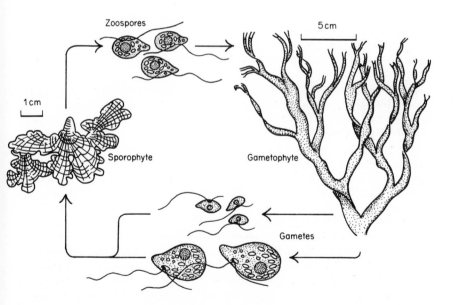

Fig. 3.19 *Cutleria multifida.* Life cycle, showing both anisogamy and the striking morphological difference between gametophyte and sporophyte. The gametophyte, although only one is shown, is dioecious. There is segregation for sex in the formation of zoospores. Gametes and zoospores diagrammatic showing relative sizes. (After Esser (1976). *Kryptogamen.* Berlin.)

The sporophyte, which displays only the prostrate component, grows as an incrustation on rocks, and was originally regarded as a distinct genus, *Aglaozonia*. The unilocular sporangia, in which, as usual, meiosis occurs, are superficial, and yield haploid biflagellate zoospores. On settling, these become attached to the substratum by a sucker-like holdfast.

The gametophytes, which arise from the attached zoospores, consist principally of an erect filamentous system. The filaments lie parallel and adhere to each other for much of their length, forming a pseudoparenchymatous thallus. For a short distance at the apex, however, the filaments are free, and here there is distinct trichothallic growth. The gametophytic phase of the Mediterranean *C. multifida* is dioecious (Fig. 3.19). The male and female gametes are both biflagellate and motile, but the male are minute compared with the female.

Laminariales

This Order, the kelps, includes *Macrocystis* and *Nereocystis*, the largest known algae, together with a number of species harvested commercially as sources of the mucopolysaccharide algin. This yields alginic acid, widely used as an emulsifying agent in the food and paint industries, and in the processing of rubber. *Macrocystis* off the coast of Southern Australia grows by as much as 1 m a day, probably the fastest growth rate in the Plant Kingdom.

The Laminariales are usually found below low water mark, but a few are regularly exposed at low tide. An example of the latter is the striking *Postelsia* of the Pacific coast of North America, a species with an arboreal habit, suggestive of a miniature coconut palm.

VEGETATIVE STRUCTURE   In *Laminaria*, which may be taken as representative of the Order, the large thallus is differentiated into holdfast, stipe and blade (Fig. 3.20a). Abrasion by tides and turbulence continually wears away the end of the blade, but the loss is made good by continued growth from a meristematic region at the base. A transverse section of the tough, pliable stipe reveals three anatomically distinct zones. Chloroplasts are confined to an outer zone (meristoderm), covered by a layer of protective mucilage. Within the meristoderm, in which some cell division persists indefinitely, is a zone of paler, elongated cells forming a well-defined cortex. Towards the centre the longitudinal walls become increasingly gelatinous, and at the centre itself is a mucilaginous matrix (medulla) containing inter-twined and branching filaments. The innermost part of the cortex and the medulla also contain columns of elongated cells broadened at each end, and hence referred to as 'trumpet hyphae'. The transverse walls of these cells are often perforated by groups of pits, recalling the sieve plates of higher plants. The stipes of some other genera of the Laminariales contain columns of cells showing an even closer resemblance to authentic sieve tubes. Like these they contain callose, and there is experimental evidence of similarity of function.

REPRODUCTION   Reproductive areas, referred to as sori, develop on the

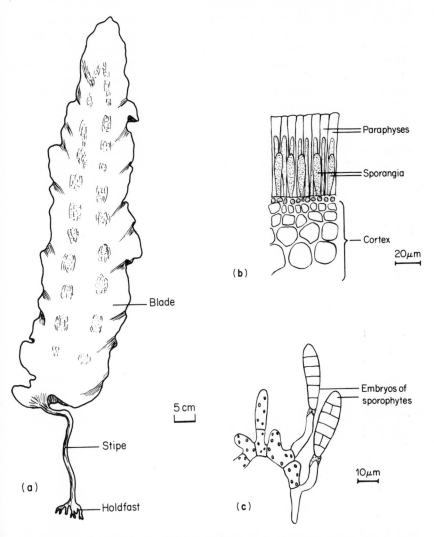

Fig. 3.20 *Laminaria.* **(a)** *L. saccharina.* Habit. **(b)** *L. cloustoni.* Transverse section of fertile region of lamina. (After Newton (1937). *A Handbook of the British Seaweeds.* British Museum Publications.) **(c)** *L. digitata.* Portion of female gametophyte. (After Sauvageau, from Fritsch (1945). *The Structure and Reproduction of the Algae; II.* Cambridge University Press.)

blades at certain times of the year. The sori consist of many unilocular sporangia interspersed with thick, sterile, protective hairs *(paraphyses)* (Fig. 3.20b). Meiosis occurs in the development of the sporangia, and they eventually yield haploid zoospores. These in turn develop into haploid gametophytic plants, much smaller than, and totally different in morphology from, the highly organized sporophytes (Fig. 3.20c). The gametophytes are dioecious, and, since it has been shown in a number of instances that spores giving rise to male and female gametophytes are produced in equal numbers in a sporangium, it appears that sex determination is genotypic.

Although the gametophytes show a tendency towards heterotrichous growth, any cell seems capable of yielding a gametangium. Oogamy is fully developed, the oogonium producing a single egg which escapes at maturity through a pore at the apex of the cell. The egg, however, does not become free, but remains seated in a cup formed by the thickened margins of the pore. The male plant produces a number of terminal antheridia, from each of which is liberated a single spermatozoid with two lateral flagella. A substance has been isolated from female gametophytes which stimulates the release of spermatozoids and subsequently acts as a pheromone. Following fertilization, the zygote secretes an external membrane and develops into a new sporophyte without any resting period. The young embryo may remain attached to the female gametophyte for a short period, but it is doubtful whether this represents anything more than purely physical adhesion.

There is no specialized asexual reproduction of either the haploid or diploid generations.

Fucales

The Fucales, the various species of which are known as 'wracks', are probably the most familiar of all seaweeds, particularly in the British Isles. The intertidal (littoral) regions of rocky shores often show a number of distinct horizontal bands, each consisting of an almost pure stand of a member of the Fucales. *Fucus* (Fig. 3.21), *Pelvetia* and *Ascophyllum* are genera frequent in these habitats. Desiccation of the thalli during exposure to air is prevented by the secretion of mucilage, and photosynthesis probably continues during low tide. A few Fucales grow in deeper water, and some (e.g. *Sargassum*) are free-floating in the warmer regions of the great oceans. Air bladders in which the proportions of oxygen and nitrogen often differ from those of normal air are frequently present in the thalli of the Fucales.

The thallus of *Fucus* is a much smaller structure than that of *Laminaria*, rarely exceeding 50 cm in length, but the differentiation into holdfast and blade is still evident. The flattened thallus is dichotomously branched and grows from apical meristems. A distinct midrib usually lies at the centre of each segment, branching in register with the thallus. The blade is generally worn away by the action of the sea until, in the region of the holdfast, only the midrib remains, giving the impression of a short stipe (Fig. 3.21).

REPRODUCTION   Both monoecious and dioecious species occur. In both, the reproductive structures form at the apices and these in turn become swollen with mucilage.

Microscopic examination reveals flask-shaped invaginations *(conceptacles)* in this swollen region, some of which contain female gametangia and others male, both interspersed with paraphyses (Figs 3.22 and

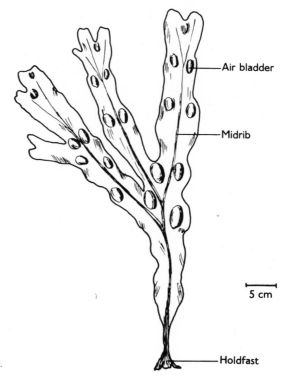

Air bladder

Midrib

5 cm

Fig. 3.21 *Fucus vesiculosus.*

Holdfast

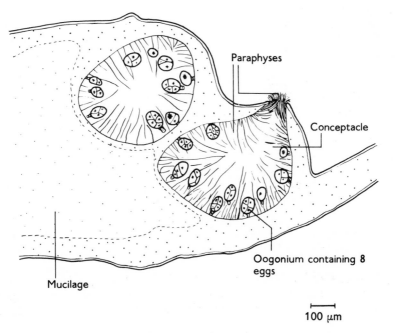

Paraphyses

Conceptacle

Oogonium containing 8 eggs

Mucilage

100 μm

Fig. 3.22 *Fucus vesiculosus.* Transverse section of thallus with female conceptacles.

3.23). Meiosis occurs during gametogenesis. The gametes are liberated at high tide.

As in the Laminariales, oogamy is fully developed but in *Fucus* the eight eggs produced in an oogonium become quite free and drift passively in the sea. The spermatozoids, each of which has two lateral flagella, are attracted to the eggs. The pheromone, named fucoserraten, is a hydrocarbon. Although hybrids have been described, fertilization in experimental conditions is species-specific and appears to depend on a glycoprotein recognition system of the surface of egg and spermatozoid. Eggs treated with fucose-binding lectins, for example, are fertilized far less readily, presumably because of masking of the binding sites.

Following fertilization the zygote continues to drift and secretes a mucilaginous envelope. It eventually settles, becomes anchored by the mucilage, and germinates. The first division of the zygote establishes the polarity of all subsequent growth, since one daughter cell gives rise to the rhizoid (and ultimately the holdfast) and the other to the blade. It has been found experimentally that if the zygote is allowed to germinate in an environment that is not uniform, such as in a gradient of light, temperature or hydrogen ion concentration, the dividing wall always forms transversely to

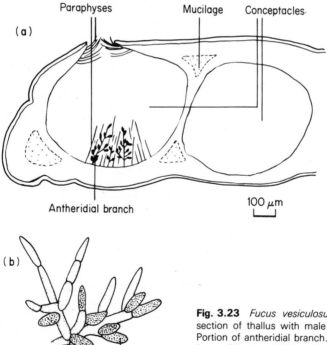

Fig. 3.23 *Fucus vesiculosus.* **(a)** Transverse section of thallus with male conceptacles. **(b)** Portion of antheridial branch. (**(b)** After Thuret, from Fritsch (1945). *The Structure and Reproduction of the Algae*, II. Cambridge University Press.)

the direction of the gradient. The subsequent behaviour of the daughter cells depends upon the nature of the gradient. With a gradient of temperature, for example, the cell on the warmer side yields the rhizoid. In Nature, of the two cells formed in the first division of the zygote, the rhizoid usually develops from that in greater contact with the substratum. It seems likely that the orientation of this first division under natural conditions is determined by a combination of the environmental gradients that have been found effective in influencing the polarity in laboratory cultures.

Once polarity is established, an electric current passes through the embryo parallel to the axis, the rhizoid pole becoming increasingly negative. This current naturally influences the distribution of ions and charged molecules within the embryo, and may promote regional differentiation. In *Pelvetia* (closely allied to *Fucus*) calcium ions move into the tip of the growing rhizoid.

Two different interpretations have been made of sexual reproduction in *Fucus*. One brings it into line with that of the Laminariales, interpreting the gametangia as homologous with unilocular sporangia, the gametophyte being reduced to nothing more than a gamete. The other draws an analogy with sexual reproduction in animals such as Man, where there is no question of an independent haploid phase.

There is no specialized asexual reproduction in the Fucales, but fragments of thalli may regenerate in favourable conditions to yield independent plants. The free-floating species of *Sargassum* reproduce solely in this manner.

Dictyotales

The thallus of *Dictyota*, a widely distributed genus, is flattened and dichotomously branched, each branch growing from a conspicuous apical cell (Fig. 3.24). The thallus is only three cells thick, lacking a midrib. The two outer layers of cells are assimilatory, and the central, consisting of large cells, may act as a storage region.

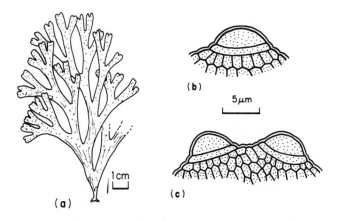

Fig. 3.24 *Dictyota.* **(a)** Habit. **(b)** Apical cell. **(c)** Incipient dichotomy of the thallus.

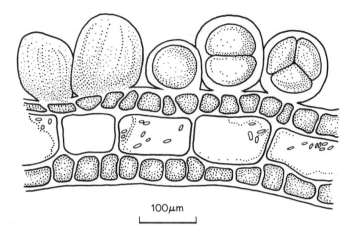

100μm

Fig. 3.25   *Dictyota*. Sporophyte producing tetrasporangia.

REPRODUCTION   Reproduction involves an isomorphic alternation of generations. Unilocular sporangia arise scattered or in groups over the surface of the diploid plant, and each yields, as a consequence of meiosis, four non-flagellate **tetraspores** (Fig. 3.25), a feature which distinguishes the Dictyotales from all other brown algae. During meiosis in *Dictyota* there is a 1:1 segregation for sex, and consequently male and female gametophytes are present in approximately equal numbers. In other genera the gametophytic phase is commonly monoecious.

The oogonia and antheridia are produced in groups (*sori*) on the surface of the thallus. Each oogonium produces a single egg which, like that of *Fucus*, drifts passively in the water when released. The antheridia are plurilocular, and the sorus is surrounded by a ring of sterile cells. The spermatozoids (antherozoids) have a single lateral flagellum (Flimmer), but electron microscopy has shown that a second is present as a rudiment enclosed in the cytoplasm, recalling the situation in some uniflagellate Chrysophyta (see p. 63). Spermatozoids are attracted to the egg by a pheromone, now known to be a complex hydrocarbon similar to ectocarpen (see p. 85). The zygote germinates very soon after fertilization.

There is no specialized asexual reproduction of either phase of the life cycle.

The relationships of the Phaeophyta

The Phaeophyta are a circumscribed Division of the algae and little can be said of their wider relationship since no forms are known simpler than the Ectocarpales. They presumably arose from some primitive flagellate ancestor, and proceeded to exploit a particular kind of pigmentation and metabolism that proved especially satisfactory in marine environments.

There is no evidence that the Phaeophyta have ever contained forms becoming adapted, as some of the Chlorophyta, to terrestrial life. The few varieties of the Fucales (e.g. *Fucus vesiculosus* var. *muscoides*), all of limited reproductive capacity, which occur in salt-marshes mixed with halophytic flowering plants appear to be instances of specialization without any far-reaching significance. It is noteworthy that photosynthesis in *Laminaria*, which often dominates a canopy, becomes saturated only at high irradiances. This is a characteristic of 'sun plants' of land vegetation, and is perhaps the consequence of an analogous specialization of the photosynthetic system.

Rhodophyta

Habitat	Aquatic (mainly marine).
Pigments	Chlorophylls *a* and *d*; α- and β-carotene, lutein, zeaxanthin; biliproteins (phycoerythrin, phycocyanin).
Food reserves	Floridean starch, compounds of sugars and glycerols.
Cell wall components	Cellulose, hemicelluloses, sulphated polysaccharides.
Reproduction	Asexual and sexual (oogamous).
Growth forms	Unicellular, filamentous, pseudoparenchymatous.
Flagella	None.

The Rhodophyta (red algae)[12] are another circumscribed group, usually easily recognized by their bright pink colour caused by the biliproteins phycoerythrin and phycocyanin. The chloroplasts are discoid or lobed and simple in structure. The thylakoids usually lie parallel, but they are not stacked, and in some forms the biliproteins are present as phycobilisome particles. These are features of the protoplasts of blue-green algae (p. 16). Floridean starch, which appears outside the chloroplasts, is chemically similar to the amylopectin of higher plant starches.

Although some species frequent rock pools, the majority live in the deep waters of warm seas, and are most commonly seen when washed up on beaches. The ability to live in the ocean depth depends on the presence of the biliproteins. Light reaching these regions lies principally at the middle of the visible spectrum, and this coincides with the maximum absorption of phycoerythrin. The energy is immediately transmitted to the chlorophyll which, at these wavelengths, experiences little direct excitation. The bright green colour of some freshwater red algae probably results from the photo-destruction of phycoerythrin in higher irradiances. An organism of this kind is the unicellular *Cyanidium* which on biochemical and structural features is probably correctly placed with the red algae. It thrives in waters issuing from volcanic springs which are often hot to the touch and may be as acid as pH 2. Some species of tropical seas develop a calcareous exoskeleton and contribute to the formation of coral reefs. A number of species, often showing reduced pigmentation, may be true parasites on parenchymatous brown algae and larger members of their own Division.

A few red algae are of economic importance. *Porphyra* (Fig. 3.26), for example, has long been used in Europe ('laver bread') and in the Far East ('nori') as a foodstuff. *Chondrus crispus* ('carrageen') is similarly used to a lesser extent in Europe. *Gelidium*, particularly from the Pacific, is the principal source of agar. The walls of the red algae in general are notably mucilaginous.

The Rhodophyta fall into the Bangioideae and the Florideae. Unicellular forms are found only in the Bangioideae, and this group also display simpler sexual reproduction. There are also cytological differences, one of the most conspicuous being the presence of pit connections between the cells in the Florideae. There are no motile forms. Even in sexual reproduction, which is uniformly oogamous, the dispersal of the male gamete is entirely passive.

Bangioideae

Apart from the unicellular forms, the simplest Bangioideae, mostly epiphytes of other marine algae, are heterotrichous with intercalary growth. *Porphyra* has a sheet-like thallus resembling *Ulva* (p. 35), and similarly one cell thick. It nevertheless develops from a filament in which the cells divide transversely and longitudinally, the divisions being confined to one plane.

REPRODUCTION   Asexual reproduction is by means of *monospores*. These are formed by simple division of vegetative cells (Fig. 3.26b) and are liberated from the surface or margins of the thallus.

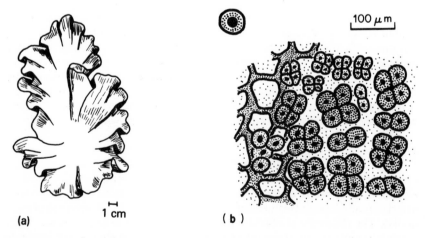

100 μm

(a)   1 cm   (b)

Fig. 3.26   *Porphyra*. **(a)** Habit. **(b)** Transverse section of portion of thallus producing clusters of monospores. (**(b)** After Esser (1976). *Kryptogamen*. Berlin.)

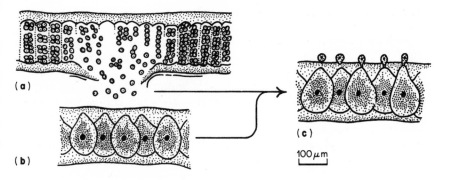

Fig. 3.27 *Porphyra.* **(a)** Portion of thallus producing spermatia. **(b)** Thallus with oogonia. **(c)** Spermatia beginning to penetrate oogonia. (All after Esser (1976). *Kryptogamen*. Berlin.)

Details of sexual reproduction in the Bangioideae are known for only a few species. In *Porphyra* male and female gametangia arise from single vegetative cells. The antheridium (spermatangium) is formed by the repeated division of a cell into 64 or 128 portions, each of which yields a male gamete (*spermatium*) (Fig. 3.27a). The oogonium (often in the Rhodophyta termed a *carpogonium*) is derived by the enlargement of a single cell, but no division occurs and only a single egg is present (Fig. 3.27b). The coming together of the spermatium and carpogonium appears to depend entirely upon water currents, but, having approached, a spermatium is then retained by the mucilage surrounding the carpogonium. Fertilization takes place by the protoplast of the spermatium first putting forth a narrow process which penetrates, possibly by means of localized lysis, the membrane of the carpogonium (Fig. 3.27c). The body of the spermatium then passes entirely into the female cell.

After fertilization, the zygote divides and releases several carpospores, each of which becomes amoeboid for a while before secreting a cell wall and germinating into a new plant. The site of meiosis in the life cycle is still uncertain. Reduction may occur at germination of the zygote, but in *Porphyra* there is also evidence for the existence of a diploid filamentous phase (*Conchocelis*) found growing on the inside of oyster and mussel shells. No similar *Conchocelis* stage is known for the freshwater *Bangia*. The life cycle thus shows little or no development of the sporophytic phase.

Florideae

Representative of the simpler Florideae is *Batrachospermum*, one of the few Rhodophyta of fresh water. The vegetative organization, an axis bearing whorls of branches, is very similar to that of the green alga *Draparnaldiopsis* (p. 41). The filaments are enveloped in copious mucilage; this together with the dark pigmentation of the cells causes colonies of the alga *in situ* in ponds and streams superficially to resemble masses of frogs' spawn. *Nemalion*

(Fig. 3.28), common in the intertidal zones of North Temperate shores, is basically filamentous, but the central filaments, which lie parallel and adhere to each other, are surrounded by an investment of short laterals enveloped in mucilage. The whole thus acquires a branching pseudoparenchymatous structure.

An example of the advanced Florideae is *Polysiphonia*, common in littoral vegetation throughout the World. The thallus, which consists of a central axis bearing freely branching laterals (Fig. 3.29a), shows well defined apical growth. This results from a dome-shaped apical cell surmounting a central column of cells discernible throughout the thallus. This axial column is surrounded from well below the summit by numerous columns of other cells produced by oblique divisions in the apical region. The structure is thus pseudoparenchymatous in origin, although in parts, especially at the nodes, it is hardly distinguishable from a truly parenchymatous condition. Even more complicated internal organization is found within those species with multi-axial thalli.

REPRODUCTION   Asexual reproduction in the Florideae takes place by a variety of spores, including monospores, all non-motile and in many instances known only in conditions of pure culture. A curious feature of the epiphytic *Centroceras* is the development of lateral branches of 4–5 cells

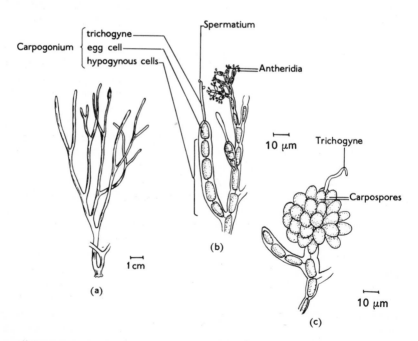

Fig. 3.28 *Nemalion multifidum.* **(a)** Habit. **(b)** Fertile lateral branch. **(c)** Formation of carpospores after fertilization. (After Newton (1937). *A Handbook of the British Seaweeds.* British Museum Publications.)

which are regularly abscised, and serve as a means of vegetative reproduction.

Even in the simpler Florideae, sexual reproduction displays a number of peculiar features not encountered in other Divisions of the algae. The male and female gametangia, in many species produced on separate plants, arise on specialized side branches (Fig. 3.28b). The antheridia are budded off from mother cells lying at the tips of short tufted branches; as many as four or five antheridia may come from one mother cell. Each antheridium liberates a single spermatium through an apical slit. The carpogonium, surmounted by a long tubular, hair-like process (the trichogyne) terminates a short carpogonial branch. When a spermatium makes contact with a trichogyne above an unfertilized egg, it becomes attached, and the intervening cell walls break down, allowing the contents of the spermatium to pass into the carpogonium. After fertilization a mucilaginous plug is secreted at the base of the trichogyne. The zygote germinates almost immediately and becomes surrounded by a tuft of short-branching filaments, often with brownish pigmentation. Carpospores are cut off from the ends of these filaments (Fig. 3.28c). In *Batrachospermum* the carpospores germinate to form a small heterotrichous *Chantransia*-stage, which may reproduce itself by monospores. Although the site of meiosis in many forms is still obscure, in *Batrachospermum* it is said to occur sporadically in cells of the *Chantransia*-stage. The cells so formed, each with a haploid nucleus, grow directly into new gametophytes. The cycle is thus diplohaplontic, but the sporophytic phase, even in *Batrachospermum*, is relatively inconspicuous.

In the advanced Florideae the gametophytes initiate sexual reproduction much as in *Batrachospermum* and *Nemalion*. Antheridia are produced in the middle region of short filaments. When mature the filaments become club-shaped with the antheridia, almost colourless and with refractive walls, densely packed at the surface.

Development of the female filament is more complex. Although initially a single file of cells, all but the tip becomes multicellular. One of the peripheral cells in the central region gives rise to a carpogonial branch. The basal cells divide sparingly giving rise to a few auxiliary cells all lying close to the carpogonium. During the development of this carpogonial branch adjacent peripheral cells divide and form an urn-shaped sheath which grows up and ultimately encloses the carpogonial branch. Only the trichogyne projects through the apical orifice. This whole structure constitutes the ***procarp***.

Following fertilization the trichogyne degenerates and the zygote makes a pit-like connection with an adjacent auxiliary cell. The diploid nucleus then migrates into this cell and its original haploid nucleus apparently degenerates. The diploid nucleus divides mitotically and forms a multicellular filamentous carposporophyte. Simultaneously, evidently stimulated by fertilization, the procarp grows into the mature ***cystocarp*** which completely encloses the carposporophyte. Ultimately the terminal cells of the carposporophyte become carposporangia. Each liberates a single densely pigmented carpospore. These escape through the apical orifice (ostiole) of the cystocarp (Fig. 3.29b).

The diploid carpospores germinate directly and give rise to plants quite similar to the gametophyte. Segments towards apices of the branches become fertile. A peripheral cell undergoes a number of divisions leading to a structure not unlike a carpogonial branch. In this branch one or more of the cells give rise to tetrasporangia. A tetrasporangium is initially uninucleate, but the nucleus undergoes meiosis and four tetraspores are formed. These escape by rupture of the sporangium (Fig. 3.29c) and germinate to form normal gametophytic plants. The cycle is thus isomorphic with the interpolation of a curious additional sporophytic phase before the tetrasporophyte.

In *Griffithsia* vegetative cells of male and female plants have been fused experimentally. The resulting binucleate heterokaryons, analogous to those of basidiomycete fungi, generate plants with diploid (tetrasporangiate) morphology showing that actual nuclear fusion is not essential for the new form of growth.

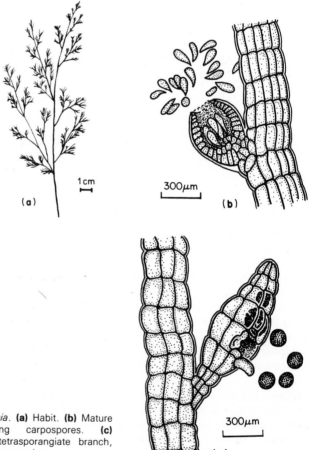

1 cm

300 μm

300 μm

(a)

(b)

(c)

Fig. 3.29 *Polysiphonia.* **(a)** Habit. **(b)** Mature cystocarp discharging carpospores. **(c)** Sporophyte with a tetrasporangiate branch, some of the tetraspores escaping.

The relationships of the Rhodophyta

As with the Phaeophyta, the assessment of the wider relationships of the Rhodophyta is rendered difficult by the absence of all but a few simple representatives of the Division. Although motile forms are generally regarded as absent, some have considered that puzzling organisms such as *Glaucocystis*, often interpreted as composite cells incorporating a blue-green alga as a symbiont, are in fact representative of a group of flagellates closely allied to the red algae. The similarities, both biochemical and structural, between the chloroplasts of the Rhodophyta and the photosynthetic cells of the Cyanophyta are certainly striking, and possibily indicate the origin of the Rhodophyta in some such source. A detailed examination at ultrastructural level of the chloroplasts during gametogenesis, when they either dedifferentiate or may even be eliminated, may give clues to their wider relationships.

There is an undoubted similarity between the reproductive process in *Polysiphonia* and that in certain fungi, but, if this resemblance is anything more than coincidental, it seems more likely that the fungi, being heterotrophic, are the derived forms.

The calcareous coral-forming Rhodophyta (e.g. *Lithothamnion*) have a fossil record that extends back to the Cretaceous. All these species belong to the more advanced Florideae.

The evolutionary trends within the algae

From the foregoing survey of the principal features and interrelationships of the algae, we can now proceed to a consideration of the evolutionary and morphological trends displayed by the algal kind of organization as a whole.

The aquatic habitat and evolutionary change

A point of general significance in relation to the evolution of the algae arises from their predominantly aquatic habitat. One of the main factors influencing evolution within the Plant Kindgom has undoubtedly been environmental change. Nevertheless, aquatic plants, particularly those that are marine, are to some extent protected from such change, at least in a catastrophic form. The volume of the sea in particular is so vast that changes in such features as salinity and temperature must of necessity be gradual. Algae, therefore, have exploited to an extent greater than that of any other component of the World's vegetation an environment which demands only comparatively slow adaptation to changing conditions. This possibly accounts for the persistence of numerous states of algal organization intermediate between the simplest unicellular and the complex multicellular, and of simple isomorphic life cycles (as in *Ulva* and *Ectocarpus*) without spores or zygotes suited to withstand unfavourable periods.

Closer inspection of individual groups reveals, of course, that continuity of structure and pattern is to some extent illusory. The Charales among the

Chlorophyta and the Florideae among the Rhodophyta provide examples of the numerous groups of living algae which lack close relatives indicating the paths along which they may have evolved.

The antiquity of the algae

Geological evidence undoubtedly points to the algae being an extremely ancient form of life. Calcareous nodules (*stromatolites*) structurally similar to those produced by some living blue-green algae (e.g. *Lithomyxa*), are known from beds of pre-Cambrian age, possibly over 2500 million years old. Filamentous remains, containing organic matter and suggestive of blue-green or even green algae, have been found in pre-Cambrian rocks not less than 2000 million years old. Discoveries of this kind, which are more frequently substantiated than contradicted as palaeobotanists investigate the older rocks, indicate that an algal form of life evolved very early. The fossil record, although admittedly still very fragmentary, also supports the view that simple unicellular and filamentous algae preceded the complex pseudoparenchymatous and parenchymatous forms we have encountered in the three major divisions: Chlorophyta, Phaeophyta and Rhodophyta.

The evolution of the vegetative thallus

If the increase in morphological complexity from flagellate unicell, through coccoid, filamentous and pseudoparenchymatous states to large parenchymatous forms such as *Macrocystis* is an evolutionary progression, the problem remains of what has caused and directed this progression. The cause presumably lies in the mutability characteristic of all life, and the direction is no doubt a consequence of natural selection. The nature of selection in an aquatic environment, with its uniformity in space and time, is however little understood. This is particularly true of the marine environment, and of planktonic algae. Consequently discussion of algal evolution involves considerable conjecture.

Nevertheless, reasoned speculation is not to be discouraged, and algae undoubtedly show many features of evolutionary significance which merit consideration. Amongst the flagellate forms, for example, it is striking that the development of motile colonies and aggregate organisms has not proceeded beyond *Volvox*. Presumably, with a diameter exceeding about 1 mm the *Volvox* system would become physically unstable, and the coordination of the thallus impossible. If the earliest forms were indeed flagellate, the evolution of a sedentary form from a motile would result in the energy that would otherwise be expended in swimming becoming available for growth and reproduction. The enhanced reproductive capacity would result in proportionately greater numbers, and provide the opportunity for the establishment of a line of sedentary organisms.

The tendency for daughter cells to remain united appears to be a basic one, and would account for the occurrence of some kind of colonial or multicellular forms in all the divisions of the algae. In the major divisions we must assume that the advantages of association, possibly again residing in metabolic and reproductive efficiency, led to the elaboration of multicellular thalli in which there gradually appeared divisions of labour amongst the cells. In a multicellular form such as *Ulothrix*, all the cells, except possibly the basal anchor, divide and liberate reproductive bodies simultaneously. There is no somatic tissue, and the individual is destroyed in reproduction. A form in which the individual persists through several reproductive phases clearly has an advantage in a situation (as might arise with prolonged turbulence) where conditions become temporarily intolerable for the reproductive bodies, but remain tolerable for the parents.

The evolution of complex thalli in which reproduction was confined to special areas or branch systems would thus be favoured. The further opportunity would then arise for various parts of the somatic tissue to become specialized, and assist either in the support or protection of the reproductive structures, as in the Florideae and many other algae, or in the exploitation of particular habitats, as with the air bladders of many Fucales.

The evolution of sexual reproduction

With regard to sexual reproduction, it seems beyond doubt that oogamy has evolved from isogamy. With forms inhabiting moving water, isogamous reproduction must be extremely wasteful, and there are evident advantages if one gamete remains relatively stationary, especially if it secretes chemotactic pheromones causing the male gamete to accumulate around it. Moreover, a zygote which begins life with a copious food reserve has a better chance of survival than one with little. Increasing size, however, severely limits motility, so again advantages can be envisaged in a situation in which one gamete, the male, remains small and motile, and the other, the female, loses motility and specializes in the laying down of food reserves.

A non-motile zygote may, of course, be disadvantageous if it settles in a situation unfavourable for the plant. This is compensated for in those algae, such as *Coleochaete*, in which the zygote germinates to produce zoospores. Another development, possibly limiting the wastage of zygotes, is shown in *Laminaria* where the zygote germinates while still attached to the gametophyte, foreshadowing a feature of the archegoniate plants.

The life cycles of algae

Sexual reproduction inevitably involves a cyclic alternation between a haploid and a diploid condition. The simplest life cycle found amongst the algae is that in which the diploid condition, generated by the fusion of morphologically identical gametes, is represented only by the zygote (Fig. 3.30a). Meiosis occurs on germination of the zygote, thereby initiating a new haploid (gametophytic) phase. A cycle of this kind, termed **haplontic**, is frequently encountered in the simpler algae, and is typical of the filamentous Chlorophyta. Here, however, it may have been retained and developed as an adaptation facilitating survival in unfavourable conditions, since only rarely does meiosis immediately follow syngamy, and the intervening diploid phase is often spent as a thick-walled, resting zygote.

Closely related to the haplontic cycle, and possibly evolved from it by a delaying of meiosis, is that in which the zygote generates a multicellular, diploid (sporophytic) phase. This eventually produces reproductive bodies, almost always zoospores, by a process involving meiosis. The haploid condition is thus restored and the cycle recommences (Fig. 3.30b). Where, as in *Ulva*, the haploid and diploid plants are morphologically similar, the alternation of phases is isomorphic. Where the phases are morphologically different, as, to take an extreme example, in *Cutleria*, the cycle is heteromorphic (see p. 85). The evidence available does not warrant any general conclusion about whether isomorphic life cycles have evolved from heteromorphic, or the converse. Life cycles in which both haploid and diploid phases are present as multicellular individuals are termed *diplohaplontic*.

Superimposed upon these two basic kinds of life cycle are various conditions of anisogamy and oogamy. The peculiar complexities of the isomorphic

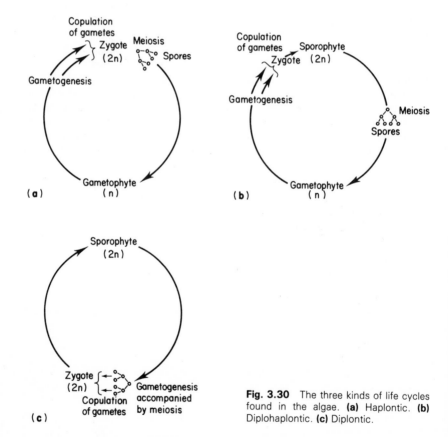

Fig. 3.30 The three kinds of life cycles found in the algae. (a) Haplontic. (b) Diplohaplontic. (c) Diplontic.

and heteromorphic life cycles of the floridean Rhodophyta have already been described.

A third kind of life cycle, similar to that of most animals, occurs in the diatoms, Fucales, and in several other isolated instances throughout the algae. The diploid condition predominates, and the haploid is represented only by the gametes, meiosis occurring during gametogenesis (Fig. 3.30c). Again the evidence does not allow any general conclusion about how this kind of cycle, termed *diplontic*, originated. As with the haplontic, it may in certain instances have selective value. In *Fucus*, for example, as compared with *Laminaria*, it is perhaps an advantage to have the gametophytic phase, possibly vulnerable to the vicissitudes of inter-tidal life, reduced to the unilocular sporangia and gametes.

The importance of the algae in the evolution of plants

Thus in the algae, the simplest of autotrophic organisms, a number of progressions, which can be regarded as representing channels of evolution, can be traced from unicellular to parenchymatous organization, from isogamy to oogamy, and from simple to elaborate life cycles. These all represent steps of fundamental importance in the evolution of plant life, and this fully justifies giving the algae considerable attention in any review of the Plant Kingdom. At its upper limit, the structural and reproductive complexity of an alga approaches that of a land plant. *Laminaria*, for example, possesses a thallus with marked morphological and anatomical differentiation, a strikingly heteromorphic life cycle, and a sporophyte attached in the early stages of its development to the gametophyte. Considered solely in terms of the level of organization, the transition from an advanced alga such as *Laminaria* to an archegoniate or even angiospermous land plant is small when compared with the evolution of that alga from a unicellular flagellate.

4
The Bryophyta
(Mosses and Liverworts)

The mosses and liverworts, although morphologically somewhat dissimilar, are classified together as the Bryophyta.[49] We treat them as comprising a Division of the Plant Kingdom, equivalent to those of the algae, possessing the following characteristics:

Bryophyta

Habitat	Mainly terrestrial.
Plastid pigments	Chlorophylls *a* and *b*, β-carotene, xanthophyll.
Food reserves	Starch, to a lesser extent fats.
Cell wall components	Cellulose, hemicelluloses.
Reproduction	Heteromorphic life cycle, the gametophytic phase normally the more conspicuous, and the sporophytic partly dependent upon it. Sex organs with a jacket of sterile cells. Spermatozoid with two whiplash flagella. Embryogeny exoscopic. Sporophyte producing non-motile, cutinized spores, in some species with heavily thickened and sculptured walls, usually all of one size (homospory). Vegetative propagation of the gametophyte by fragmentation or specialized gemmae.
Growth forms of gametophyte	Thallus flattened, with some internal differentiation, or consisting of a main axis with leafy appendages.

The Bryophyta are the simplest terrestrial plants, although in some parts of the world, such as the bogs of temperate regions and the mist forests of tropical mountains, they are a dominant part of the vegetation. Some species form dense communities submerged in Antarctic lakes. Vast bogs in the Northern hemisphere have been built up largely by the growth of the moss *Sphagnum*. The dead stems and leaves accumulating below the growing surface become consolidated to peat, often several metres in depth. In some places peat is an important fuel, and in granulated form is widely used in horticulture as a source of humus.

The largest bryophyte, *Dawsonia*, is a tufted moss of swampy places in south-east Asia and Australia. Individual stems of this genus may reach or even exceed a metre in length, but dimensions of this order are quite atypical of bryophytes. Most are lowly plants, many of them inconspicuous and not easily seen without a hand lens. The cellular differentiation within the larger bryophytes is greater than in the algae, but lacks the complexity found in vascular plants. Amongst the mosses it reaches its maximum in *Polytrichum*, and amongst the liverworts in *Symphogyna*. In both genera groups of thickened and elongated cells occur in the central region of the stem. They approach in form the tracheids of higher plants, and lignin may even be present, but detailed patterns of thickening are absent.

Although a cuticle has been demonstrated in some bryophytes, in general they are little able to resist desiccation and are consequently found principally in damp and humid localities. The exceptions, such as species of the liverwort *Metzgeria* and of the moss *Orthotrichum*, common on tree-trunks, seem to produce sufficient cellular colloids to retain the water necessary for the maintenance of life in even prolonged dryness.

The morphology of the bryophytes is on the whole more complex than that of the algae. In the mosses, for example, the mature thallus regularly takes the form of a stem bearing leaves. On the other hand, the immature gametophyte (*protonema*) of many mosses closely resembles a heterotrichous green alga (see p. 39), and may, as, for example, in the moss *Pogonatum aloides*, persist for many months. Like the algae, the bryophytes produce no roots, although both the mature and immature forms of the thallus bear rhizoids. In a few forms, such as the liverwort *Symphogyna* and the moss *Polytrichum*, the aerial parts of the plant are continuous with fine subterranean creeping axes, superficially resembling the filiform rhizomes of the smaller filmy ferns. So far as is known, only one bryophyte (*Cryptothallus mirabilis*, a thallose liverwort) is subterranean in habit and heterotrophic.

The bryophyte life cycle

Of the two kinds of plant in the bryophyte life cycle, the haploid gametophyte is the more persistent. The sporophyte (which consists of little more than a capsule, a stalk (*seta*) and a basal foot) grows upon it, and is wholly or partly parasitic and usually of limited life span. All bryophytes rely on free water for the dispersal of the spermatozoids and for fertilization.

Classification

The Bryophyta fall naturally into three Classes, the Hepaticae (liverworts), the Anthocerotae (hornworts), and the Musci (mosses). The features used in this classification are the nature of the thallus, and (where present) of the leaves, the extent of the development of the juvenile phase of the gametophyte (protonema), and the presence or absence of an opening mechanism in the capsule.

Hepaticae

Despite the diversity of the liverworts, there is little doubt that they form a natural group. The protonemal phase of the gametophyte is usually ill-defined and the mature thallus almost always shows recognizable dorsiventrality. A characteristic feature of many liverworts is the presence of oil bodies in the cells, thought by some to render the tissues unpalatable to grazing insects. The antheridia break open irregularly, instead of by a distinct cap cell. The capsule of the sporophyte matures before the elongation of its stalk, the converse of the situation in the mosses.

Of the seven Orders of Hepaticae, the common Marchantiales, Jungermanniales and Metzgeriales will be considered in some detail, and the small Orders Sphaerocarpales and Calobryales mentioned on account of special features which claim attention.

Marchantiales

VEGETATIVE STRUCTURE   The Marchantiales are exclusively thalloid. Although some species are simple in appearance, internal organization more complex than that found in any other thalloid liverwort is encountered in this Order.

The thallus of *Marchantia* itself (Fig. 4.1), found growing in damp places and areas of burnt ground, is dichotomously branched, with a thickened central rib and the surface divided into hexagonal areas visible with the naked eye. On examination with a hand-lens, a pore can be observed at the centre of

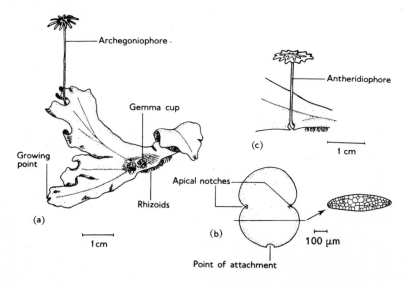

Fig. 4.1 *Marchantia polymorpha*. (a) Habit of female plant. (b) Structure of gemma. (c) Antheridiophore.

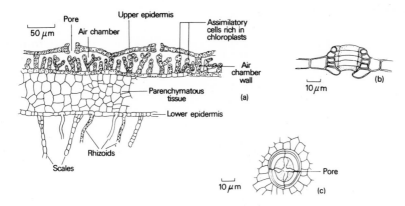

Fig. 4.2 *Marchantia polymorpha.* **(a)** Transverse section of thallus. **(b)** Transverse section of pore. **(c)** Surface view of pore.

each hexagonal area, which in transverse section is seen to consist of an air chamber containing photosynthetic tissue (Fig. 4.2). The pore, like the stoma of a higher plant, probably allows aeration of the thallus with the minimum dehydration, but is incapable of significant change in its aperture. Below the chlorophyllous tissue is a compact body of cells largely lacking chloroplasts. The lower side of the thallus bears up to eight rows of scales and unicellular rhizoids, the walls of some of which bear peg-like invaginations ('peg rhizoids'). The growth of the thallus is sensitive to photoperiod, and ceases in long days. This is accompanied by an accumulation of lunularic acid, an endogenous growth regulator possibly found in liverworts generally.

Amongst other members of the Marchantiales showing a chambered thallus are *Conocephalum*, the thallus of which yields a characteristic fragrance when crushed, and *Preissia*, in which the aperture of the pore shows some variation in response to atmospheric humidity. This feature may account for *Preissia* being able to tolerate drier habitats than *Marchantia*. The midrib of the thallus of *Preissia* also contains conspicuous elongated fibrous cells.

Riccia represents the simplest kind of structure found in the Order. The lower part of the thallus is again a compact colourless tissue, but the upper part consists of columns of chlorophyllous cells, separated by narrow air channels. The upper cells of the columns are colourless and fit closely together, leaving no distinguishable pores. In *Riccia fluitans* (Fig. 4.9) the narrow, dichotomously branched, floating thallus is divided almost entirely into air chambers separated by partitions one cell thick.

SEXUAL REPRODUCTION In *Marchantia*, which is dioecious, sexual reproduction is induced by increasing day length. The male and female gametes are produced on upright, umbrella-shaped structures termed antheridiophores and archegoniophores respectively (and gametangiophores collectively) (Fig. 4.1a and c). Both structures develop from one half of a

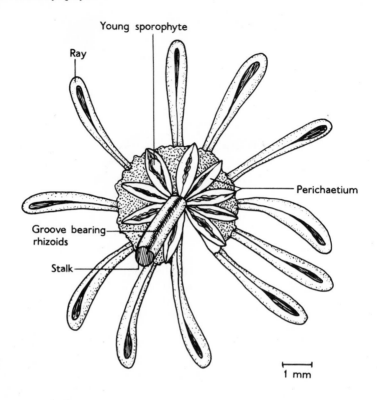

Fig. 4.3   *Marchantia polymorpha*. Archegoniophore seen from below.

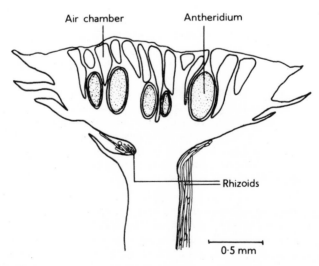

Fig. 4.4   *Marchantia polymorpha*. Vertical section of antheridiophore.

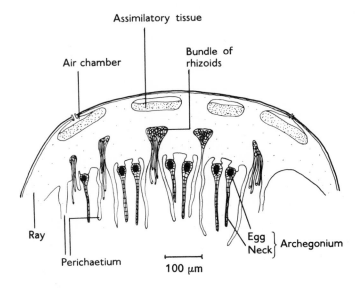

Fig. 4.5 *Marchantia polymorpha*. Tangential section of an almost mature archegoniophore.

dichotomy and are therefore homologous with a bifurcation of the thallus. Their morphological nature is clearly demonstrated by the rhizoids which grow down grooves in the stalks (Figs 4.3 and 4.4) and by the characteristic photosynthetic chambers which develop in the caps of the mature game-tangiophores. The female organs (archegonia) arise in radial rows on the upper surface of the cap. During the maturation of the archegonia, the cap grows more above than below, with the result that the archegonia become transferred to the lower surface. In the mature archegoniophore (Fig. 4.5) each row of archegonia is separated from its neighbours by a curtain-like out-growth, termed a **perichaetium**. In addition, sterile processes emerge radially from the upper surface of the cap between the rows of archegonia, giving the whole its familiar stellate appearance (Fig. 4.3).

The archegonium, as always in the bryophytes, is formed in its upper parts by a single layer of cells, and has a strikingly long neck (Fig. 4.6). The egg lies at the dilated base of the ventral canal and, when mature, appears to be suspended in fluid. It is surmounted by a ventral canal cell, and a number of neck canal cells. These degenerate at maturity, and their products, when hydrated, give rise to a mucilage through which the spermatozoids swim to reach the egg.

The antheridiophore lacks the complexity of the archegoniophore, being merely an elevated cap (Fig. 4.4), with the antheridia on the upper surface. Although superficial in origin, the mature antheridia are sunk in pits, each opening to the exterior by a narrow pore. Each antheridium is borne on a short stalk and bounded by a single layer of sterile jacket cells. When mature it contains a mass of small cubical cells (**spermatocytes** or antherocytes) in

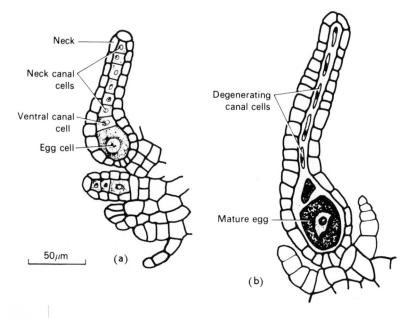

Neck

Neck canal cells

Ventral canal cell

Egg cell

Degenerating canal cells

Mature egg

50 μm   (a)

(b)

Fig. 4.6 *Marchantia polymorpha.* **(a)** Before breakdown of neck canal cells. **(b)** Mature archegonium.

each of which differentiates a biflagellate spermatozoid. This is in the form of a single gyre of a helix and consists of a head piece, an elongated nucleus, and a cytoplasmic tail (Fig. 4.7). The head piece includes the two posteriorly directed flagella, one inserted behind the other, and a large mitochondrion. Close to this is a lamellate body, the multilayered structure. A ribbon of microtubules extends from the surface of this body along the outside of the nucleus and enters the tail. The tail contains a large plastid, parts of which extend as a flap over the end of the nucleus, and a mitochondrion lying in a depression in the plastid. The differentiation of these highly specialized gametes from cells which are initially more or less isodiametric presents many problems of gene activation and cell mechanics. Particularly interesting is the state of the chromatin. Although fully condensed in the mature gamete, intermediate stages show coarse fibrils becoming oriented parallel to the longitudinal axis of the nucleus.

The mature antheridia open in moist conditions and the spermatocytes are dispersed. After a short period these in turn break open and release the spermatozoids. For fertilization to be possible the male and female plants must be growing together. It seems likely that the gametangia become mature, and fertilization occurs, before elongation of the gametangiophores.

DEVELOPMENT OF THE SPOROPHYTE OF *MARCHANTIA*   Although detailed observations are few, germination of the zygote probably begins within 48 hours of fertilization. The first division is by a horizontal wall, transverse to

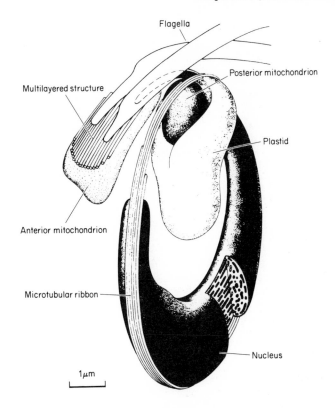

Flagella

Posterior mitochondrion

Multilayered structure

Plastid

Anterior mitochondrion

Microtubular ribbon

Nucleus

1μm

Fig. 4.7 *Marchantia polymorpha*. Diagram showing the disposition of the organelles in an almost fully differentiated spermatozoid. The nucleus is cut open to show the condensing chromatin. (After Carothers (1975). *Biological Journal of the Linnean Society*, **7**, Supplement 1, 71. Scale approximate.)

the longitudinal axis of the archegonium. Since it is from the outer cell that the apex of the sporophyte arises, embryogenesis is said to be *exoscopic*. The products of the inner cell form the foot, by which the sporophyte remains anchored in the gametophyte. Continued growth and differentiation, which are dependent upon nutrients drawn from the gametophyte, lead to an embryonic sporophyte consisting of three distinct regions. At the summit is the immature capsule containing the sporogenous cells, below this is a short seta, and at the base the foot (Fig. 4.8).

At this stage, the young sporophyte is not only enclosed by the proliferated jacket cells of the archegonium, which form a *calyptra*, but is also surrounded by a further tubular outgrowth of the gametophyte called a *pseudoperianth* or *perigynium*. Division of the sporogenous tissue (*archesporium*) inside the capsule is mitotic until spore mother cells are produced, when meiosis takes place, giving rise to tetrads of spores. These separate in the capsule, become rounded in outline, and develop thickened

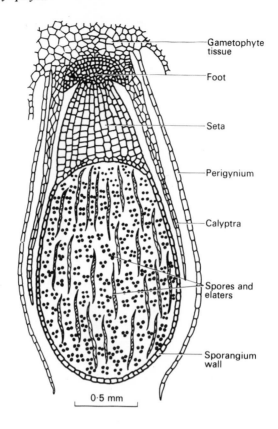

Gametophyte tissue

Foot

Seta

Perigynium

Calyptra

Spores and elaters

Sporangium wall

0·5 mm

Fig. 4.8 *Marchantia polymorpha*. Longitudinal section of sporophyte rupturing the calyptra. Note the lines of the elaters. (After Parihar (1967). *Bryophyta*. Allahabad.)

walls. Not all the cells inside the capsule become spores; some (referred to as *elaters* because of their subsequent behaviour) elongate and lay down spiral thickenings.

Elongation of the cells of the seta eventually causes the calyptra to rupture, and, once exposed to air, the single layer of cells surrounding the capsule soon bursts, so revealing the mass of yellow, haploid, spores. The loosening of this mass and the dispersal of the spores are now assisted by the contortions of the elaters. These contortions are caused by the spiral bands in their walls, presumably consisting of cellulose microfibrils, altering their curvature and pitch as they dry. In response to the strains generated in this way the cell as a whole makes jerky twisting movements.

The spores germinate rapidly on a damp surface, giving rise to short, alga-like, filaments of cells. Division of the apical cell then ceases to be confined to one plane and subsequent growth leads to the mature form of the gametophyte.

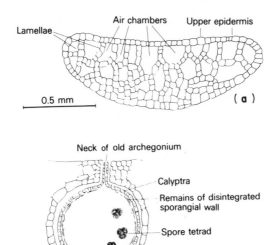

Fig. 4.9 *Riccia fluitans*. **(a)** Transverse section of thallus. **(b)** Thallus containing mature sporophyte.

SEXUAL REPRODUCTION IN OTHER MARCHANTIALES   No other genus has gametangiophores as elaborate as those of *Marchantia*. In *Conocephalum*, for example, the archegoniophore is a simple cap without emergent rays. In *Riccia* gametangiophores are entirely absent, both archegonia and antheridia merely lying at maturity in pits in the dorsal surface of the thallus. The sporophyte generation is again dependent on the gametophyte for nutrition, but at maturity it consists of only a sac of spores, with no seta or foot (Fig. 4.9). By the time the spores are mature no living diploid tissue remains, and dispersal must await the decay of the gametophyte.

ASEXUAL REPRODUCTION   Proliferation of the vegetative plant often follows from bifurcation of the thallus being accompanied by progressive decay of the older posterior region. In this way, an area becomes quite rapidly colonized by many seemingly individual plants. Additional to this, *Marchantia* has a notably elegant means of asexual reproduction. Multicellular bodies, called *gemmae*, develop inside cup-like growths on the upper surface of the thallus (Fig. 4.1a). Each gemma is slightly biconvex, with two diametrically opposed, marginal notches, each containing a small meristem (Fig. 4.1b). When mature, the gemmae become detached from the short stalk on which they are borne, and are readily dispersed. Experiments have shown that the newly detached gemmae have no innate dorsiventral symmetry. This becomes fixed at germination by gradients of light, temperature and other factors in the immediate environment. Each meristem grows out to form a new thallus and, finally, two individuals result from decay of the central portion.

Culture experiments have shown that short days promote the production of gemmae. The germination of the gemmae while in the cup is inhibited by growth-regulating substances diffusing basipetally from the apical meristem of the parent thallus.

The evolution of the Marchantiales

Marchantia seems to represent the highest level of organization achieved by a wholly thalloid gametophyte. Are we therefore to regard the simple *Riccia* as a primitive Marchantialean plant, and *Marchantia* as an advanced form? Although this would appear plausible, some striking breeding experiments with *Marchantia* point in the other direction. A number of mutants were raised from species of *Marchantia* in culture and hybridized in various ways, with the result that a whole series of forms was obtained which reproduced features found in other genera of the Order. The thallus of the var. *dumortieroides*, for example, lacks air chambers and resembles that of *Dumortiera*, a genus which, except for this feature, is close to *Marchantia*. Similarly the var. *riccioides* resembles *Riccia* in its narrow branching and the immersion of sex organs in the prostrate thallus. This reservoir of variation in *Marchantia* suggests that its evolutionary antecedents may have yielded the other genera of the Order by a process of simplification. *Riccia* would then be regarded as a reduced form.

On the other hand, perhaps both *Marchantia* and *Riccia* should be regarded as evolved forms. The archegoniophores of *Marchantia*, which elevate the seta-less capsule and facilitate the wide dispersal of the thin-walled spores in air currents, can reasonably be regarded as an advantageous development. In *Riccia*, however, the spores are thick-walled and long-lived. Despite its apparent rudimentary sporophyte, *Riccia* is probably no less well adapted than *Marchantia*, but to a different, Mediterranean environment. Clearly identifiable remains of both *Marchantia*- and *Riccia*-like Marchantiales have been found in Upper Triassic coals of Sweden.

Jungermanniales and Metzgeriales

VEGETATIVE STRUCTURE   Both thalloid and leafy states are represented in these Orders, the largest among the Hepaticae. Most genera achieve only a small size in temperate regions, but species reaching several centimetres in length are common in the humid Tropics, where they are frequently epiphytic. The thalloid genera are typified by the common *Pellia*, which grows dichotomously as *Marchantia*, but differs in outward appearance. A poorly defined midrib of elongated cells extends to each apical region. Examination under the microscope shows that the thallus is indeed simple, having none of the specialized photosynthetic tissue of the Marchantiales. Some genera have much more distinct midribs (*Pallavicinia*), while others, in which the thallus is regularly dissected (*Fossombronia*), begin to resemble the leafy forms. Morphological differentiation in the thalloid forms reaches its peak in some of the tropical representatives. In *Symphogyna*, for example, a filiform underground rhizome gives rise, in a sympodial fashion, to a sequence of erect aerial thalli. These, to which the photosynthesis is confined, and on which the sex organs are borne, may reach a height of 2 cm.

The leafy liverworts also show a wide range of vegetative morphology, but here principally in the form of the leaves. These may be simple, more or less circular, plates of cells, as in the common *Odontoschisma sphagni*, or more often twice or several times lobed. Sometimes the leaves achieve great delicacy. In *Blepharostoma*, for example, the lobes consist only of a single file of cells. In *Frullania*, the leaf has two lobes, the lower of which is shaped like a minute helmet, and possibly serves as a water sac.

In most leafy liverworts the stem is inclined or prostrate, and its symmetry clearly dorsiventral. Although the leaves are usually in three ranks, only those on the dorsal side are fully developed. Those of the third, ventral, row (termed **amphigastria** or under leaves) remain small, and are often shed a short distance behind the apex.

Classification of the leafy liverworts is based largely on the features of the leaves, including the orientation of their insertions on the stem. When the anterior margins of the leaves lie regularly beneath the posterior of those in front, the arrangement is said to be **succubous** (Fig. 4.10b) and when the converse **incubous**. The leaves of liverworts regularly lack nerves of the kind seen in mosses, but the lower lobes of the bilobed leaves of *Diplophyllum* possess a

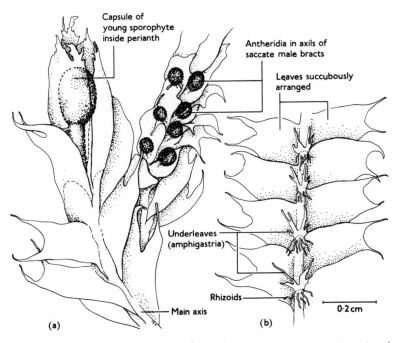

Fig. 4.10 *Lophocolea cuspidata.* **(a)** Fertile shoot seen from above. **(b)** Ventral surface showing amphigastria and the succubous arrangement of the leaves (see text).

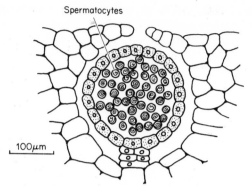

Fig. 4.11 *Pellia*. Vertical section through antheridium.

conspicuous central row of elongated cells. Conduction of water along the stems of leafy liverworts is probably largely by surface capillarity. Experiments have shown that such transport is more rapid with succubous arrangements of leaves than with incubous.

SEXUAL REPRODUCTION Reproduction is essentially similar to that described for the Marchantiales, except that specialized gametangiophores are never produced. The antheridia, superficial in origin, usually occur singly, and either come to lie in a cavity in the upper surface of the thallus (e.g. *Pellia*, Fig. 4.11), or, in leafy forms, lie in the axils of leaves of special branches of limited growth (Fig. 4.10a). The spermatozoids are basically

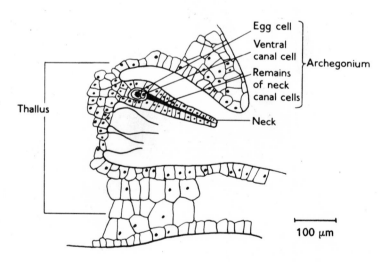

Fig. 4.12 *Pellia*. Vertical section of thallus showing archegonia.

similar to those of *Marchantia*, but in at least one species of *Pellia* they are considerably bigger, the nucleus extending for several gyres. Also in *Pellia* the microtubular ribbon, as seen in transverse section, is not closely applied to the nucleus across its whole width, but is inclined at an angle of about 45° to its surface. The archegonia are usually grouped, and are produced either laterally, as in *Pellia* (Fig. 4.12), or at the tip of the main shoot, as in most leafy liverworts. When archegonia are apical, they, and ultimately, the sporophyte, terminate the growth of the main shoot, so that vegetative growth is continued by a resulting lateral, sympodial branching. Both monoecious and dioecious forms occur, sometimes in the same genus. The common *Pellia epiphylla*, for example, is monoecious, but *P. fabbroniana*, frequent in calcareous districts, is dioecious.

The sporophyte of the Jungermanniales and Metzgeriales (Fig. 4.13) has a higher proportion of sterile tissue than that of the Marchantiales. The sporo-phytes are green when young, and capable of appreciable photosynthesis. Minerals are probably largely transmitted from the gametophyte, but some may be absorbed directly. The cells immediately adjacent to the foot of the sporophyte often have highly convoluted walls ('transfer cells'). Although often regarded as a device facilitating transport, the significance of the labyrinthine structure of the wall probably lies more in the accumulation of materials at the boundary, and the consequent disturbance of the cell wall metabolism, than in their transmission into the sporophyte. The capsule develops while still enclosed in the calyptra and the ultimate extension of the seta, which may reach 1 cm or more, is extremely rapid (rates of 1 mm per hour have been recorded in *Pellia*). Although this extension is solely by cell elongation, phototropism is commonly observed. In *Pellia* experiments have shown that curvature occurs only in the regions illuminated. There is no

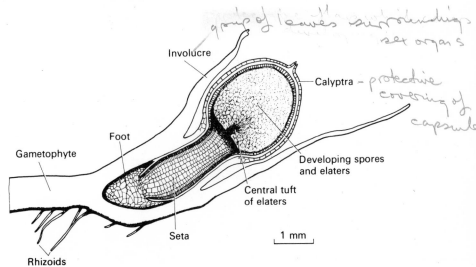

Fig. 4.13 *Pellia*. Longitudinal section of maturing capsule.

transmission of the stimulus, a conspicuous feature of phototropism in coleoptiles and seedlings, and this difference is presumably related to the absence in the sporophyte of *Pellia* of any localized site of growth.

The mature capsule of the Jungermanniales and Metzgeriales contains a mixture of spores and elaters. In *Pellia* the spores undergo several divisions before being shed (Fig. 4.14), and remains of the spore mother cells are visible between them. The wall of the mature capsule usually dehisces into four valves which, as a consequence of differential thickenings in the wall, become sharply reflexed. The dispersal of the spores is again aided by the elaters (Fig. 4.14). In *Pellia* the elaters, some of which remain as a brush attached to the top of the seta (Fig. 4.13), are similar to those of *Marchantia*. In some genera, however, the elaters are 'explosive'. In *Cephalozia bicuspidata*, for example, the elaters are loosely attached at one end of the valves of the capsule. After the capsule opens they begin to dry and in consequence to twist. Suddenly they violently untwist, hurling both the elater and its adhering spores into the air. The sudden expansion of the elater is believed to be caused by the shrinking column of fluid in the drying cell being put under such tension that it eventually spontaneously vaporizes, thus increasing its volume many times.

ASEXUAL REPRODUCTION   Vegetative multiplication of many species takes place by regeneration from fragments of mature plants. Multicellular gemmae are not uncommon and are sometimes conspicuous, as the clusters

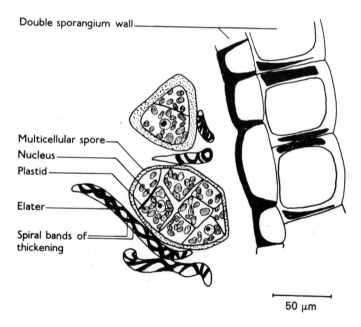

50 μm

Fig. 4.14 *Pellia*. Portion of capsule showing detail of the wall, multicellular spores and elaters.

of reddish, two-celled, gemmae on the margins of the upper leaves of *Sphenolobus exsectiformis*, a plant frequent on rotting wood. In *Blasia pusilla*, a thalloid form, multicellular gemmae are produced in remarkable flask-shaped receptacles on the dorsal side of the thallus.

Sphaerocarpales

The three genera of this Order are similar to the simpler Marchantiales, such as *Riccia*, in habit, but the thallus has frilly margins suggestive of leaves. The antheridia and archegonia occur in small clusters, each cluster enclosed in a distinctive involucral sheath. As in *Riccia*, the sporophyte is little more than a capsule.

Sphaerocarpus has been much used in experiments. It was found, for example, that the wave length of ultraviolet light most effective in producing mutations in nuclei of immature spermatozoids was that most strongly absorbed by nucleic acids, providing the first evidence for the special role of these acids in heredity. *Sphaerocarpus* was also the first plant in which sex chromosomes were demonstrated. The female gametophyte possesses a large X chromosome and the male a small Y, in each instance accompanied by seven autosomes. The usefulness of the plant for genetical work derives from the tendency of the spores in many species to adhere in their original tetrads. The products of each meiosis can then be separated and cultured individually, allowing a direct demonstration of any genetic recombination ('tetrad analysis'). The spores are also of interest in relation to the inheritance of the pattern of thickening of the wall, a feature used to discriminate species. The spores in hybrid capsules show the thickening of the female parent, suggesting that the determining factors are inherited in the cytoplasm.

Naiadita, a fossil liverwort from uppermost Triassic rocks, is believed to belong to the Sphaerocarpales.

Calobryales

This Order, though small, is highly distinctive and the two genera it contains have a similar growth form. The upright stems, which are radially symmetrical and bear three ranks of leaves, rise from a creeping rhizome-like axis lacking rhizoids. The archegonia are effectively terminal. There are no involucral leaves protecting the young sporophyte, but the calyptra is particularly conspicuous. Some bryologists have considered the sexual reproductive structures of the Calobryales to be the most primitive amongst the bryophytes as a whole. Possibly allied to the Calobryales is *Takakia*, a curious liverwort of Japan and the Pacific Northwest with an upright stem bearing linear appendages ('phyllids'), usually in groups of two or three. The archegonia are terminal, but the sporophytes are still unknown.

Anthocerotae

This Class contains the single Order, Anthocerotales. Although formerly included with the Hepaticae, the Anthocerotales are now usually placed in a separate Class, mainly on account of their unique sporophyte. *Anthoceros* is representative of its Class.

VEGETATIVE STRUCTURE   The gametophyte of *Anthoceros* recalls *Pellia* in external morphology except that there is neither regular dichotomous growth nor a midrib (Fig. 4.15). The thallus is undifferentiated, apart from internal cavities which contain mucilage and occasionally the blue-green alga *Nostoc*, a genus known to fix atmospheric nitrogen. In most species a single chloroplast, containing a complex pyrenoid, occurs in each cell, a situation unknown elsewhere in the Bryophyta or in higher plants (with the exception of some species of *Selaginella*; see p. 157) but common in the algae. This has led to the suggestion that the Anthocerotae are closer to an algal ancestry than other Bryophyta.

Some species of *Anthoceros* (such as those of the Mediterranean region) regularly form tubers, enabling them to tide over a dry season unfavourable for growth.

SEXUAL REPRODUCTION AND THE FORM OF THE SPOROPHYTE   In *Anthoceros*, as in *Marchantia*, the formation of the sex organs has been

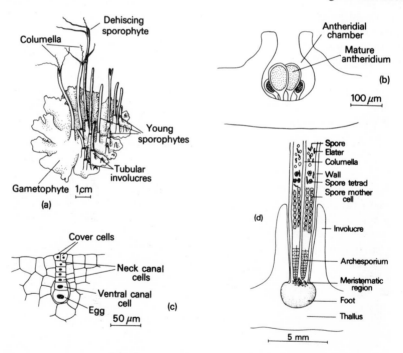

Fig. 4.15   *Anthoceros laevis*. **(a)** Female thallus with sporophyte. **(b)** Section of antheridial chamber. **(c)** Section of nearly mature archegonium. **(d)** Diagrammatic longitudinal section of a sporophyte showing the different regions.

shown to depend upon photoperiod. Here however gametogenesis in most species is initiated by diminishing day length, so that fertilization occurs during winter.

The antheridia arise from a cell beneath the surface, one to several antheridia (depending upon species) coming to lie in a closed chamber, the roof remaining intact until the antheridia are mature (Fig. 4.15b). Archegonia arise superficially, but the wall of the archegonium is continuous with the thallus, the neck opening at the surface (Fig. 4.15c). This resembles the situation in some vascular archegoniates (see, for example, Fig. 5.12). Development of the diploid zygote leads to a slender cylindrical sporophyte (Fig. 4.16) with a relatively small proportion of fertile tissue. The sporophyte remains inserted into the gametophyte by a conspicuous lobed foot, and the basal part is surrounded by an upgrowth of the thallus, the involucre.

The development of the archesporium begins as a dome-like layer within the summit of the sporophyte. The sporophyte continues to grow for several weeks from a meristem close to its base. During this growth the archesporium differentiates basipetally as a hollow cylinder, the centre of which is occupied by a sterile columella. Multicellular elaters (which lack distinct spiral thickenings and are often referred to as pseudo-elaters) differentiate amongst the sporogenous cells. When the spores at the top of the capsule are ripe, the capsule dehisces basipetally (Fig. 4.15a) along two longitudinal slits, the opening beginning near the tip. As the upper part of the capsule dries, the valves separate completely above, and they begin to twist longitudinally. The consequent contortions of the valves expose the spores and elaters adhering

Fig. 4.16 *Anthoceros laevis*. Female plants bearing young sporophytes. (Approx. × 0.67)

to the central column, and dispersal begins. The separation of the valves continues downwards as the spores mature, and meanwhile the basal meristem generates new sporophytic tissue at about the same rate. Consequently a single sporophyte continues to yield spores over a considerable period. The spores have conspicuously thickened and sculptured walls.

Anthoceros and its allies further differ from other liverworts, but resemble the mosses, in possessing photosynthetic tissue in the outer layers of the extending sporophyte. Stomata are also present, as in the capsules of some mosses. The sporophyte is thus not entirely dependent on the gametophyte for nutrition, but, since the sporophyte will still mature even if it is covered with a tinfoil cap, it seems likely that it enjoys a considerable amount of translocation from the parent gametophyte.

The evolutionary position of the Anthocerotales

The Anthocerotales are remarkable amongst the liverworts in recalling the features of both the algae (the presence of the pyrenoid in the chloroplast) and the mosses and higher plants (the presence of stomata and the continued growth of the sporophyte). They are consequently thought by some to stand close to the line of evolution leading from the algae to terrestrial vegetation. The intermediate position of the Anthocerotales extends even to their ultrastructure. The electron microscope confirms that the chloroplasts of *Anthoceros* resembles those of the algae, but the chloroplasts of *Megaceros*, where there are several in each cell, possess irregular grana and are more like those of other archegoniate plants.

Another feature of the Anthocerotales which has excited much attention is the axial form of the sporophyte. Since the sporophyte is often long-lived and may even persist for a time after the death of the parent gametophyte, it may indicate how simple axial plants, such as the psilophytes of the Silurian and early Devonian (p. 146), have evolved. Alternatively the *Anthoceros* condition may be derived, the sporophyte having become reduced and almost deprived of its independence.

Although evidence relating to either possibility is lacking, it seems quite plausible that a growth form similar to that shown by the living Anthocerotales did play some part in the evolution of land plants from algal ancestors. Unfortunately the only fossils so far known attributable to the Anthocerotales are spores from the Tertiary.

Musci

The mosses are a Class much greater in number and more widely distributed than the liverworts, occurring in almost every habitat supporting life. Apart from being the dominant vegetation in acid bogs, and alpine and Arctic regions, they are familiar features in woodlands and hedgerows, while some species even survive the polluted atmosphere of urban areas, often forming dark green cushions between paving stones and in other damp crevices.

The protonema of mosses is usually conspicuous, and is commonly filamentous and heterotrichous. In some species the prostrate system becomes cytologically distinguishable from the erect, the cells having elongated chloroplasts and being yellowish instead of green (*caulonema*). The buds which initiate the mature form of the gametophyte arise on the prostrate filaments.

In addition to the development of the gametophyte, other features which distinguish mosses from liverworts are the multicellular rhizoids, the growth of the sporophyte from an apical cell, the complex opening mechanisms of the capsules, and the absence of sterile elaters amongst the spores. The three Orders, Sphagnales, Andreaeales and Bryales, differ principally in the nature of the protonema and the structure of the capsule.

Sphagnales

The Sphagnales, represented by a single genus, *Sphagnum*, are confined to acid waterlogged habitats. They are the principal component of peat bogs, where they form a more or less continuous spongy layer.

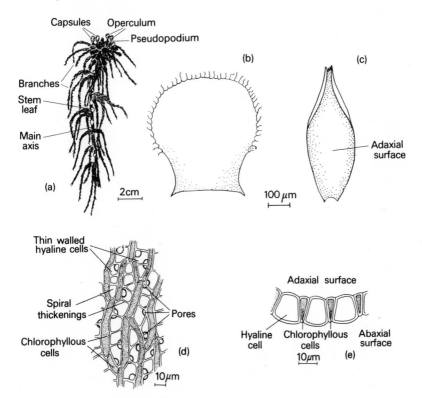

Fig. 4.17 *Sphagnum fimbriatum*. **(a)** Upper portion of shoot with sporophytes. **(b)** Stem leaf. **(c)** Branch leaf. **(d)** Arrangement of cells in leaf, adaxial surface. **(e)** Section of leaf.

VEGETATIVE STRUCTURE   The adult gametophyte comprises an upright main axis from which whorls of branches arise at regular intervals (Fig. 4.17). The leaves, which are closely inserted, have a peculiar structure which is diagnostic of the genus, and also, in its finer details, of the many species (Fig. 4.17c and d). When first formed, the leaves are made up of many diamond-shaped cells. These then cut off narrow daughter cells, but on two sides only. The daughter cells develop chloroplasts, while the mother cell remains colourless, often becomes spirally thickened, and eventually dies. These dead cells also develop pores during differentiation, enabling them to act as reservoirs. This peculiar leaf structure accounts for the ability of the *Sphagnum* plant to retain large quantities of water, and consequently for its outstanding bog-building properties. Well preserved leaves very similar to those of living *Sphagnum* have been found in Permian deposits in the USSR, indicating beyond doubt the great antiquity of this kind of construction.

REPRODUCTION   Reproduction of *Sphagnum* is probably principally vegetative, the decay of the older parts eventually causing branches to separate, and thus to become new individuals. Mature plants do, however, produce sex organs in favourable situations, and both monoecious and dioecious species occur. The antheridia, each of which begins its development from a single apical cell, lie in the axils of leaves towards the tips of small upper branches (Fig. 4.18). These antheridial branches are often strongly pigmented and clustered in a conspicuous comal tuft. The female inflorescence consists of a bud-like aggregate of archegonia and bracts borne laterally near the summit of the main stem.

After fertilization the zygote yields a sporophyte (Fig. 4.17a) consisting principally of a capsule, containing a dome-shaped archesporium, and a foot. The seta remains inconspicuous, and the function of elevating the

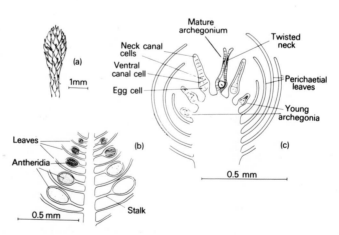

Fig. 4.18   *Sphagnum* sp. **(a)** Antheridial branch. **(b)** Longitudinal section of antheridial branch. **(c)** Longitudinal section of archegonial branch.

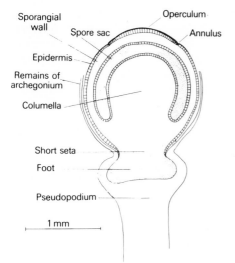

Fig. 4.19 *Sphagum* sp. Longitudinal section of nearly mature sporophyte.

capsule is taken on by the base of the female inflorescence which, as the capsule matures, grows up as a leafless axis, or ***pseudopodium*** (Fig. 4.19). Release of the spores is brought about by air pressure which builds up in the lower half of the capsule as it dries. Eventually this pressure is sufficient to dislodge the clearly differentiated lid (***operculum***) with explosive force, and the spores are effectively dispersed.

The spores germinate to form a filament, but this is rapidly replaced by a small thallose protonema. This in turn gives rise to a bud which develops into the familiar leafy gametophyte, the protonema meanwhile becoming moribund and disappearing.

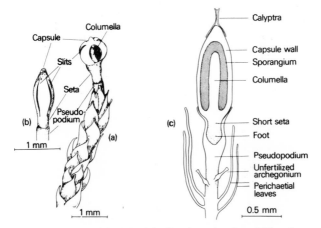

Fig. 4.20 *Andreaea nivalis*. **(a)** Habit of fertile plant showing dehisced capsule in dry condition. **(b)** Dehisced capsule in wet condition. **(c)** Longitudinal section of mature sporophyte.

Andreaeales

The Andreaeales are another Order containing only a single genus, distinguished by its peculiar capsule.

The leafy gametophyte of *Andreaea* (Fig. 4.20) rarely exceeds 1 cm in height. It is usually found growing on rock, chiefly in cold, exposed and relatively dry regions. The leaves are olive-brown in colour, composed of rounded cells, and in most species showing no distinct midrib.

Sex organs are formed apically. The sporophyte resembles that of *Sphagnum* in having a domed archesporium (Fig. 4.20c), and in being borne on a pseudopodium at maturity. Dehiscence of the capsule takes place by four longitudinal slits which do not meet at the tip (Fig. 4.20b). The hygrosopic properties of the wall cause the slits to close in damp conditions, and to open again in dry (Fig. 4.20a).

The protonema of *Andreaea* is similar to that of *Sphagnum*.

Bryales

The 600 or so genera of the Bryales form a well-defined Order. Although there is a common basic morphology and life cycle within the Order, the variation in size, detailed structure and habitat preferences is considerable. Many mosses are confined to permanently damp situations in woodlands and

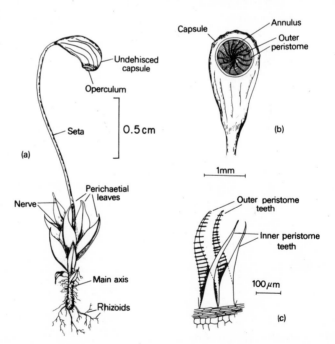

Fig. 4.21 *Funaria hygrometrica*. **(a)** Habit of fertile plant. **(b)** Mature capsule showing intact peristome. **(c)** Portion of peristome viewed from the inside.

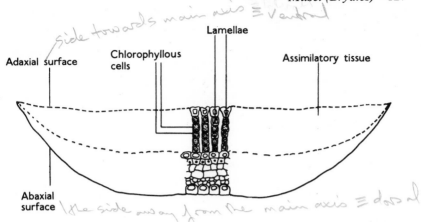

Side towards main axis ≡ ventral

Abaxial surface \the side away from the main axis ≡ dorsal

Fig. 4.22 *Polytrichum commune*. Transverse section of leaf showing the assimilatory lamellae.

by springs, but others, for example *Tortula ruraliformis*, are able to survive periods of drought in sand dunes and other arid habitats. The cells of these species appear to have acquired the capacity to continue metabolism at a reduced rate while partially dehydrated. At the other extreme are a few sub-aquatic species, such as *Fontinalis antipyretica*.

VEGETATIVE STRUCTURE   The most conspicuous form of the moss plant is the adult gametophyte (Figs 4.21 and 4.26). This consists of a main axis bearing leaves which, although usually spirally inserted, may in some forms come to lie on one plane, giving the shoot a complanate appearance (e.g. *Neckera*, common in banks and rocks). In a few species (e.g. *Fissidens*) the leaves are equitant and arranged in two ranks. The leaves of most mosses consist of a single sheet of cells, although the central region may be thickened and contain a well-defined midrib (often referred to as a 'nerve'), sometimes excurrent in a hyaline point. The most complex leaf is found in *Polytrichum* and its allies. Here a number of parallel longitudinal lamellae grow up from the upper surface (Fig. 4.22) and the chloroplasts occur principally in these cells. The shape of the leaf, and the nature and development of the midrib and of the cells at the margin of the leaf are important features in the taxonomy of the mosses.

Anatomically, moss gametophytes offer little that is remarkable, the most complex differentiation being found, as already mentioned, in the stem of *Polytrichum* (Fig. 4.23). Not only is there an approach here to the development of tracheids, but there is also a clear radially symmetrical zonation in structure, recalling that of the axes of some of the smaller ferns. Surrounding a central core of tracheid-like cells (sclereids), containing scattered thin-walls (hydroids), is a zone of cells conspicuously large in transverse section, regarded as an approach to phloem (leptoids). This central region, legitimately regarded as a rudimentary vascular tissue, is surrounded by a sheath of parenchymatous cells containing starch (hydrome sheath), and this in turn

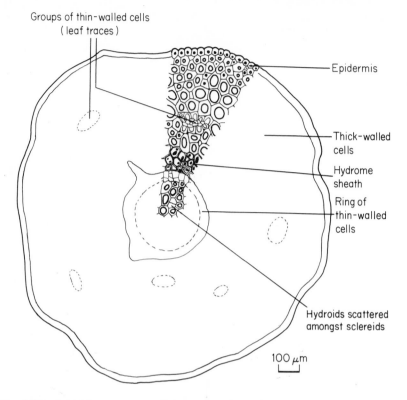

Groups of thin-walled cells
(leaf traces)

Epidermis

Thick-walled cells

Hydrome sheath

Ring of thin-walled cells

Hydroids scattered amongst sclereids

100 μm

Fig. 4.23   *Polytrichum commune*. Transverse section of stem.

by a sclerenchymatous cortex. Experiments have shown that where moss stems contain a central strand with thin-walled cells these are the principal site of internal conduction. In most mosses however there is also substantial external conduction through matted rhizoids, and adpressed leaf bases. The significance of the internal conduction in the physiology of the mosses is not yet clear.

SEXUAL REPRODUCTION   In sexual reproduction, the Bryales show every possible arrangement of the archegonia and antheridia. Both monoecious and dioecious species occur, and amongst the monoecious species, the game-tangia may be either mixed together in a bud-like inflorescence, or separate. Whatever the arrangement, the antheridia and archegonia are often numerous and interspersed with sterile hairs or paraphyses (Figs 4.24 and 4.25), the whole cluster of sex organs commonly being surrounded by a whorl of closely adpressed leaves (perichaetium). The archegonia usually have long necks, each consisting of several tiers of cells, and the central canal may contain as many as 10 cells. The antheridia are stalked, and one or more cells at the apex usually form a distinct lid at maturity, opening as if on a hinge

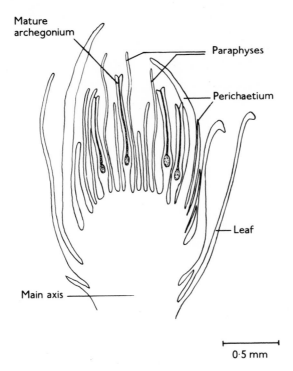

Fig. 4.24 *Mnium* sp. Longitudinal section of female head.

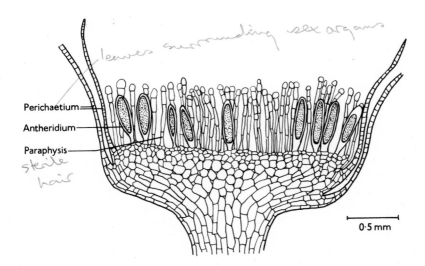

Fig. 4.25 *Mnium hornum*. Longitudinal section of male head.

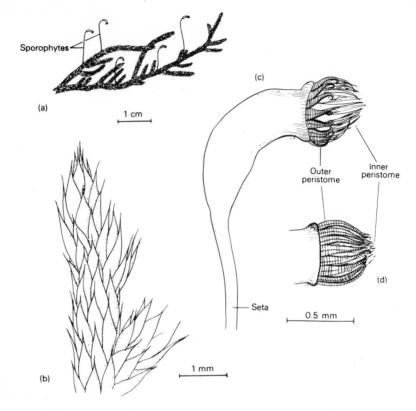

Fig. 4.26 *Hypnum cupressiforme*. **(a)** Fertile shoot system. **(b)** A portion of the shoot showing the closely inserted leaves. **(c)** Capsule, showing peristome in dry state. **(d)** Peristome in wet condition.

while the mass of spermatocytes is discharged.

There is a striking correlation in the Bryales between the position on the plant where the sex organs are produced, and the growth habit. Where the reproductive organs terminate the main axis, and growth is consequently sympodial, the main axis is almost invariably upright. These are the **acrocarpous** mosses. In the remainder, where the sex organs are produced laterally (the **pleurocarpous** mosses), the main axis is usually creeping (Fig. 4.26). With only a few exceptions, the tufted mosses are acrocarpous. Photoperiod and temperature affect the onset of the sexual phase in many mosses, and these factors are probably responsible for the annual reproductive cycle seen in many temperate species.

The development of the sporophyte begins immediately after fertilization, and the regions of the embryo yielding the foot, seta and capsule are soon distinguishable. The venter of the archegonium is also stimulated into growth by germination of the zygote and a cap-like calyptra is formed over the young sporophyte. Extension of the seta tears the calyptra away from the main body

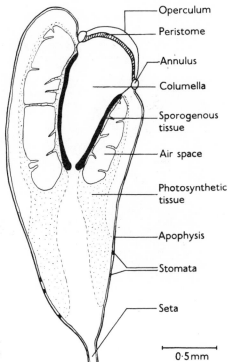

Operculum

Peristome

Annulus

Columella

Sporogenous tissue

Air space

Photosynthetic tissue

Apophysis

Stomata

Seta

0·5mm

Fig. 4.27 *Funaria hygrometrica*. Median longitudinal section of immature capsule.

of the gametophyte, but it continues to ensheathe the developing capsule and exert formative effects on the upper part of the sporophyte. Premature removal results in anomalous growth. The influences emanating from the calyptra seem to be partly physical and partly chemical. Compared with that of a typical liverwort, the moss sporophyte develops slowly, and the extension of the seta begins well before the capsule is mature.

The capsule itself (Figs 4.27 and 4.28) is a complex organ, but its differentiation follows a regular radial pattern, and two concentric regions of tissue can be recognized which follow distinct developmental paths. An inner region, termed the **endothecium**, gives rise to the archesporium, which in the Bryales is never domed, but is always a cylinder, often with a central sterile columella. Outside the endothecium is the **amphithecium** which, in most Bryales, differentiates a ring of remarkable tooth-like structures, the **peristome**. This remains as a fringe around the mouth of the opened capsule (Figs 4.21b and 4.26c). The peristome is developed from three layers of cells. Thickenings are laid down on both sides of the tangential walls bounding the middle layer. Only these thickenings remain at maturity. They then fall into a number of columns, each of which yields a peristome tooth. Because of the ordered sequence of mitoses in the differentiation of the amphithecium, and the regular spacing of the cells giving rise to the peristome, the number of teeth is constant in any given species and is always a power of two.

Fig. 4.28   *Bryum pallens*. Fruiting condition. (Approx. × 2.5)

Chlorophyllous tissue occurs in the immature capsule, particularly in the basal region (*apophysis*), where there are also stomata in the epidermis. The sporophyte is thus to some extent autotrophic. In some species (e.g. *Funaria hygrometrica*), air spaces occur between the archesporium and the wall of the capsule. This conspicuous aeration has perhaps been developed in relation to the respiratory demands of the developing archesporium.

Meiotic division of the spore mother cells heralds the last phase in the maturation of the capsule. The columella usually breaks down at this stage, so that the centre of the capsule is occupied solely by spores. The operculum ultimately drops off, exposing the peristome (Fig. 4.21b,c), which now begins to play an important role in the dispersal of the spores. The poly-saccharide material forming the peristome teeth is hygroscopic and, since the macromolecular orientations of the thickenings in the two columns of cells giving rise to a tooth differ, tensions are generated in the tooth with changes in its hydration. These are released by sharp twisting and bending move-ments. The peristome thus forms a very effective scattering mechanism, activated by changes in atmospheric humidity. The exact nature of the move-ments of the teeth varies with the species. In some mosses, for example *Funaria hygrometrica*, the peristome is incurved when wet and recurved when dry, but in others the movements are less regular. The number of spores in a capsule varies from a few thousand in species with small capsules to a million or more in those with large capsules. In a few mosses (e.g. *Macromitrium* of the Southern hemisphere) the ripe capsule contains spores of two sizes, the smaller giving rise to diminutive male plants. The four spores

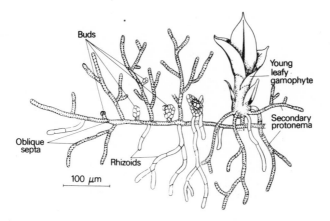

Fig. 4.29 *Funaria hygrometrica.* Development of buds on protonema.

in each tetrad are initially similar, but subsequently two become distinctly larger. Evidently there is segregation of sex at meiosis.

The spores of the Bryales have thin walls, and germinate rapidly on a damp surface. The protonema is usually well developed, resembling a hetero-trichous green alga, but distinguishable by the obliquely transverse walls of the prostrate filaments. In spore cultures growing on agar the buds giving rise to the mature plants (Fig. 4.29) frequently arise in a number of concentric zones. Mutants of *Physcomitrella* are known which will not produce buds unless cytokinin is added to the medium.[1] The existence of mutants of this kind facilitates the investigation of the genetical and physiological factors bringing about the change in form of growth, but the results of experiments *in vitro* must always be interpreted with caution. When *Funaria*, for example, is grown in pure culture on medium containing activated charcoal the protonema produces buds as readily as on soil, but in the absence of charcoal they are delayed. It seems likely that the charcoal absorbs substances secreted by the protonema into the medium which would otherwise accumulate and have morphogenetic effects.

ASEXUAL REPRODUCTION   Vegetative propagation undoubtedly plays a large part in the reproduction of the Bryales. Almost any part of the gametophyte – leaf, stem or even rhizoid – is capable of regeneration, either directly or by the production of gemmae (Fig. 4.30), and giving rise to a new individual. Some species are hardly known in the sporophytic condition, but are nevertheless widely distributed. These must be dispersed almost entirely by vegetative means.

The sporophyte is also capable of regeneration under experimental condi-tions: for example, if segments of the seta are placed on a mineral-agar medium. Almost always, however, growth is of the gametophytic form, an example of *apospory*. In this way diploid gametophytes, and ultimately tetraploid sporophytes, can be obtained. There is some evidence that

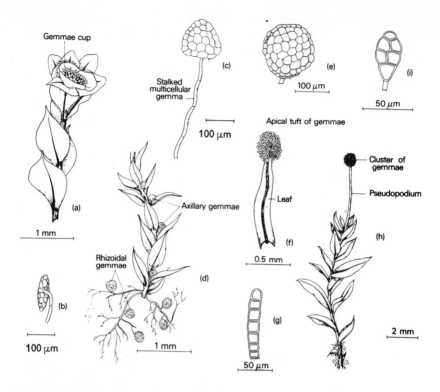

Fig. 4.30 Asexual reproduction in mosses. **(a-c)** *Tetraphis pellucida*: **(a)** habit of gemmiferous plant; **(b,c)** mature gemma, front and side views. **(d,e)** *Bryum rubens*: **(d)** habit; **(e)** gemma. **(f,g)** *Ulota phyllantha*: **(f)** young leaf with apical tuft of gemmae; **(g)** gemma. **(h,i)** *Aulacomnium androgynum*: **(h)** habit; **(i)** gemma.

autodiploid gametophytes exist in Nature, and they may have arisen in this way.

The mosses provide ideal material for the experimental study of the life cycle. In *Physcomitrium*, for example, protonema developing secondarily from leaves will give rise to vegetative buds, or in the presence of sucrose to sporophytes (*apogamy*).[27] In such a well defined situation it will ultimately be possible to identify the factors which are responsible for the switch from gametophytic to sporophytic growth.

The relationships of the Bryophyta

The origin of the bryophytes

There are so many similarities, particularly in the protonemal phase, between the bryophytes and the algae, that it seems beyond doubt that the mosses and liverworts had their origin in some algal form. Further, it is clear that the Bryophyta share many more basic features, such as the nature of the photosynthetic pigments, cell wall components and food reserves, with the Chlorophyta than with any other algae. The flagella of the spermatozoids are also of the same kind, and the multilayered structure underlying the flagellar bases recalls similar structures in the motile stages of the Chaetophorales. These algae also have well developed oogamy and notably heterotrichous gametophytes. Nevertheless, it is also clear that the Bryophyta are considerably more highly organized than any Chlorophyta. This is shown by their terrestrial habit, differentiated thallus, regular phasic alternation of gametophyte and sporophyte, and the production of aerial spores. The regular enclosing of the reproductive organs in a wall of sterile cells also contrasts with the naked gametangia and sporangia of the Chlorophyta. Although the Chaetophorales and Charales indicate how archegonia and antheridia may have evolved, little information is available about the origin of the bryophytes themselves. No fossil record of the transitional forms has yet been discovered or recognized, and they have left no evident descendants. The protonema of many mosses suggest that the intermediates may have been heterotrichous (p. 39). The thallose liverworts, for example, could have been derived from an heterotrichous ancestor by progressive reduction and ultimate elimination of the aerial component, coupled with development of the prostrate part of the plant. Heteromorphic life cycles would have been no novelty since they are well represented in the algae, but examples of non-motile spores, such as those produced by *Dictyota*, are few. The evolution of non-motile spores with a well developed exine, an essential for terrestrial life, thus probably accompanied the transmigration, and those forms which remained as algae have never developed this feature.

The fossil record gives little information about the origin of the bryophytes. Possible transitional forms from the oldest rocks are usually so ill-preserved that their exact status is uncertain. The most promising candidate is perhaps *Sporogonites* from the Lower Devonian, a form in which a number of capsules with stalks a few centimetres in height seem to be arising from a creeping thallus. Unfortunately no detail can be discerned and the bryophytic nature of the fossil must remain conjectural. So far the fossil record is wholly silent on how the ancestral organisms gave rise to the sporophytes characteristic of bryophytes.

The morphological and reproductive differences between the mosses and liverworts appear to extend as far back as the Carboniferous, since the general classification of the fossil bryophytes from these ancient rocks is readily apparent. This strengthens the view that mosses and liverworts have been independent evolutionary lines from a very early period, and it is even possible that they had independent origins from transitional archegoniate forms.

Apart from their archegoniate reproduction, bryophytes have no obvious relationships with even the simplest vascular plants, living or fossil. A number of fossils from the Rhynie Chert may have belonged to plants intermediate between Bryophyta and Tracheophyta,[34] but the evidence is not compelling. The Bryophyta appear never to have been a major component of the Earth's vegetation. They have probably remained isolated from the main line of evolution of land plants, changing only slowly, and exploiting a relatively circumscribed ecological niche. The failure to reach the morphological complexity characteristic of the remainder of the land flora may have been a consequence of the restriction of independence to the gametophytic phase in the life cycle. A haploid organism has no possibility of carrying a reservoir of variability in the form of recessive genes, capable of being advantageously expressed in future chance recombinations. Despite this impediment to evolutionary change a number of anatomical and morphological trends have been strikingly similar in bryophytes and vascular plants. Bryophytes, for example, have acquired both leafiness and, in the stem of *Polytrichum*, a rudimentary vascular strand. This consists of elements which resemble structurally, and very possibly also functionally, xylem and phloem. Gametophytic and sporophytic organisms have evidently responded in a similar manner to common environmental factors. Nevertheless the achievement of the haploid plants has in comparison been modest.

Evolutionary relationships within the bryophytes

The evolutionary relationships within the mosses and liverworts themselves are also obscure. The liverworts, for example, show a whole series of forms from the creeping thallus (e.g. *Pellia*), to thalloid with two rows of ventral scales (e.g. *Blasia*), thalloid in which the margin of the thallus is so deeply crenulate that the lobes resemble leaves (e.g. *Fossombronia*), and ultimately leafy forms with upright stems and radial symmetry (e.g. *Haplomitrium* of the Calobryales). There has been much argument about whether this series represented a phylogenetic advance, or whether the first liverworts resembled *Haplomitrium*, the other forms being derived. The view that the radially symmetrical leafy form is primitive is strengthened by the surprisingly complex features present in many of the thalloid forms (e.g. in *Marchantia* and *Anthoceros*). Also a thalloid form such as *Pellia* is so outstandingly simple that it stands under the suspicion of being reduced and specialized.

Similar arguments apply to the mosses, although here the relative uniformity of the group makes comparative morphology even less informative. Few would regard those species whose capsules possess only rudimentary peristomes, or even lack them altogether, as anything other than reduced. The moss capsule with its clearly differentiated operculum and peristome must therefore have been an early and distinguishing feature of the Class. We have no direct evidence of how the peristome evolved, but it is possible that the 'pepper-pot' mechanism present at the mouth of the capsule of *Polytrichum* (Fig. 4.31) and the bristle-like peristome found in a few other genera, indicate steps in a developmental pathway that culminated in the typical peristome of the Bryales (p. 131).

Although we must regard the basic morphological features of the bryophytes as having arisen very early in the evolution of land plants, evolution of a more superficial nature has no doubt continued in the Division, probably

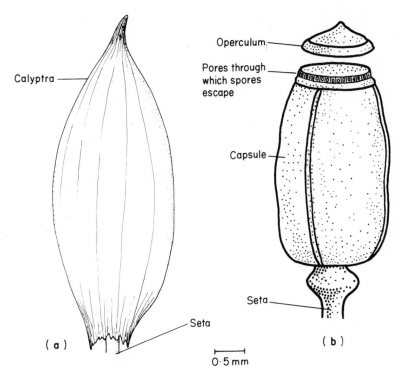

Calyptra

Operculum

Pores through which spores escape

Capsule

Seta

(a)

(b)

Seta

0·5 mm

Fig. 4.31 *Polytrichum juniperinum*. Capsule. **(a)** Before removal of calyptra. **(b)** Operculum removed showing 'pepper-pot' mechanism.

being influenced by evolution of vegetation as a whole. The restriction on variability imposed by the haploid nucleus may have been partly overcome by widespread polyploidy, a notable feature of the living bryophytes. We can envisage that the rise of the angiospermous forests in the Cretaceous provided many new surfaces well suited for colonization. A renewed burst of evolution, consisting principally of diversification of established morphological forms, may have occurred at this time leading to the substantial epiphytic element in the existing bryophyte flora.

5
The Tracheophyta, I
(Psilopsida, Lycopsida, Sphenopsida)

Under the system of classification we have adopted, all the vascular plants (i.e. those possessing the lignified conducting tissue, xylem) are placed in a single Division, the Tracheophyta. Discussion of the many and diverse Orders which are combined in the Tracheophyta, and their interrelationships, will occupy the remaining chapters.

The general characteristics of the Tracheophyta may be defined as follows:

Tracheophyta

Habitat	Predominantly terrestrial or epiphytic.
Plastid pigments	Chlorophylls *a* and *b*, carotenoids (principally β-carotene), xanthophylls (usually principally lutein).
Food reserves	Starch; to a lesser extent fats, inulin and other polysaccharides. Proteins.
Cell wall components	Cellulose, hemicelluloses, lignin.
Reproduction	Heteromorphic life cycle; sporophyte the conspicuous phase. Sex organs with or without a jacket of sterile cells. Male gametes in some flagellate. Embryogeny various. Spores rarely green, usually with well defined wall (exine) impregnated with sporopollenin. Often of two sizes, produced in different sporangia, the larger (megaspores) female and smaller (microspores) male (heterospory). Specialized vegetative reproduction of the sporophyte infrequent.
Growth forms	Predominantly axial.

The Tracheophyta can be conveniently sub-divided into the Psilopsida (psilopsids), Lycopsida (lycopods), Sphenopsida (horsetails) and Pteropsida (ferns, gymnosperms and angiosperms). The spore- (as opposed to the seed-) bearing Tracheophyta are often referred to collectively as the Pteridophyta.[41]

The angiosperms, with about 200 000 living species, are by far the largest component of living vegetation. They are nevertheless the most recent Tracheophyta to have been evolved, and their dominance is comparatively recent. The remainder of Tracheophyta, with the exception of a few gymnosperms, are all archegoniate and they have a relatively rich fossil record, extending back in some instances well into the Palaeozoic. Taken as a whole, the archegoniate Tracheophyta show a progressive ability to exploit terrestrial habitats. The sexes become separated in the gametophytic phase, and the male gametophytes progressively reduced. Ultimately the male gamete ceases to be motile and fluid is no longer necessary for fertilization. These reproductive changes are accompanied by increasing protection of the plant body from unfavourable climatic conditions, thus widely extending the range of environments available to plant life.

Although the simpler and, on the basis of the fossil evidence, more primitive vascular plants are archegoniate, they are, nevertheless, very different from the mosses and liverworts. In contradistinction to the situation in the Bryophyta, in the archegoniate Tracheophyta it is the sporophyte which is the conspicuous phase, being longer-lived and possessing considerably greater anatomical complexity than the gametophyte. There are no living plants clearly intermediate between bryophytes and tracheophytes. Even if intermediates existed in the Devonian,[34] the common ancestors were undoubtedly remote and perhaps were present only at the time of the beginning of the colonization of the land. If such ancestors possessed little or no lignified tissue, they would of course stand small chance of preservation as macrofossils.

A feature common to all living Tracheophyta is the presence of cells surrounding the sporogenous tissue which break down during maturation of the sporangium. The materials so liberated are utilized in the final development of the spores as the tetrads are released from the remains of the spore mother cells. These specialized cells, which often lie in one or more concentric layers, form the so-called *tapetum*.

We shall begin our consideration of the Tracheophyta with the Psilopsida, since the evidence of palaeobotany and comparative morphology points to their being amongst the most primitive of vascular plants.

Psilopsida

Sporophyte consisting of more or less dichotomously branching axes, often with small leaf-like appendages. Roots absent, the subterranean axes bearing rhizoids. Vascular tissue consisting of tracheids and ill-defined phloem. Sporangia terminal, homosporous. Gametophyte (in living forms) subterranean, sometimes with vascular tissue, resembling portions of the sporophyte rhizome. Spermatozoids flagellate. Embryogeny exoscopic.

Only two genera of living psilopsids are known, *Psilotum* (Fig. 5.1a) and *Tmesipteris* (Fig. 5.1b). The former is pantropical and not uncommon, but the latter, although locally abundant, is confined to Australasia and Polynesia. A number of Palaeozoic fossil forms are known from various

Fig. 5.1 **(a)** *Psilotum nudum*. Part of the fertile region. The trilocular synangia are subtended by small forked bracts. ($\times$ 6.6) **(b)** *Tmesipteris tannensis*. Part of the fertile region. The bilocular synangia are attached at the forks of conspicuous bifid bracts. ($\times$ 5)

parts of the world, principally from rocks of Devonian age. The living forms are included in the Psilotales, and the fossil in the Psilophytales. Inclusion within the same Sub-Division should not be taken to imply a phylogenetic relationship between the two Orders (see p. 149).

Psilotales

The sporophyte of *Psilotum* (Fig. 5.2), which may be either terrestrial or

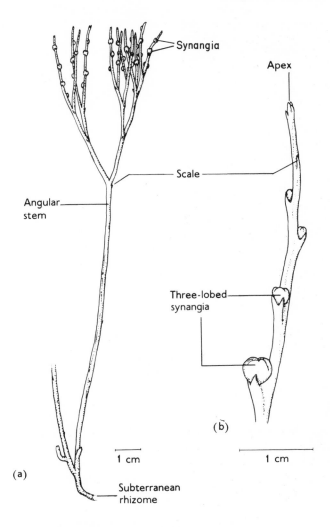

Fig. 5.2 *Psilotum nudum*. **(a)** Habit. **(b)** Fertile shoot.

epiphytic, consists of upright (or, in one epiphytic species, pendulous), dichotomously branching axes arising from a horizontal system of similarly branching rhizomes. The rhizomes bear rhizoids, and contain an endophytic fungus, probably in symbiotic association (*mycorrhiza*). Small scales, which are at first green but soon become scarious, occur at irregular intervals on the stem. They possess no vascular strand, but in one species a strand does approach the base of the leaf but stops below the insertion. *Psilotum*, like all vascular plants, possesses stomata, but here they are confined to the epidermis of the stem.

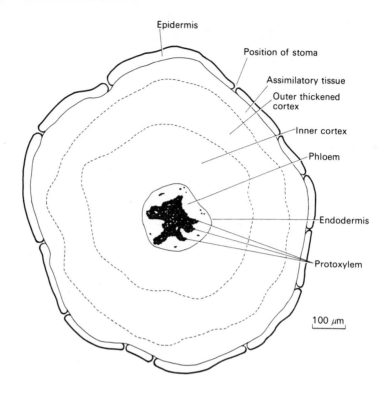

Fig. 5.3   *Psilotum nudum.* Transverse section of lower part of aerial branch.

THE ANATOMY OF THE AXES   The stem contains a simple stele (***protostele***), frequently enclosing in the upper regions a central parenchymatous medulla. The xylem, consisting solely of tracheids, is often stellate in transverse section, the protoxylem lying at the tips of the arms which often subtend poorly defined ribs at the exterior of the stem (Fig. 5.3). Phloem surrounds the xylem, but, apart from the lateral sieve areas, the sieve cells are little different from elongated parenchyma cells. An endodermis, separated from the phloem by a narrow zone of pericyclic parenchyma, marks the boundary of the stele. Phlobaphene, a condensation product of tannin, is often deposited in considerable quantities in the cells of the inner cortex.

The branches of both the aerial and terrestrial systems of the sporophyte grow indefinitely from single or small groups of apical cells. There is no evidence that the dichotomy of the axes follows median longitudinal division of a single apical cell, as in certain algae (e.g. the brown alga, *Dictyota dichotoma* (p. 91).

REPRODUCTION   Spores are produced in the upper region of the sporophyte (Fig. 5.2). The spore-bearing organs, which are distant from each other, are three-lobed, and each is subtended by a bifid appendage (Fig. 5.4). The lobes

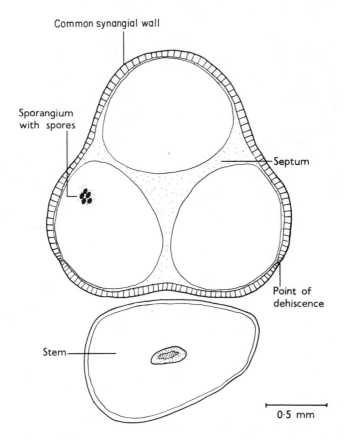

Common synangial wall

Sporangium
with spores

Septum

Point of
dehiscence

Stem

0·5 mm

Fig. 5.4 *Psilotum nudum.* Transverse section of synangium.

correspond to three internal chambers, separated by septa, and each filled
with spores. It is still not clear whether this spore-bearing organ is to be interpreted as a trilocular sporangium, or as a synangium formed by the fusion of
three sporangia. Three primordia are, however, visible early in the ontogeny
of the organ, perhaps indicating its synangial nature. Since a distinct vascular
strand extends into the base of the synangium, it is usually regarded as
terminating a lateral axis, rather than arising in association with a
sporophyll. This view is strengthened by the existence of a cultivated form,
'Bunryu-zan', in which the synangia terminate vertical axes.

Each of the three groups of archesporial cells in the synangium arises from
several cells, and is thus *eusporangiate* in origin. Part of the peripheral
archesporium, although not regularly layered, functions as a tapetum. The
remainder yields sporogenous cells. Meiosis leads to tetrads of spores, each
bilaterally symmetrical with a single linear scar (*monolete* spores). Such
spores can be considered as formed by the bisection of two conjoined hemi-

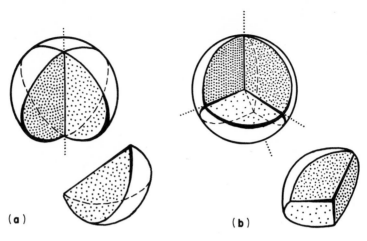

Fig. 5.5   Diagrammatic representation of manner of formation of **(a)** monolete and **(b)** trilete spores.

spheres by a plane perpendicular to the equator (Fig. 5.5). The linear scar thus represents the line of common contact between the four segments. The exine is lightly reticulate with no very distinct ornamentation. The massive wall of the sporangium, some five cells thick at maturity, dehisces at three sites symmetrically placed with respect to the loculi.

The gametophyte of *Psilotum* is a subterranean axial structure, dichotomously branching, and resembling short lengths of the sporophytic rhizome. The similarity extends to the anatomy, the finer axes being wholly parenchymatous, and the broader containing a central vascular strand. The peripheral cells, like those of the rhizome, house an endophytic fungus. Both the gametophyte and the sporophytic rhizome produce globular multicellular gemmae, a means of vegetative propagation.

Antheridia and archegonia arise from superficial cells in the region of the growing points of the gametophyte (Fig. 5.6). The antheridia, depending upon their size, liberate up to 250 spermatozoids. The archegonium has four tiers of neck cells, but at maturity all but the lower one or two tiers degenerate. Fertilization, which depends upon the presence of a film of water, is brought about by spirally coiled, multiflagellate spermatozoids.

The first division of the zygote is in a plane transverse to the longitudinal axis of the archegonium, as in the bryophytes. The outer (or epibasal) cell yields the apex of the embryo, and the inner (or hypobasal) the foot. Psilotales provide one of the few examples of such exoscopic embryogeny amongst the archegoniate Tracheophyta.

Tmesipteris

Tmesipteris is frequently an epiphyte with trailing stems (Fig. 5.7). The gross morphology is similar to that of *Psilotum*, but branching is much less

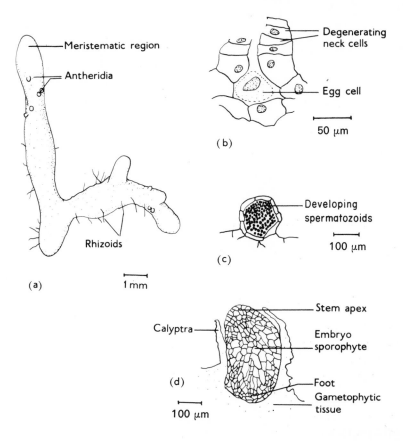

Fig. 5.6 *Psilotum nudum.* **(a)** Gametophyte. **(b)** Mature archegonium. **(c)** Antheridium. **(d)** Young sporophyte. (All after Bierhorst (1953). *American Journal of Botany,* **40**, 469; and Bierhorst (1954). *American Journal of Botany,* **41**, 274.)

frequent. The appendages are larger and more leaf-like, remaining green and possessing stomata, and also frequently a vascular strand. The insertion of the appendages is however, peculiar, being longitudinal instead of transverse, so that they appear more as flange-like outgrowths of the axis than as normal foliage leaves. Spore-bearing organs occur in the upper parts of some of the shoots, each subtended by a bifid appendage. These organs are again regarded as synangia terminating very short lateral branches, but in *Tmesipteris* each consists of only two fused sporangia.

The reproduction of *Tmesipteris* is very similar to that of *Psilotum*. The foot of the young embryo of *Tmesipteris* is lobed, and the whole bears a striking resemblance to the young sporophyte of *Anthoceros*. It is doubtful, however, whether this bears the phylogenetic significance that some have claimed.

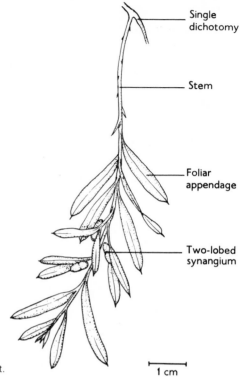

Single
dichotomy

Stem

Foliar
appendage

Two-lobed
synangium

1 cm

Fig. 5.7   *Tmesipteris tannensis.* Habit.

Psilophytales

The Psilophytales take their name from *Psilophyton*, a Devonian fossil first discovered in sandstones in eastern Canada in 1859 and subsequently reported from rocks of the same age in many parts of the world. The original '*Psilophyton*' was a mixture of the remains of several different plants, but the name *Psilophyton* is retained for forms with dichotomously branching main axes bearing vegetative lateral branches in irregular spirals (Fig. 5.8). The vegetative regions were interspersed with fertile ones in which the branches, after a series of dichotomies, ended in fusiform sporangia. *Pertica* is a related plant from later Lower Devonian beds in North America. Here there was a distinct main axis with lateral branches inserted in four ranks. After a series of dichotomies the upper branches ended in tassels of 50 or more elliptical sporangia.

Also attributed to the psilophytes in the wider sense are plants from the Rhynie Chert, a siliceous rock of Lower Devonian age in Scotland. Except for *Asteroxylon* (see p. 165) the Rhynie plants had a general resemblance to *Psilotum. Rhynia* (Fig. 5.9), for example, consisted of aerial axes arising from a horizontal rhizome. The aerial axes, the thicker of which reached a

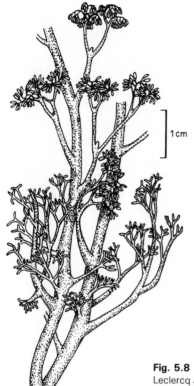

1 cm

Fig. 5.8 *Psilophyton dawsonii.* (After Banks, Leclercq and Hueber (1975). *Palaeontographica Americana*, **8**, 75.)

diameter of 6 mm, branched dichotomously and ascended to about 50 cm. A strand of xylem, endarch and wholly tracheidal, extended throughout the plant, and stomata were present in the epidermis of the aerial branches. Some branches terminated in fusiform sporangia, about 12 mm long, containing cutinized spores. In *Horneophyton* the rhizome was tuberous, and the archesporium of the capsule domed, as in some bryophytes. The Rhynie plants, probably growing in a bog petrified *in situ* by seepage from volcanic ash, are so beautifully preserved that detailed anatomical investigation is possible. Even germinating spores have been clearly seen in some preparations.

Increasing knowledge of the psilophytes has brought suggestions for reclassification. Those with a clear indication of a main axis are now often referred to the Trimerophytopsida, and those with solely dichotomous branching to the Rhyniopsida.

The life cycles of the psilophytes are still not certainly known. The spore-bearing plants were probably diploid since the spores were in tetrads. The nature of the gametophytes is disputed. It is possible that small forms of *Rhynia* did not bear sporangia and were the gametophytes of larger spore-

Fig. 5.9   *Rhynia.* (After Delevoryas (1962). *Morphology and Evolution of Fossil Plants.* Holt, Rinehart and Winston, New York, from a reconstruction in the Chicago Natural History Museum.)

bearing plants. Portions of subterranean axes are also known bearing bud-like protuberances, identified by some as embryos. It still has to be demonstrated that these axes, which certainly recall gametophytes of *Psilotum*, actually bore antheridia and archegonia. If the gametophytes of the psilophytes were in fact axial it would be strong evidence for the homologous theory of the relationship of the two phases of the life cycle. This holds that in the transitional forms the two phases were morphologically similar, but subsequently diverged.

The evolution of the Psilopsida

The Psilophytales do not show any relationship to any known alga, nor do they resemble (with the possible exception of the sporophyte of *Anthoceros* (p. 121)) any of the Bryophyta. They may have arisen independently from some algal stock, but it is clear that, as present in the Devonian, they are already far removed from any transitional form. Even *Cooksonia*, which appears indisputably Upper Silurian in age, possessed tracheids, and a *Rhynia*-like form. Cutinized spores, generally regarded as characteristic of land plants, are however found even earlier in the Silurian. Although the plants producing them are hitherto unknown, they may have been primordial psilophytes with algal features.

The predominantly axial morphology is a striking feature of both the Psilophytales and Psilotales, but there may be no direct phylogenetic continuity, since *Psilotum*-like remains are not identified with certainty from intervening geological periods. The origins of the Psilotales are in fact obscure. Their generally high chromosome numbers (52 to 210 in the gametophytic phase) may indicate a complex polyploid series developed with little evolutionary change over geological time, but evidence of this kind is inconclusive. A remarkable similarity, both in sporophyte and gametophyte, between *Psilotum* and certain New Caledonian ferns (see p. 223) suggests that the affinities of the Psilotales may lie with the Filicales rather than with the Psilophytales.

Lycopsida

Sporophyte consisting of more or less dichotomously branching axes, but differentiated into root and shoot. Shoot bearing microphylls, each containing a single vein, but leaf trace leaving no gap in the stele. Vascular tissue consisting of tracheids and phloem. Sporangia in or near axils of microphylls, homo- or heterosporous. Gametophyte (in living forms) terrestrial or subterranean. Spermatozoids flagellate. Embryogeny endoscopic (see p. 154).

Only five living genera are included in the Lycopsida, which, although more numerous than the Psilopsida, are again a minor component of contemporary vegetation. The fossil record, however, shows that the lycopsids were abundant in the Carboniferous period, many then being represented by substantial trees. Only herbaceous forms have persisted until the present day. Five Orders of Lycopsida are recognized, three containing the living genera and their fossil relatives, and the other two only extinct forms.

Lycopodiales

Members of this Order, which includes the living *Lycopodium* (club moss) and *Phylloglossum*, are distinguished by their eligulate leaves and homospory.

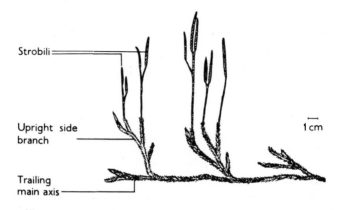

Strobili

Upright side
branch

Trailing
main axis

1 cm

Fig. 5.10   *Lycopodium clavatum*. Habit.

Lycopodium (Fig. 5.10), with about 200 species, is distributed throughout the world from Arctic to tropical regions, with a related range in the growth form of the sporophyte. The colder climates favour species with short, erect stems, or creeping stems giving rise to short upright side branches, while those in the Tropics have much laxer growth, are sometimes stoloniferous, and are often epiphytes. Taxonomists are now inclined to assign species of *Lycopodium* to a number of smaller genera. For present purposes however the concept of *Lycopodium* will be retained, and alternative nomenclature will be indicated where appropriate.

The stem is surrounded by microphylls, a kind of leaf which is typically small, simple in outline, and with a single median vascular strand. Some species of *Lycopodium*, especially the epiphytic, are heterophyllous, the lateral rows of leaves being expanded in the plane of the shoot system and the upper and lower rows adpressed and smaller. The fertile leaves (*sporophylls*) may be similar to the sterile, as in *L.* (*Huperzia*) *selago*, and the fertile regions not clearly set off from the sterile along the axis. More usually, however, the sporophylls differ from the sterile leaves in size, shape and the extent of the chlorophyllous tissue, and are often grouped together, as in *L. clavatum*, in distinct cones (strobili) of determinate growth.

Roots arise endogenously, emerging from the underside of the stem in the prostrate species, or from near the base of the stem in upright species. In some of the latter, initiation of the roots occurs near the shoot apex, but instead of emerging there the roots grow down inside the cortex, and break out only when they reach the level of the soil.

GENERAL FEATURES OF THE ANATOMY   The stem, which grows from a group of initial cells (Fig. 5.11), contains a central stele. Although basically a protostele, the detailed anatomy shows considerable variation with species. In its simplest form the stele consists of a core of tracheids, more or less stellate in section, with phloem lying between the arms (*actinostele*). In other

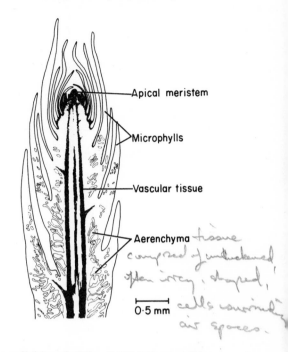

Apical meristem

Microphylls

Vascular tissue

Aerenchyma tissue composed of unthickened often irregularly shaped, cells surrounding air spaces.

0·5 mm

Fig. 5.11 *Lycopodium (Diphasiastrum) alpinum.* Longitudinal section of apical region of vegetative shoot.

species, the xylem and phloem form parallel bands (Fig. 5.12) (*plectostele*), or the phloem and xylem may be intermingled, anastomosing strands of sieve cells being scattered amongst the tracheids. In every instance differentiation of the xylem begins at the exterior and then proceeds centripetally, leaving no undifferentiated tissue at the centre. The xylem is thus uniformly exarch. In the phloem the sieve plates occur on both the lateral and tapering end walls. The regular presence of callose has not been confirmed. The vascular tissue, entirely primary, is usually surrounded by a narrow zone of parenchyma, and this in turn by an endodermis, the cells of which have a distinct Casparian strip (see p. 295).

The leaves, the chief site of photosynthesis, are structurally simple. There are abundant stomata on both surfaces, and internally numerous intercellular spaces. There is, however, no clearly differentiated mesophyll.

REPRODUCTION Sporangia, borne singly in the axil of a sporophyll or close to its insertion (Fig. 5.13), develop from a group of initial cells, and are hence termed eusporangiate. Continued cell division within the primordium leads to a central mass of spore mother cells surrounded by a wall several cells thick. The inner layers of the wall function as a tapetum, breaking down to provide food materials which nourish the maturing sporocytes. Following meiosis the four young spores arrange themselves in a tetrahedral fashion so that each spore has three triangular faces at its proximal pole where it is in contact with its fellows. As the spores separate these areas become less thickened than elsewhere, and the edges between them conspicuous, often

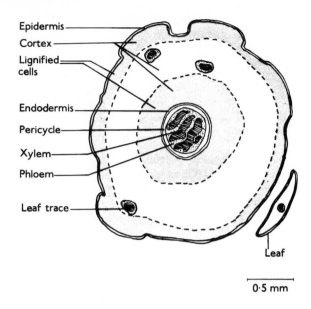

Epidermis

Cortex

Lignified cells

Endodermis

Pericycle

Xylem

Phloem

Leaf trace

Leaf

0·5 mm

Fig. 5.12 *Lycopodium clavatum.* Transverse section of stem.

appearing as a prominent triradiate scar in the mature spore (*trilete* spores) (Fig. 5.5). The sporangia dehisce transversely along a line of thin-walled cells (*stomium*) and hence the minute spores (each about 50 μm in diameter) are distributed by wind.

In some species of *Lycopodium* the exine of the spore has quite elaborate reticulate ornamentation, while in others it is almost smooth. The rough spores are not easily wetted and their germination may be delayed for several years, probably until weathering and attrition have rendered the coat permeable. The process may be simulated in the laboratory by immersing such spores in concentrated sulphuric acid. Following this treatment the spores germinate freely in pure culture. In natural conditions initially unwettable spores may be washed deep into the soil before germination occurs, and this is reflected in the nature of the gametophyte. For example, in *L. clavatum*, the spores of which have a pronounced reticulate relief, the gametophyte is a subterranean saucer-shaped structure, growing saprophytically and persisting for several seasons. The sex organs are produced on a cushion in the central region (Fig. 5.14). In the tropical *L.* (*Lycopodiella*) *cernuum*, where the spores are smooth, germination and development occur rapidly, producing lobed, cup-shaped gametophytes which possess chlorophyll and last for little more than a single season. In *L.* (*Huperzia*) *selago*, in which the spores are somewhat intermediate in the development of the wall, there is a corresponding ambivalence in the habitat of the gametophyte. In all instances the gametophyte of this species is a small carrot-shaped body growing saprophytically and producing sex organs on the

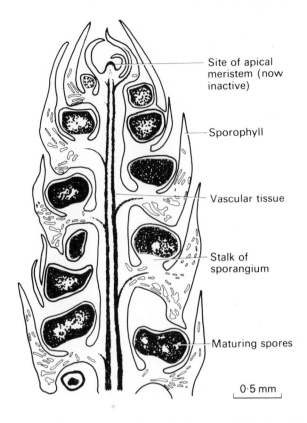

Site of apical
meristem (now
inactive)

Sporophyll

Vascular tissue

Stalk of
sporangium

Maturing spores

0·5 mm

Fig. 5.13   *Lycopodium clavatum*. Longitudinal section of upper region of young strobilus.

upper cushion. It may, however, be either buried, or at the surface, and if the latter the upper part becomes pale green. There is thus a general relationship between the nature of the spore wall and the form and duration of the gametophytic plant.

The sex organs of *Lycopodium* are similar to those of *Psilotum*, except that in many species the archegonia have more conspicuous necks, and both the male and female gametangia are enclosed to a greater extent by vegetative tissue. The spermatozoids are biflagellate, about 8 µm long, and spindle-shaped. The ultrastructure is basically similar to that of bryophyte spermatozoids, but the nucleus is less elongated and is not coiled. Fertilization is probably facilitated by chemotactic attraction of spermatozoids to archegonia. The zygote divides by a wall transverse to the axis of the archegonium. The outer cell, termed the *suspensor*, divides no further, but the inner remains meristematic and gives rise to two regions of cells. The central region becomes the foot, while the inner differentiates into the root, first leaf, and stem apex of the embryo proper. This development results in

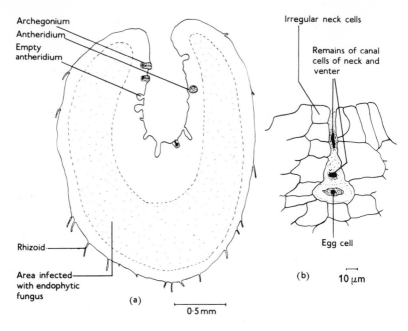

Archegonium
Antheridium
Empty
antheridium

Irregular neck cells

Remains of canal
cells of neck and
venter

Rhizoid

Area infected
with endophytic
fungus

Egg cell

(a)

(b)

0·5 mm

10 μm

Fig. 5.14   *Lycopodium volubile*. **(a)** Vertical section of subterranean gametophyte. **(b)** Mature archegonium.

the embryo being directed inwards, and the embryogeny is consequently said to be ***endoscopic***.

At first the apex of the embryo points downwards, but expansion of the foot region pushes it to one side and the young sporophyte finally breaks out of the surface of the gametophyte. Before it becomes fully established, the young plant depends upon food materials absorbed from the gametophyte, probably through the foot. In some species of *Lycopodium* (e.g. in *L.* (*Lycopodiella*) *cernuum*) the differentiation of the embryo is delayed and it emerges as a parenchymatous protuberance, termed a ***protocorm***. This eventually gives rise to one or more growing points, each of which yields a normal plant. Apogamous sexual reproduction occurs in *L. cernuum* in culture, but it has not been observed in Nature. *L. selago* commonly produces bud-like gemmae in the axils of the upper leaves.

Phylloglossum

Phylloglossum, confined to Australasia, is the only other living genus of the Lycopodiales. Its single species, *P. drummondii*, consists of an upright sporophyte, reaching 5 cm or less, with a basal whorl of leaves and a pedunculate strobilus (Fig. 5.15). During growth a lateral axis arises near the base of the plant and extends down into the soil, its tip eventually becoming transformed into a tuber. This tuber forms an organ of perennation, persisting

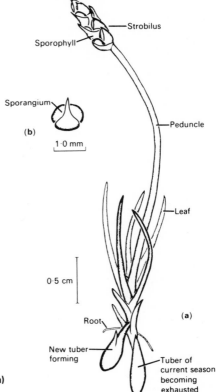

Fig. 5.15 *Phylloglossum drummondii.* **(a)**
Habit. **(b)** Sporophyll and sporangium.

through the dry period (when the remainder of the plant perishes) and giving
rise to the following year's growth. The gametophyte of *Phylloglossum* is
similar to that of *Lycopodium (Lycopodiella) cernuum*, and there is an
interesting analogy between the protocorm stage in the development of the
sporophyte of this species and the perennating tuber of *Phylloglossum*.
Although strikingly different from *Lycopodium* in habit, *Phylloglossum* is
clearly not distant in the basic features of anatomy and reproduction. It is no
doubt a form that has become specialized in relation to a particular kind of
habitat.

Selaginellales

In this Order, which includes a single living genus *Selaginella*, the habit
resembles that of the Lycopodiales, but the microphylls are ligulate, the
sporophylls are always aggregated into distinct strobili, and the spores are of
two kinds, differing in size and in the sexes of the gametes they eventually
produce.

There are more than 700 species of *Selaginella*, most of which are tropical, ranging from small epiphytes to large climbing plants. A few species of arid areas are capable of flexing their branches and, in a rolled-up form, resisting prolonged desiccation. The basic cell structure and enzymes are conserved in the dehydrated cells. Normal form and metabolism are rapidly resumed on moistening ('resurrection plants'). The stems are commonly much more branched than in *Lycopodium*, the branches and sub-branches often growing in the same plane and forming fern-like fronds. In many species, leafless rhizophores originate from the stem at points of branching and grow down towards the soil, dichotomizing as they approach its surface. Contact apparently acts as a stimulus, causing roots to arise endogenously and penetrate the substratum. The leaves are usually spirally inserted, and in some species they radiate around the stem, as in *S. selaginoides*. On the lateral branches of other species, however, the leaves become coplanar with the branch system, so increasing the frond-like effect of the whole. This often involves heterophylly, the two lower rows of leaves becoming expanded in the plane of the branch system, and the upper leaves remaining small and adpressed. This is well seen in *S. kraussiana* (Fig. 5.16), now becoming common in some parts of southern England. The leaves of *Selaginella* resemble those of *Lycopodium*, but a minute, tongue-like ligule is inserted into the upper side of the leaf close to the axis. This organ, of uncertain function, appears early in the life of the leaf, but remains colourless since the plastids fail to develop internal lamellae. The prominent Golgi bodies in the cells may be the source of the mucilaginous investment of the young ligule. The cells soon die and as the leaf expands the ligule shrivels. In some species

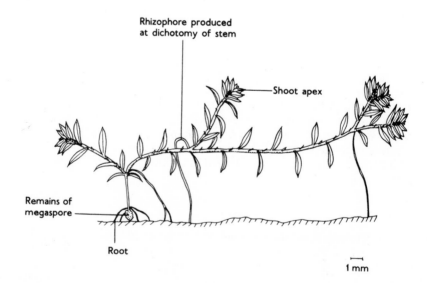

Fig. 5.16   *Selaginella kraussiana.* Young sporophyte.

of *Selaginella* the upper cells of the leaf contain a single cup-shaped chloroplast, similar to that of *Anthoceros* (p. 120), but lacking a pyrenoid.

GENERAL FEATURES OF THE ANATOMY   The anatomy of *Selaginella* differs little from that of *Lycopodium*, although occasionally the end walls of the tracheids fail to differentiate, thus giving rise in places to continuous channels simulating vessels. The endodermis, of a kind unknown elsewhere in the Plant Kingdom, consists of elongated hypha-like cells which suspend the stele in the central cavity (Fig. 5.17). The formation of this trabecular endodermis must reflect a greater expansion of the cortical region than of the vascular, so generating the intervening space. The endodermal cells remain alive and, as is usual in the endodermis, the plasmalemma forms a tight junction with the inner surface of the Casparian strip (see p. 295). The stele of *Selaginella* is basically a protostele lacking internal parenchyma, but it is often ribbon-like instead of cylindrical, and, especially in aerial axes, several steles may ascend the stem together (polystely). In the rhizomes of a few

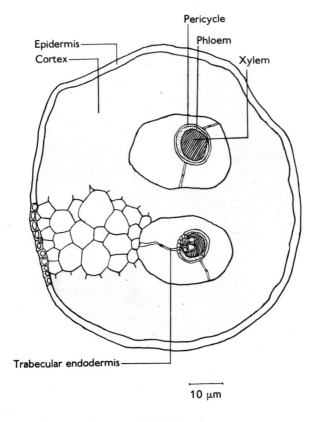

Fig. 5.17   *Selaginella* sp. Transverse section of stem.

species the xylem is in the form of a hollow cylinder lined on both surfaces by phloem. An endodermis also occurs both externally and internally. Such a stele, which is said to be **amphiphloic,** and is referred to as a **solenostele,** clearly exhibits a more complicated pattern of differentiation than a protostele, but the factors controlling differentiation of this kind are still little known.

Rhizophores develop from meristematic areas often lying below the points of branching. These primordia seem to be to some extent indeterminate, and if a branch is removed the adjacent primordium will grow out to form a shoot instead of a rhizophore. However, if the branch is replaced by a source of the growth-regulating substance indole-3-acetic acid (auxin), the primordium behaves normally. This was one of the first and most striking demonstrations of how growth-regulating substances, diffusing from one area to another in a plant, maintain the familiar pattern of morphogenesis. The form of growth in *Selaginella* seems to be particularly labile. The apices of roots in pure culture in the absence of auxin may begin to produce leaf primordia and generate a shoot.

REPRODUCTION   In the strobili, which terminate the main or lateral axes, the sporophylls are usually in four ranks. Each sporangium lies in the axil of a sporophyll, between the ligule and the axis (Fig. 5.18). There are two kinds of sporangia, producing **mega-** and **microspores** respectively, located in different regions of the cone. The initial development of each kind of sporangium is the same, and closely resembles that of the sporangia of *Lycopodium*. Development diverges with the formation of the spore mother

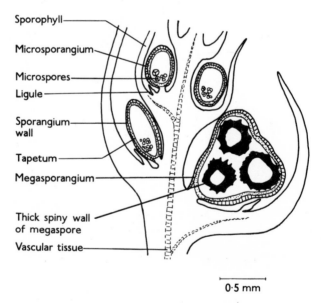

Sporophyll
Microsporangium
Microspores
Ligule
Sporangium wall
Tapetum
Megasporangium
Thick spiny wall of megaspore
Vascular tissue

0·5 mm

Fig. 5.18   *Selaginella kraussiana.* Longitudinal section of strobilus.

cells. In the microsporangia all the mother cells undergo meiosis and form microspores, but in the megasporangia not all the sporogenous cells complete their development. Only one or a very few mother cells yield spores, and not all of these may persist. The consequence is that the number of spores reaching maturity is very small and ranges, depending upon species, from one (e.g. *S. sulcata*) to about 12. The fully differentiated megaspores are conspicuous for their size, their store of food materials built up at the expense of the resorbed abortive sporogenous tissue and spores, and the thickening and ornamentation of their walls. They are in consequence some of the most remarkable spores in the Plant Kingdom; those of *S. exaltata*, for example, may exceed 1 mm in diameter.

Some nuclear divisions occur in the spores while they are still in the sporangia, but growth of the gametophytes, leading to rupture of the spore wall, does not resume until after the spores are shed. In the microspore an unequal mitosis produces a large antheridial cell and a small cell, termed a prothallial cell, which represents the sole development of the somatic tissue of the gametophyte. The antheridial cell continues to divide and yields a normal antheridium, from which 128 or 256 biflagellate spermatozoids are eventually liberated. Each is about 25 μm long, with a narrow elongated

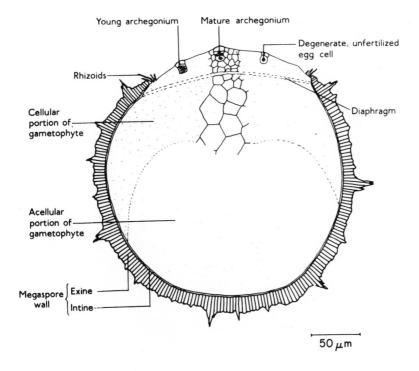

Fig. 5.19 *Selaginella kraussiana.* Vertical section of megaspore with endosporic gametophyte.

nucleus in which longitudinal strands, probably of chromatin, become apparent during differentiation. In the megaspore the food material comes to occupy a central position, and free nuclear division occurs at its periphery. Subsequently cell formation begins beneath the triradiate scar, which is eventually forced open by the general swelling, exposing a cap of gametophytic tissue (Fig. 5.19). The somatic tissue of the female gametophyte is thus more extensive than that of the male, although chlorophyll is quite absent from both. There is some specific variation in the extent to which the cellular portion of the female gametophyte is delimited from the food supply below. In some species the boundary is imprecise, and cell formation gradually extends down into the lower region, but in other species a distinct diaphragm separates the upper cellular region from a largely acellular food reserve.

The female gametophyte, once exposed, protrudes as an irregular cushion bearing rhizoids at its margins and in the central region archegonia. The necks at the archegonia are very short, consisting of no more than two tiers of cells (Fig. 5.20). Fertilization necessarily depends upon a microspore falling close to a megaspore, and a film of water being present when the gametangia are mature. The rhizoids around the female gametophyte may in these conditions serve to retain a 'fertilization drop' above the archegonia in which the spermatozoids congregate.

EMBRYOLOGY   The embryogeny of *Selaginella* is like that of *Lycopodium* in being endoscopic, but differs in the greater development of the suspensor, and the wide variation in detail between species. In *S. selaginoides*, for example, elongation of the suspensor pushes the developing embryo down into the food reserve. In *S. kraussiana*, however, in which the food reserve is

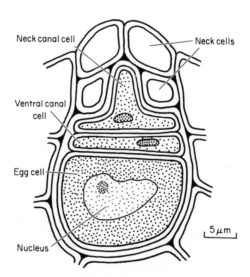

Fig. 5.20 *Selaginella kraussiana*. Mature archegonium (from an electron micrograph).

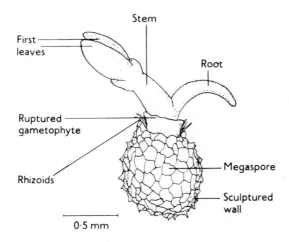

Fig. 5.21   *Selaginella kraussiana*. Sporophyte emerging from gametophyte.

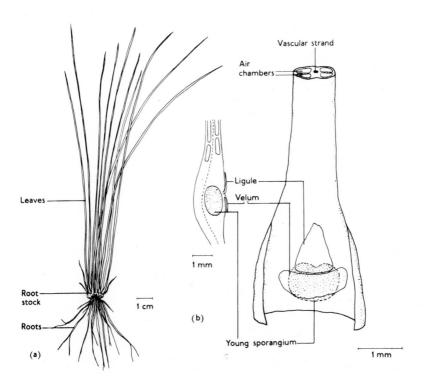

Fig. 5.22   *Isoetes echinospora*. **(a)** Habit. **(b)** Leaf base with young sporangium in face view and longitudinal section.

cut off by a diaphragm, the development of the suspensor is markedly less, but a curious downward extension of the archegonial canal carries the embryo through the diaphragm into the centre of the gametophyte. Apart from these features, there is also variation in the development of the foot region of the embryo, and in the relative positions in which the various parts of the embryo arise. In all species the embryo eventually emerges from the upper surface of the gametophyte (Fig. 5.21).

The substantial thickening of the megaspore wall in *Selaginella* probably protects the spore for considerable periods from desiccation and decay. Once conditions are favourable for germination, development is rapid and, because of the considerable food reserve of the megaspore, independent of an external supply of nutrients. Compared with the life cycle of the homosporous *Lycopodium*, there is considerably less time spent in the gametophytic phase. Since this, in view of the delicacy of the gametophytic tissues and their dependence on the maintenance of humid conditions, is the most vulnerable phase of the life cycle, the modifications that lead to its curtailment no doubt confer a considerable selective advantage on *Selaginella*. This is perhaps reflected in its numerous species.

In some species of *Selaginella* (e.g. *S. rupestris*) young plants emerge from the female regions of strobili. This has been regarded as following from the lodging of microspores between the megasporophylls and the fertilization of an egg while the megaspore was still *in situ*, thus simulating an early step in the evolution of a seed. It is doubtful, however, whether this can be substantiated. In *S. rupestris*, at least, reproduction is apogamous, the megaspore producing an embryo directly without fertilization.

Isoetales

The members of this Order have a remarkable, rush-like habit, quite unlike that of any other lycopsid. They have short, fleshy, upright rootstocks, occasionally showing one or two dichotomies, and bearing a tuft of quill-like microphylls (Fig. 5.22). The microphylls are ligulate, and reproduction is heterosporous. All living representatives, contained in the two genera *Isoetes* and *Stylites*, are aquatic, or plants of situations subject to periodic or seasonal inundation.

GROWTH HABIT AND ANATOMY   *Isoetes* is widely distributed, three species occurring in Britain. The rootstock in all species is mostly below the level of the substratum, and is rarely branched. The leaves, which in some aquatic species may reach a length of 70 cm, are arranged, at first distichously, but subsequently in a dense spiral about a depressed apical meristem at the upper end of the rootstock. The roots arise from the lower end of the stock where the meristem is again depressed, but here lies extended along a transverse cleft. The roots are initiated at the base of the cleft (Fig. 5.23). Accompanying this singular morphology is an equally remarkable manner of growth. A cambial zone arises around the small amount of primary vascular tissue in the stock, but it contributes more to the cortex than to the stele. The activity of

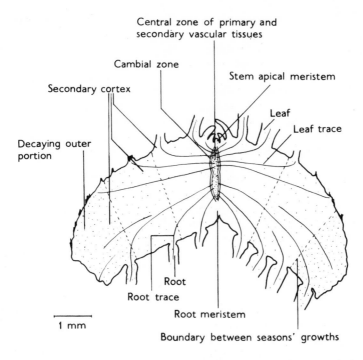

Central zone of primary and
secondary vascular tissues

Cambial zone

Stem apical meristem

Secondary cortex

Leaf

Leaf trace

Decaying outer
portion

Root

Root trace

Root meristem

1 mm

Boundary between seasons' growths

Fig. 5.23 *Isoetes echinospora.* Median longitudinal section of rootstock in plane perpendicular to that of the basal cleft.

this cambium, in temperate species at least, is seasonal. In step with the addition of new material within, a girdle of outer tissue, complete with its decaying leaves above and roots beneath, sloughs away. Consequently, having reached its mature diameter (which may exceed its length), the stock remains more or less the same size, the new leaves and roots being carried up on to the shoulders of their respective meristems by the expansion of the products of the anomalous cambium.

The leaves contain a single vascular strand, often very tenuous, surrounded by four air canals, interrupted at intervals by transverse septa. These canals are especially striking in the aquatic species. The broadened leaf bases lack chlorophyll and overlap widely, forming a tight comal tuft. The anatomy of the stock presents a number of peculiar features. The primary xylem consists of more or less isodiametric tracheids, and at the base of the stele they are arranged in an anchor-like bifurcation lying in the same plane as the basal cleft of the stock. The tissue produced on the inside of the anomalous cambium differentiates as a mixture of tracheids, sieve cells and parenchyma.

The remainder of the tissue in the stock is parenchymatous, and no recognizable endodermis delimits the vascular tissue. The roots possess a single vascular strand surrounded by a cortex of two distinct zones: an outer fairly resistant to decay, and an inner of more delicate tissue with numerous air spaces.

In an American mat-forming species of *Isoetes* the leaf arrangement remains distichous, and the rootstock extends laterally in the plane of the leaves. Adventitious buds arise at the extremities of these extensions and give rise to additional plants. *Stylites* closely resembles *Isoetes*, but the rootstock is dichotomously branched and may reach a height of 15 cm. The roots are confined to a single furrow which runs along the side of the stock. The plant has been found only in the High Andes of Peru, where it forms dense cushions by the sides of glacial lakes.

REPRODUCTION   Mature plants of *Isoetes* are usually abundantly fertile. The leaves first formed in a season's growth bear megasporangia, and those later microsporangia, although the sporangia frequently abort on the last formed leaves. The sporangium is initiated much as in *Selaginella* between the ligule of the sporophyll and the axis, but distinctive features emerge as development proceeds. Part of the central tissue, for example, remains sterile and differentiates as trabeculae which divide the mature sporangium into a number of compartments (Fig. 5.22b). Also, the ripe sporangium becomes enclosed in a thin envelope (velum). This orginates just below the ligule and grows down over the sporangium, leaving a central pore (foramen). Spore formation in *Isoetes* and *Stylites* is also peculiar in that the microspores are monolete and the megaspores are trilete. Not only are these genera unique in the living lycopsids in producing monolete spores, but they are also among the few plants in which monolete and trilete spores are produced by the same individual. The spores are liberated by the decay of the sporophylls. Their subsequent germination and development are similar to those of the micro- and megaspores of *Selaginella*. In *Isoetes*, however, the male gametophyte is wholly endosporic. The single antheridium contains only four spermatozoids, differing from those of *Lycopodium* and *Selaginella* in being multi-flagellate. They are released by rupture of the microspore wall. The female gametophyte is similar to that of *Selaginella*. Cell formation usually extends down into the body of the spore, and the diaphragm seen in many species of *Selaginella* is absent. Although initially endosporic, the expanding gametophyte ruptures the megaspore at the site of the triradiate scar, but, as in *Selaginella*, fails to develop chlorophyll. The archegonia are also similar, but the necks consist of four tiers of cells in place of two.

The first division of the zygote is slightly oblique. No suspensor is formed, but the embryogeny can still be termed endoscopic since the outer cell gives rise to the foot and the remainder of the embryo comes from the products of the inner cell. Differential growth causes the embryo to turn round so that it becomes directed towards the upper surface of the gametophyte. It eventually breaks through, but the young plant remains for some time partially enclosed by a sheath of gametophytic tissue. Apogamy does not appear to have been detected in *Isoetes*, but the production of a bud in place of a sporangium is not uncommon.

Although *Isoetes* and *Stylites* are like the remainder of the Lycopsida in

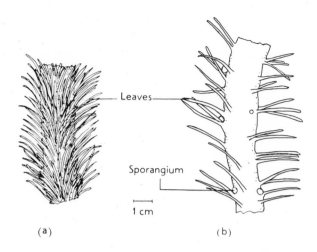

Leaves

Sporangium

1 cm

(a)   (b)

Fig. 5.24 *Baragwanathia.* **(a)** Vegetative shoot. **(b)** Shoot bearing sporangia. (After Lang and Cookson (1935). *Philosophical Transactions of the Royal Society,* **224B**, 421.)

essentials, they share a number of remarkable features unrepresented else-where in archegoniate plants. They appear to be the products of a line of lycopsid evolution that has been independent for a considerable period.

The early Lycopsida

The early recorded Lycopsida, the exact classification of which remains diffi-cult, lived in the warm, humid climate of the Lower Devonian. An example is provided by the remarkable *Baragwanathia* (Fig. 5.24) from Australian rocks. This consisted of dichotomizing, aerial axes, some of which reached a diameter of 6.5 cm, bearing numerous overlapping, needle-like leaves about 4 cm long. The stem contained a core of tracheids, from which slender strands ascended into the leaves. Some of the shoots had fertile regions in which reniform sporangia, containing cutinized spores, lay amongst the leaves. Although the precise attachment of the sporangia is still unknown, the general resemblance of *Baragwanathia* to a lycopod is so striking that an affinity is clear. A similar plant with mesarch xylem (*Asteroxylon*) occurs fossilized in the Rhynie Chert. *Protolepidodendron* (Fig. 5.25), from the Middle Devonian, apparently grew in a manner similar to *Lycopodium clavatum*. A prostrate, dichotomously branching stem gave rise to upright branching shoots, the whole system being clad in microphylls forked at the tip. The leaves of some of the erect branches bore unprotected sporangia on their adaxial surfaces. In the herbaceous lycopods, living and fossil, there is a correlation between exarch protoxylem and the lateral position of the sporangia, contrasting with the endarch protoxylem and terminal sporangia of *Rhynia*.

Carboniferous rocks yield an abundance of lycopsid fossils, both of

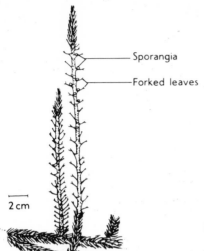

Sporangia

Forked leaves

2 cm

Fig. 5.25 *Protolepidodendron*. Habit. (After Kräusel and Weyland, from Delevoryas (1962). *Morphology and Evolution of Fossil Plants*. Holt, Rinehart and Winston, New York.)

vegetative and reproductive structures and of isolated spores. Of the many forms present, some were evidently herbaceous. Those having a general resemblance to living *Lycopodium* are placed in *Lycopodites*, but others, in which heterospory has been demonstrated, are believed to have been more like *Selaginella* and are placed in *Selaginellites*. The most impressive Lycopsida of the Carboniferous, however, were undoubtedly the arborescent Lepidodendrales, some of which achieved a height of 30 m. *Lepidodendron* (Fig. 5.26), for example, consisted of a trunk, 1 m or more in diameter at its base, which rose as a single column until it broke up by numerous dichotomies into the dense crown of branchlets. The upper parts of the tree bore simple ligulate microphylls, up to 20 cm long, triangular in cross-section, and arranged in regular spirals.

The trunk and lower branches of *Lepidodendron* and its relatives retained a characteristic pattern of diamond-shaped leaf scars (Fig. 5.27). In each of these occurred a pit, which indicated the site of the ligule, and two lateral softer areas, one on each side of the vascular bundle. These are the remains of strands of aerenchyma (**parichnos**) which ran from the cortex of the stem to the leaf. The trunk was anchored at ground level by four radiating arms which, since they were first found detached and not immediately recognized, were named *Stigmaria*. The Stigmarian axes dichotomized freely and the smaller bore rootlets, anatomically similar to those of *Isoetes*, in spiral sequence.

Despite the girth of *Lepidodendron*, the anatomy of the trunk was comparatively simple, and its manner of growth consequently puzzling. The primary vascular tissue is often well preserved, and even sieve cells have been seen in sufficient detail to reveal a close resemblance to those of *Lycopodium*. A vascular cambium evidently arose at an early stage, but it added only a

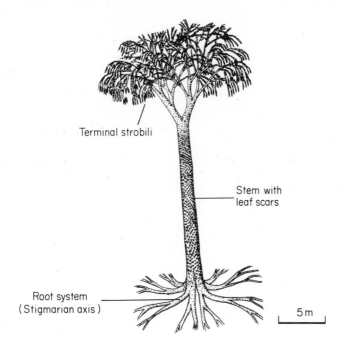

Fig. 5.26 *Lepidodendron*. Habit. (After Hirmer (1927). *Handbuch der Palaeobotanik*. Oldenbourg, Munich.)

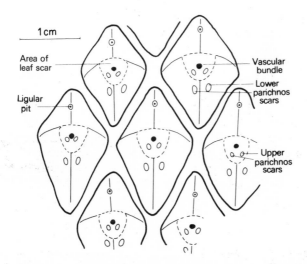

Fig. 5.27 *Lepidodendron*. Surface of cast of stem showing leaf bases. The thin-walled tissue of the parichnos strands has collapsed in the leaf cushions, producing the depressions referred to as secondary parichnos strands below the leaf scars.

narrow zone of secondary tissue to the primary stele, and the formation of secondary phloem is doubtful. Additional secondary activity occurred in the outer cortex, resulting in a hard sclerotic periderm which undoubtedly provided the principal mechancial support to the trunk. The inner cortex contained elongated cells, a so-called 'secretory tissue', which may have been a primitive form of phloem. The central zone of cortex, usually fragmentary or missing in the fossilized material, probably consisted of thin-walled aerenchyma, continuous with that in the Stigmarian axes and leaves. The curious anatomy, particularly the small amount of secondary xylem, has led to the view that the growth of these trees did not continue indefinitely, but was determinate. It is envisaged that the plant first generated a massive apical meristem, the activity of which then produced an axis of considerable height before dichotomy began. At each dichotomy the apices became smaller, until eventually they ceased to be active.

In most Lepidodendrales the strobili, varying from 1 to 3.5 cm in width and 5 to 40 cm in length, were terminal on the branchlets. A few homosporous cones are known, but most, even some of the earliest (e.g. *Cyclostigma* from the Upper Devonian), were heterosporous, resembling the cones of *Selaginella* in general features and the placement of the sporangia in relation to the ligule. Germinating megaspores and microspores of the Lepidodendrales have occasionally been found in petrified material. Some megaspores contained largely endoscopic gametophytes with rhizoids and archegonia, and possible mitotic figures have been seen in germinating microspores. It is clear that reproduction was basically as in *Selaginella* and *Isoetes*. A feature unrepresented in living lycopsids is seen in *Lepidocarpon* (Fig. 5.28), a form in which the megasporangium was enclosed in an involu-

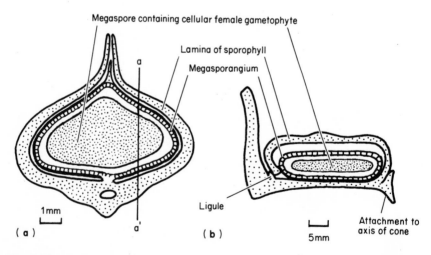

Fig. 5.28 *Lepidocarpon*. **(a)** Transverse section showing megasporangium enclosed in folded sporophyll. (After Arnold (1947). *An Introduction to Paleobotany*. McGraw-Hill, New York.) **(b)** Vertical section of region indicated by a . . . a' in **(a)**. (After Hoskins and Cross, from Arnold (1947). *In Introduction to Paleobotany*. McGraw-Hill, New York.)

tion of the sporophyll, with a micropyle-like opening at the distal end. The female gametophyte was evidently retained in this structure, providing evidence that in the Carboniferous period some lycopsids produced a female organ approaching a rudimentary ovule. Many specimens of *Lepidocarpon* have been found containing embryos. The detached sporophyll with the megasporangium and embryo may have floated, and provided a means of distribution of the plant in the Carboniferous swamps.

Pleuromeia (Fig. 5.29), the sole representative of the Pleuromeiales, is a little known plant found in Triassic sandstone. Its interest lies in its being morphologically intermediate between the Lepidodendrales and the Isoetales. A single trunk, little more than 1 m high, with spirally arranged leaf scars, rose from a rounded or lobed base which produced roots in the same manner as Stigmarian axes. It terminated above in a crown of narrow ligulate leaves, and when fertile a single erect cone with curious obtuse sporophylls. *Pleuromeia* was heterosporous and may occasionally have been dioecious. Ecologically it seems to have been maritime, some species possibly forming coastal thickets.

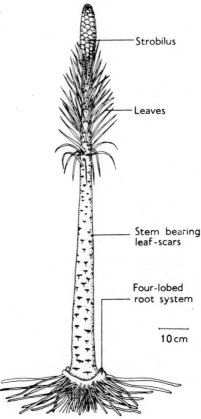

Fig. 5.29 *Pleuromeia sternbergi*. Reconstruction. (After Hirmer, from Andrews (1961). *Studies in Paleobotany*. Wiley, New York.)

Strobilus

Leaves

Stem bearing leaf-scars

Four-lobed root system

10 cm

The Jurassic and Cretaceous periods yield evidence of herbaceous forms very similar to *Lycopodium* and *Selaginella*. *Selaginellites hallei*, for example, was a small, heterophyllous plant, with leaves faintly denticulate at the margin, like those of many species of *Selaginella*. There were four megaspores in each megasporangium, and each megaspore reached a diameter of about 500 μm, about 10 times the size of the microspores. *Nathorstiana*, from Cretaceous rocks of Germany, recalls at once the Isoetales. An upright stock, made irregular by leaf scars, reached a height of about 12 cm and bore a crown of needle-like leaves. The base was divided into a number of vertical lobes from which the roots emerged. Nothing is yet known of the sporophylls.

The phylogeny of the Lycopsida

It is clear from the fossil record that the living Lycopsida are the relicts of a component of the Plant Kingdom which reached the apogee of its morphological complexity and floristic success in the Palaeozoic, about 200 million years ago. Besides the antiquity of the lycopsid kind of construction, the fossil record also indicates that the Lycopsida always had a distinctly heteromorphic life cycle. As in the living representatives, there is a conspicuous absence of morphological and anatomical similarities between the sporophyte and gametophyte phases, a situation which contrasts strongly with that in the Psilopsida. In the Lycopsida the gametophytes appear never to have been other than wholly parenchymatous plants of lowly status.

The reduction of the gametophytes in the heterosporous Lycopsida to short-lived, almost wholly endosporic thalli can be regarded as a major step in the direction of becoming independent of humid conditions for the survival of this vulnerable phase. Had the Lycopsida acquired this valuable adaptation to terrestrial life earlier, and proceeded to the evolution of seeds, their spectacular decline at the end of the Palaeozoic might never have occurred.

Sphenopsida

Sporophyte consisting of a monopodial branch system, some axes rhizomatous and bearing roots. Leaves microphyllous, borne in whorls. Vascular tissue of tracheids and phloem. Sporangia borne on sporangiophores, aggregated in terminal strobili. Gametophyte (in living forms) terrestrial. Spermatozoids multiflagellate. Embryogeny exoscopic.

The only living sphenopsid is *Equisetum*, but the Sub-Division has a rich fossil record. Four Orders are recognized: the Equisetales (containing *Equisetum* and a number of fossil genera), and the extinct Calamitales, Sphenophyllales and Pseudoborniales.

Equisetales

The striking feature of this Order is the jointed structure of the stem, and, in the regions of uniform diameter, the regular alternations of the microphylls in the successive whorls. The stems are often conspicuously ridged, each ridge lying beneath a leaf. Consequently, provided the number of leaves in a series of whorls remain the same, the ridges also show regular alternation from one internode to the next.

Equisetum (Fig. 5.30), the horsetail, is a familiar sight in parts of the North Temperate zone, but is rarer in the Tropics and southern hemisphere, being absent altogether from Australia and New Zealand. About 15 species are now living, but others are known as Mesozoic fossils. The genus, or a form very closely similar, was widespread in Cretaceous times, and some species appear to have formed dense stands at the fringes of Cretaceous lakes. Moist habitats, such as river banks and marshy ground, are also favoured by most living species, but some are able to thrive in much drier places.

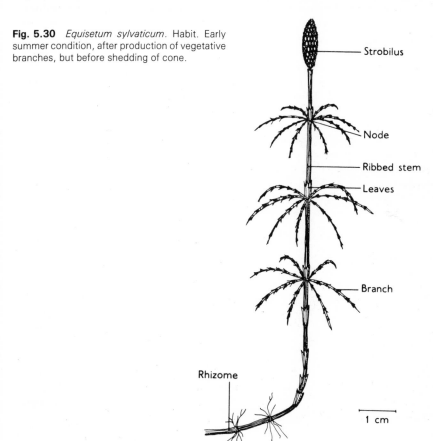

Fig. 5.30 *Equisetum sylvaticum*. Habit. Early summer condition, after production of vegetative branches, but before shedding of cone.

Strobilus

Node

Ribbed stem

Leaves

Branch

Rhizome

1 cm

E. arvense, for example, often flourishes on well-drained railway embankments. All species have a similar growth form. A perennial underground rhizome gives rise to green aerial shoots, and occasionally also to perennating tubers packed with starch. In temperate and arctic regions the aerial shoots die back at the end of the growing season and new shoots emerge in the following spring.

MORPHOLOGY AND ANATOMY OF *EQUISETUM*   Although the height of the aerial system varies from a few centimetres in arctic and alpine species to as much as 10 m in the tropical *E. giganteum*, its morphology is strikingly uniform. Branches appear only at the nodes, and where several are present they too are whorled. The branch primordia are not, however, axillary, but they arise between the microphylls and eventually break through the sheath formed by their congenitally fused bases. In subterranean axes roots emerge from directly below the sites of branch primordia. Branch and root primordia are in fact present at every node, but they develop only in appropriate environmental conditions, sometimes reproducible in the laboratory. In *E. arvense*, for example, green branches can be made to grow from the nodes of an etiolated unbranched fertile shoot if they are enclosed in a moist chamber.

The structure of the axes of *Equisetum* also shows little variation. The mature stems have a large central cavity, surrounded by a ring of vascular bundles (Fig. 5.31). These bundles are of the same number as the ribs on the

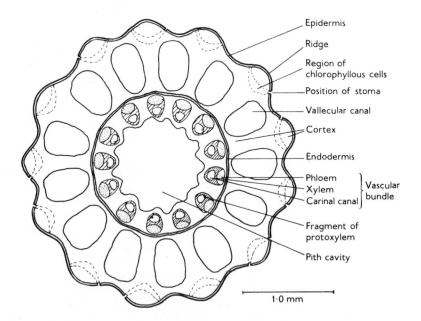

Epidermis

Ridge

Region of chlorophyllous cells

Position of stoma

Vallecular canal

Cortex

Endodermis

Phloem
Xylem   } Vascular
Carinal canal } bundle

Fragment of protoxylem

Pith cavity

1·0 mm

Fig. 5.31   *Equisetum arvense*. Transverse section of young stem.

outside of the internode (and hence as the leaves of the node above), and are also co-radial with them. An endodermis can usually be distinguished, either encircling the stele on the outside alone (as in *E. arvense*), or forming two continuous cylinders, one inside and one outside the ring of bundles (as in *E. hyemale*), or surrounding each bundle individually (as in *E. fluviatile*). In some species the position of the endodermis in the rhizome is different from that in the aerial stem. Alternating with the vascular bundles, and lying between the endodermis and the periphery, are large longitudinal air chambers known as *vallecular canals*.

The vascular bundles themselves contain very little lignified tissue. In a differentiated internode the protoxylem is represented solely by fragments of tracheids adhering to the sides of a cavity, termed the *carinal canal*, present on the adaxial side of each bundle. The metaxylem differentiates as two groups of tracheids, one placed tangentially on each side of the phloem. The vascular anatomy of the nodes is highly peculiar. Here the tracheids, resembling those of the metaxylem, run horizontally and form a ring linking the bundles of the adjacent internodes. In *E. hyemale* the phloem close to the node is notable in lacking well-defined sieve cells. The root, branch and leaf traces also originate at the level of the node, the leaf trace departing immediately above the entry of the bundle from the internode below.

The strength of the *Equisetum* stem depends principally upon the cortical ridges. These consist of sclerenchymatous cells reinforced by deposits of silica. The support that this rather tenuous outer framework can provide is, of course, limited. The taller species of *Equisetum* usually grow in groves and, since the rough siliceous stems and branches do not readily slide over each other, the plants hold each other up.

The microphylls of *Equisetum* soon become scarious, and photosynthesis takes place predominantly in the surface layers of the stem. Apart from a few curious 'water stomata' or *hydathodes* in the adaxial epidermis of the tip of the microphyll, the stomata are confined to the valleys of the internodes, and thus lie above the vallecular canals. The stomata are often deeply immersed, merely a pore being visible externally. Each guard cell is flanked by a subsidiary cell, and each subsidiary cell bears transverse bars of cutin on its exposed surfaces (Fig. 5.32).

THE GROWTH OF THE AXES   The growth of the axis of *Equisetum* provides a striking example of co-ordinated differentiation. The axis is surmounted by a single tetrahedral apical cell. The extent to which this particular cell undergoes division is doubtful, but adjacent to its three posterior faces daughter cells are cut off in a regular, clockwise sequence. While the stem is growing there is no pause in the meristematic activity of the apical initials and their products are recognizable as three tiers of cells in the extreme apex. The primordium of a whorl of leaves first becomes visible as a ring at the base of the apical cone. While this ring is being superseded by another, leaf teeth initials become visible around the upper margin of the first. These initials are of limited growth and give rise to the free part of the microphylls, the basal ring meanwhile forming the sheath of fused bases.

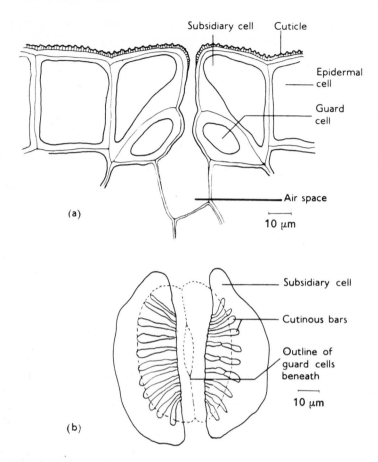

Fig. 5.32   *Equisetum* sp. Structure of stomata. **(a)** In vertical section. **(b)** In surface view partially macerated, showing the bars of cutin on the subsidiary cells. (After Hauke (1957). *Bulletin of the Torrey Botanical Club*, **84**, 178.)

The upper three or four nodal primordia remain close together, but the cells between them become organized as an intercalary meristem which surrounds, but does not cut across, the procambial strands. This meristem becomes active between the fourth and fifth nodes, and cuts off cells acropetally which go to form the internode. The activity continues while four or five new nodes are initiated at the apex so that the mature length of the internode is reached by about the ninth or tenth node. The secretion of silica on to the surface of the internode begins at the top and proceeds downwards, becoming completed as the intercalary growth ceases. The metabolic pathway of the silica and the manner of its secretion are not yet understood.

The procambial tissue, the disposition of which foreshadows that of the vascular bundles in the mature axis, advances continuously, in step with the advance of the apex, at the level of the uppermost leaf whorl primordia. Differentiation of the procambial strand is not, however, a simple acropetal process. Protoxylem begins to appear at about the fourth node, and differentiation extends acropetally into the leaf and basipetally into the node below. Since this differentiation occurs during the time of maximum extension, most of the tracheids first formed in the internode are ruptured, and the area of weakness provided by the differentiating cells is pulled apart by the radial and tangential expansion of the stem, so yielding the carinal canal. Metaxylem begins to appear at the fifth node, and also differentiates basipetally. The rate of differentiation is such that the descending strands do not fuse with the metaxylem of the node beneath until the tenth node, when extension of the internodes is ceasing. The internodal metaxylem thus escapes rupture, although occasional small lacunae indicate that it is subjected to some stress during the concluding phases of differentiation. The differentiation of the phloem follows more or less the same course as that of the metaxylem.

In temperate and arctic habitats the aerial branches perish in the winter, and in this instance the growth of the shoot is a little different from that just described. Almost the whole of the next year's shoot overwinters in primordial form as a subterranean bud. In spring elongation of the internodes and further differentiation proceeds acropetally, so generating an aerial system of limited growth.

REPRODUCTION   In all species of *Equisetum* the sporangiophores are aggregated into a terminal strobilus (Fig. 5.33) which terminates either a vegetative axis (as in *E. palustre*) or a specialized axis lacking pigmentation which appears early in the growth season (as in *E. arvense*). An intermediate state is shown by *E. sylvaticum*, in which the fertile shoots are at first colourless, but after release of the spores become green and branch. The vascular system inside the cone recalls that of a node, since it consists of a cylinder of metaxylem. Fine traces depart to the sporangiophores, and the cylinder is broken here and there by parenchymatous perforations which bear no evident relation to the departing traces. The sporangiophores, which are not necessarily arranged in whorls, are peltate, and are tightly packed so that heads acquire a polygonal outline. About 20 sporangia are pendent from the margin of the head, and so lie more or less radially in the intact cone.

A sporangium develops from a group of cells, in which a central archesporial tissue, surrounded by a wall several layers thick, can soon be distinguished. The inner layer of the wall functions as a tapetum, together with about a third of the archesporial tissue, the remainder being sporogenous. While the spores are maturing the spore coat becomes differentiated into several membranes, some of the components probably coming from the tapetum. The innermost cellulose membranes are simple, but the outer part of the wall takes the form of an X, the arms of which remain tightly wrapped around the spore as it lies in the sporangium. On release the arms separate and respond to changes of humidity in the same way as the peristome teeth of the Bryales. They are thus referred to as elaters (or *haptera*), and their quite violent movements assist distribution of the spores. The sporangium

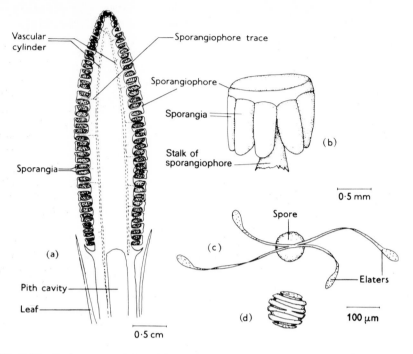

Fig. 5.33 *Equisetum maximum.* **(a)** Median longitudinal section of strobilus. **(b)** Peltate sporangiophore. **(c)** and **(d)** Spores with elaters in dry and moist condition respectively.

opens along a longitudinal stomium as the result of tensions arising during the drying out of the indurated and spirally thickened cells elsewhere in the wall.

Equisetum is usually considered to be homosporous. In *E. arvense* a large sample of spores falls into two classes whose mean diameters differ by about 10 μm, but this is an effect of drying. The spores, which contain chlorophyll, soon lose their viability if stored. On germination the cellulose wall of the spore is continuous with that of the filament. It has been shown that at the site of germination there is a large concentration of potassium ions, probably leading to a localized increased plasticity of the wall, since potassium ions are known to reduce the amount of cross-linking in cellulose microfibrils. The filament which emerges from the spore is transformed by division in a number of planes into a cushion of cells anchored by rhizoids. Subsequent growth is from a marginal meristem which forms a number of obliquely ascending lobes (Fig. 5.34), on the upper surface of which the sex organs appear. Growth in size is, however, limited, and the gametophytes rarely exceed 1 cm in diameter and 3 mm in height, although much larger forms have sometimes been found.

Culture experiments have shown that there are two different kinds of

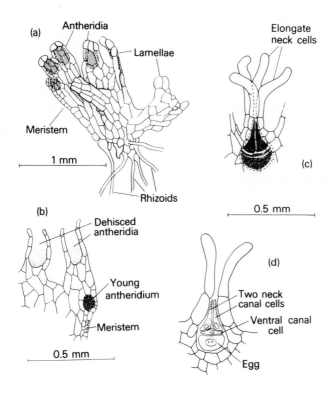

Fig. 5.34 **(a,b)** *Equisetum sylvaticum*. **(a)** Young male gametophyte. **(b)** Section through male branch. **(c,d)** *E. palustre*, archegonia. **(c)** General view. **(d)** Longitudinal section. (After Duckett (1968). Ph.D. Dissertation, Cambridge University.)

gametophytes. About half the gametophytes of a mass sowing of spores remain small and produce only antheridia, dying soon afterwards. The remainder are larger and longer lived. They first produce archegonia, and then, if none is fertilized, a crop of antheridia, followed by another of archegonia. Such gametophytes may last at least two years and in favourable circumstances produce several sporophytes. There is thus some evidence that *Equisetum* is heterothallic, but, since the proportions of the two kinds of gametophyte are related to the density of sowing and other cultural conditions, the differences in behaviour cannot have a simple Mendelian basis. Chance variations in the cytoplasmic complements of the spores, leading to differences in competitiveness, may influence the growth and subsequent gametogenesis.

The antheridia of *Equisetum*, like those of the Lycopsida, contain numerous spermatocytes. They open by the parting of the cover cells and the spermatocytes are discharged. Each is furnished with a number of long fibrils, an interesting parallel to the elaters attached to the spore. The flagella of the enclosed spermatozoids soon become active and the spermatocytes break open, releasing the gametes. Each is about 20 μm long and is twisted into a left-handed screw of three gyres, with an apical crown of about 100 flagella. The ultrastructure is similar to that of other archegoniate male gametes, and a ribbon of microtubules adjacent to the nucleus is a prominent feature.[14] The spermatozoid swims at about 300 μm per second in a helicoid path. The archegonia are formed mostly between the aerial lobes of the gametophyte. The necks project, and at maturity the four distal cells become elongated and reflexed. Fertilization, dependent, as in the Lycopsida, upon the presence of a film of water, is followed by the division of the zygote in a plane perpendicular to the axis of the archegonium. The subsequent embryology is exoscopic, and the products of the outer cell give rise to the stem apex, the primordium of the first whorl of leaves, and, in a variable lateral position, a root apex.

THE DEVELOPMENT OF THE SPOROPHYTE   The development of the young sporophyte follows a curious course (Fig. 5.35), without parallel in other living archegoniate plants. The first axis, which contains a simple protostele, is of determinate growth and never increases in diameter. In *E. arvense*, for example, it produces about six whorls, each of about three leaves. As this shoot ceases to grow, a bud grows out from below the first whorl of leaves.

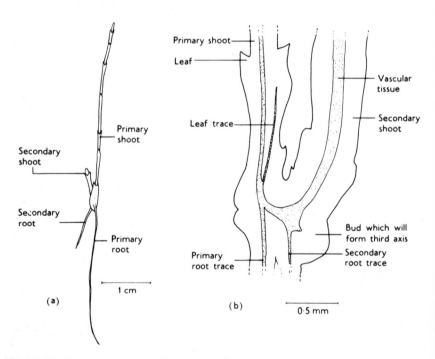

Fig. 5.35   *Equisetum arvense.* **(a)** Sporeling. **(b)** Development of mature form of plant. (After Barratt (1920). *Annals of Botany*, **34**, 201.)

This also produces an upright axis, of slightly greater diameter than the first and containing a protostele that shows a tendency towards medullation. This process is repeated two or three times until the axis reaches the diameter and structure characteristic of the mature plant. The rhizomatous growth habit is then initiated. The primary root of the embryo persists only a short time, and roots are produced freely from the nodes at and below the soil surface as the young plant becomes established.

The early Sphenopsida

The fossil record of the Sphenopsida parallels that of the Lycopsida, beginning early in the Palaeozoic and expanding in the Carboniferous, when the Sphenopsida must have formed a large part of the Earth's vegetation. From the end of the Palaeozoic to the present time they have been of diminishing importance until today only a single genus remains.

The fossil sphenopsids which show the closest resemblance to *Equisetum* are the Calamitales. These occurred predominantly in the Carboniferous, reached the proportions of trees and were probably one of the main competitors of the arborescent lycopsids. The leaves were whorled, and, although in some of the early forms forked (e.g. *Asterocalamites*), in the latter they were simple. The stems were conspicuously ridged, the ridges in some forms alternating from node to node, and in others lying superposed. The stomata of *Calamites* resembled those of *Equisetum*, transverse bars of cutin being present on the subsidiary cells (see Fig. 5.32). Petrifactions also reveal that the vascular system of the Calamitales was basically similar to that of *Equisetum*. In the internodal region a pith cavity was present, and the protoxylem was associated with a canal. A cambium, however, arose between the primary xylem and phloem and contributed a considerable amount of secondary vascular tissue to the stem. Probably as a consequence of this radial expansion, no air canals were present in the cortex.

The strobili of the Calamitales, which terminated lateral branches, consisted of alternate whorls of sporangiophores and bracts, although the cone of one early form appears to have contained peltate sporangiophores alone. Since cones are often found detached, they are placed in *form genera*, defined by the relative arrangements of the sporangiophores and bracts. The two form genera most widely represented are *Calamostachys* (Fig. 5.36) and *Palaeostachys*. The cones assigned to these form genera may, of course, have been produced by plants differing widely vegetatively. But homosporous and heterosporous cones have been described, and in at least one (*Elaterites*) the spores were clearly furnished with elaters.

Apart from the Cretaceous plants closely resembling living *Equisetum*, referred to earlier, there is evidence of herbaceous forms having also existed in the Palaeozoic. *Equisetites hemingwayi*, for example, from the Upper Carboniferous, is the remains of a fertile shoot very like *Equisetum* in the structure of the cone and order of size.

The Sphenophyllales, another Order of the sphenopsids, appeared in the late Devonian and were prominent in Carboniferous floras. They share little

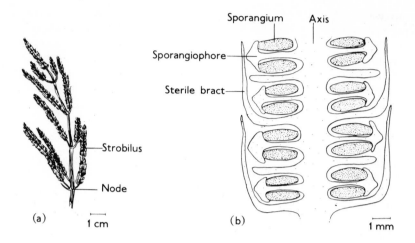

Fig. 5.36   *Calamostachys.* **(a)** *C. ludwigi.* Fertile shoot (from a compression). (After Weiss, from Andrews, Jr. (1961). *Studies in Paleobotany.* Wiley, New York.). **(b)** *C. binneyana.* Longitudinal section of cone. (After Andrews, Jr. (1961). *Studies in Paleobotany*, Wiley, New York.)

with the Equisetales and Calamitales except the whorled arrangement of the leaves. They were probably scrambling plants, supporting themselves on other vegetation in the manner of the familiar *Galium aparine* (cleavers or goose-grass). The leaves were wedge-shaped, with dichotomously branching venation, and were usually borne in multiples of three (Fig. 5.37). The stems contained a solid core of primary xylem, triangular in transverse section, with

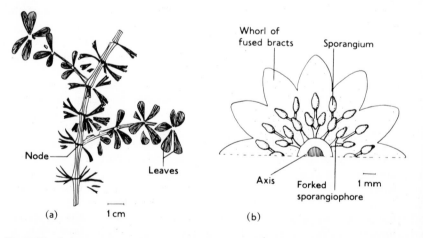

Fig. 5.37   *Sphenophyllum.* **(a)** *S. verticillatum.* Habit of vegetative shoot (from a compression). (After Potonié (1899). *Lehrbuch der Pflanzenpalaeontologie.* Dümmler, Berlin.) **(b)** *S. (Bowmanites) dawsoni.* Fertile whorl viewed from above. (After Hirmer (1927). *Handbuch der Palaeobotanik.* Oldenbourg, Munich.)

protoxylem at the vertices. The primary xylem was surrounded by secondary in radial files, resulting in a vascular system strikingly reminiscent of that of a root. These stems of *Sphenophyllum* provide the finest example in the Palaeozoic of a spore-bearing plant in which well defined secondary vascular tissue, both xylem and phloem, was regularly produced from a bifacial cambium. The strobili were terminal and frequently consisted of whorls of sterile bracts, each forming a cup-like sheath around the axis, on the adaxial side of which were attached the sporangiophores (Fig. 5.37b). The distal portion of the sporangiophore, which was free, branched and recurved, bore a number of sporangia. In other forms of cone the sterile part was less well developed, and the sporangiophore branching more complex. Most of the cones were homosporous, but distinct heterospory has been detected in one form. The Sphenophyllales disappeared at the beginning of the Triassic period.

The Pseudoborniales are based upon remains from the Upper Devonian of Bear Island. They appear to have borne their leaves, each of which forked once or twice, in whorls of four, and to have produced loose, terminal strobili, 30 cm or more in length. The sporangiophores were whorled, and each included a prominent sterile portion. Some of the stems reached a diameter of 10 cm, indicating that the Pseudoborniales were probably bushes or small trees.

The origin of the Sphenopsids

The bearing of the leaves or branches in whorls, and the articulate structure of the stem, although characteristic features of the Sphenopsida, are not of course confined to this Sub-Division of the Plant Kingdom. They are present in some of the algae (e.g. *Draparnaldia* (p..41), *Batrachospermum* (p. 95)) and in the flowering plants (e.g. *Galium, Casuarina*), and have clearly arisen a number of times in plant evolution. Nevertheless, the Sphenopsida appear to have specialized in this organization of the plant body from their beginnings, and they may even have had a common origin in the remote past.

Unfortunately no fossils are known which show convincingly how the Sphenopsids might have arisen. *Calamophyton* and *Hyenia*, both Middle Devonian in age, and consisting of axes bearing forking appendages, were once thought to be articulate plants close to the inception of sphenopsid evolution. This interpretation is now largely discarded; the transverse markings on the axes, formerly taken as an indication of nodes, appear to be nothing more than regular cross-fractures developing during fossilization. In the absence of any firm evidence we may, however, reasonably speculate that the origin of the sphenopsids probably lay in a plant with a simple axial form like that of the psilopsids, but bearing lateral appendages. These appendages may have forked once or twice, those towards the tips of some of the shoots terminating in sporangia. We must suppose a tendency for these lateral appendages to be produced in whorls, followed by the consolidation of this pattern of development, with corresponding changes in vascular anatomy, in the genetic constitution, leading eventually to plants with a well-defined articulate morphology.

Although in some of the Carboniferous sphenopsids sporangia were produced in association with bracts, these complexes seem to have no affinity with the lycopsid sporophyll, a reproductive structure, as we have seen (p. 165), well established in early Devonian times. It appears unlikely, therefore, that the origin of the sphenopsids is to be sought amongst these plants.

Although the Sphenopsida themselves may have had a common origin, the relationships between the Calamitales and Equisetales are clearly closer than between the other Orders. There seems to have been a general tendency towards the sporangiophore becoming peltate, and this perhaps hindered the evolution of seed-like structures, of which there are only rare indications in the sphenopsids. The almost total elimination of the sphenopsids points to their limitations, which possibly lay principally in the reproductive mechanism. It is difficult to account for the survival of the Equisetales in preference to the other Orders. It may have been because only this Order contained herbaceous forms sufficiently adaptable to meet the fluctuating conditions of the Mesozoic.

6
The Tracheophyta, II
(Pteropsida: Filicinae)

In the classification we have adopted the fourth Sub-Division of the Tracheophyta, referred to as the Pteropsida, encompasses the diverse range of vascular plants known as the ferns, gymnosperms and angiosperms. This grouping, which gives the Pteropsida the same taxonomic status as, for example, the Sphenopsida, is justified by a number of basic features common to the higher archegoniate and flowering plants, but absent from the Sphenopsida, Lycopsida and Psilopsida. The general features of the Pteropsida may be summarized as follows:

Pteropsida

Sporophyte usually branching freely, the branches bearing a distinct, often axillary, relationship to the leaves. The leaves often with an elaborately branched vascular system, the leaf trace commonly compound and leaving a parenchymatous gap at its departure from the stele. Vascular system consisting of tracheids or tracheids and vessels, and phloem. Position of sporangia various, sometimes enclosed in sporophytic tissue. Heterospory frequent, the female gametophyte in many forms contained within the sporophyte. Male gametes flagellate, or without any specialized locomotory apparatus. Embryogeny largely endoscopic, the embryo often retained and distributed within a seed.

The Pteropsida fall naturally into three Classes, namely the Filicinae, Gymnospermae and Angiospermae. The first two Classes have a rich fossil record, but that of the angiosperms is meagre and mostly recent. The Filicinae are thus some of the most ancient of the Pteropsida, and they remain far less specialized in their organization than flowering plants. A slice containing the apical region of a fern, for example, no more than 0.25 mm in thickness, will generate a normal stem in culture. With a flowering plant the same result can be achieved only with a substantially greater explant. Further it is clear that the cause of the morphological alternation in the life cycle of the land plants between a diminutive gametophyte and a conspicuous sporophyte will come from continued study of the reproduction of the homosporous Pteropsida. Reproduction of the flowering plants is so overlain with later evolutionary change that the essential principles can be identified only by those with a sound knowledge of the homosporous plants.

Filicinae[3,15]

Sporophyte herbaceous, leaves often large and resembling branch systems. Vascular system consisting of tracheids and phloem, usually lacking clearly defined secondary tissue. Sporangia borne on the leaves, but never on the adaxial surface of a microphyll. Heterospory rare. Gametophyte (of living species) usually autotrophic, spermatozoids multiflagellate. Embryogeny mostly endoscopic.

Homosporous plants whose axes lacked secondary thickening, and which appear referable to the ferns, are found in rocks as old as the Devonian. The ferns were undoubtedly an important component of the succeeding Carboniferous vegetation, and, although many of these early forms died out with the Palaeozoic, others were able to adapt themselves to changing conditions. Consequently, the ferns, with about 10 000 living species, have remained an important and, particularly in humid regions, a conspicuous element of the World's vegetation. Apart from this relative floristic success, the ferns show the greatest range of growth forms amongst the archegoniate plants, and the greatest diversity in the morphologies of individual organs. The leaves, however, remain in most forms relatively large and branched structures, branching also being characteristic of their vascular systems. This kind of leaf (often called a frond), strikingly different from the microphylls of the Lycopsida, is termed a ***megaphyll***.

The ferns are conveniently considered as five Orders, namely the Cladoxylales, Coenopteridales, Ophioglossales, Marattiales and Filicales. The first two Orders are wholly extinct.

Cladoxylales

The Cladoxylales are based upon a small number of peculiar, probably homosporous, plants from the Devonian and Lower Carboniferous. The axes (Fig. 6.1), the bigger reaching diameters of the order of 1.5 cm, branched irregularly, and bore small dichotomously forking leaves, rarely reaching 2 cm in length. Some of the terminal shoots were fertile, and here small fan-shaped sporangiophores took the place of leaves, each segment of the sporangiophore terminating in a sporangium.

The vascular system of the stem consisted of ascending, anastomosing plates of xylem, each in life probably surrounded by phloem and endodermis. Since the system has the appearance of falling into a number of independent sub-systems, it is said to be polystelic. The vascular supply of the branches was compound in origin, departing from several of the adjacent ascending plates. A curious anatomical feature was the presence of small parenchymatous areas within the xylem, especially towards the periphery of the

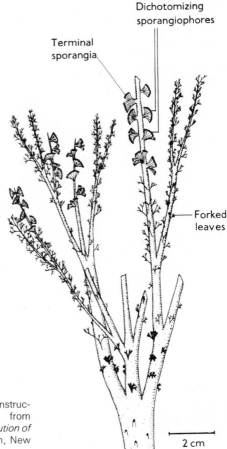

Dichotomizing sporangiophores

Terminal sporangia

Forked leaves

2 cm

Fig. 6.1 *Cladoxylon scoparium*. Reconstruction. (After Kräusel and Weyland, from Delevoryas (1962). *Morphology and Evolution of Fossil Plants*. Holt, Rinehart and Winston, New York.)

stem. Some species of *Cladoxylon* appear to have had rudimentary secondary thickening.

Coenopteridales

The Coenopteridales, which were undoubtedly ferns, flourished in the later Palaeozoic, but apparently did not persist into the Mesozoic. They showed a wide range of habitat, from tree ferns reaching a height of three or more metres to comparatively small epiphytes. The leaves were large and spreading, often resembling sprays of branchlets (Fig. 6.2). The branching was not always in one plane, and in one group, the zygopterids, the branchlets were in a distinct quadriseriate arrangement. The morphological similarity of leaf and stem was sometimes reinforced by large decumbent fronds giving rise here and there to roots and shoot buds. An Upper Carboniferous species of

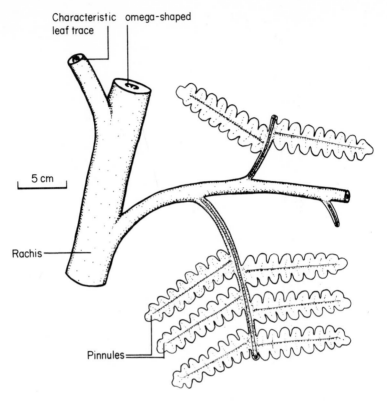

Characteristic omega-shaped
leaf trace

5 cm

Rachis

Pinnules

Fig. 6.2 *Botryopteris*. Reconstruction of portion of frond. (After Delevoryas and Morgan, from Delevoryas (1962). *Morphology and Evolution of Fossil Plants*. Holt, Rinehart and Winston, New York.)

Botryopteris, which must have formed low sprawling clumps, showed this kind of growth particularly well. In another form, *Metaclepsydropsis* (Fig. 6.3) from the Lower Carboniferous, the stem was rhizomatous and bushy fronds, elaborately branched in a quadriseriate manner, arose from it at intervals. That the complexity of the fronds appeared early in the phylogeny of this group of ferns is shown by equally remarkable fossils from the Devonian period. Notable amongst these very early coenopterids was *Rhacophyton* from the Upper Devonian. This had a stem up to 2 cm in diameter, possibly scrambling, bearing fronds in a crowded helix. The vegetative fronds were pinnately branched (except possibly in the ultimate segmentation), but the complex fertile fronds were branched in a zygopterid manner.

THE VASCULAR ANATOMY The vascular system of the stem of the coenopterids was commonly a protostele, often with a parenchymatous medulla, and lacking (except in *Zygopteris*) any secondary thickening. The

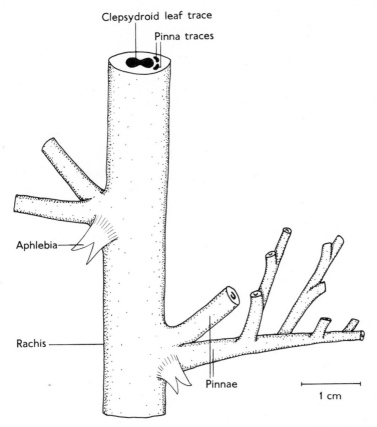

Clepsydroid leaf trace

Pinna traces

Aphlebia

Rachis

Pinnae

1 cm

Fig. 6.3 *Metaclepsydropsis.* Reconstruction of portion of frond showing quadriseriate arrangement of pinnae.

leaves, which were usually in a recognizable phyllotactic spiral, received a single vascular trace which assumed in the petiole a definite and characteristic symmetry. Consequently the genera and species of these ferns are largely based upon the profile of the leaf trace in transverse section. In *Metaclepsydropsis*, for example, the section was hour-glass shaped (clepsydroid). The protoxylem lay towards each pole, each group associated with an island of parenchyma. Pinna traces arose from each pole in alternate pairs, in register with the quadriseriate branching. In *Stauropteris*, where the frond was similarly constructed, the petiole contained four groups of tracheids ascending in parallel. The petiolar trace of *Tubicaulis*, which produced complanate fronds, formed an arc in transverse section, with the convex surface adaxial. In early species of *Botryopteris* the petiolar trace was ovate in section, but in later species had the form of a ω, the curvature being abaxial and protoxylem lying at the tips of the adaxial extensions. Well-defined phloem with elliptical sieve areas has been seen in several of these petioles. In many of the coenopterid fronds there were small branched emergences, each with an exiguous vascular supply, at the base of the rachis and at the sites of branching of the fronds. These are referred to as ***aphlebiae***, and occur in some living ferns (e.g. *Hemitelia* and other tree ferns).

THE REPRODUCTIVE ORGANS   In general in the coenopterids the sporangia terminated branchlets, either a basal or a terminal region of the frond being fertile. The sporangia of the Devonian *Rhacophyton* lacked any special opening mechanism, but in later forms the sporangia commonly possessed an annulus of thick-walled cells interrupted by a thin-walled stomium. The sporangia of some of these ancient ferns were surprisingly modern in appearance and position. In some species of *Botryopteris*, for example, sporangia, 255–340 μm in diameter, were attached to the abaxial surfaces of fronds and had a wall only one cell thick with a lateral annulus. The resemblance to the sporangia of living *Osmunda* is striking. Homospory was the predominant condition in the coenopterids, but a well-established example of heterospory is provided by the Carboniferous *Stauropteris burntislandica*. The megasporangia (Fig. 6.4) of this species, which were parenchymatous at

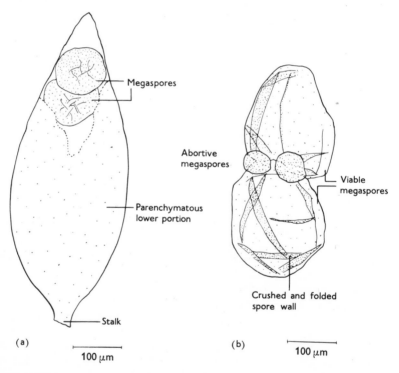

(a)   (b)

100 μm   100 μm

Fig. 6.4 *Stauropteris burntislandica*. **(a)** Megasporangium (from a specimen in the Oliver collection). **(b)** An isolated tetrad (*Didymosporites*). (After Chaloner (1958). *Annals of Bontany, New Series*, **22**, 197.)

the base, produced only one tetrad, consisting of two large and presumably functional spores about 200 μm in diameter, with two abortive spores lying between them. In a related species, believed to be homosporous, spores germinating in a manner typical of living ferns have been found petrified within sporangia. Apart from a few isolated instances of this kind, nothing further is known of the gametophytic phase of the coenopterids.

Ophioglossales

The Ophioglossales form a small and morphologically peculiar Order of living ferns which, since they have no well-established fossil record, are of obscure origin. In all members the fertile region of the frond takes the form of a spike or pinnately branched structure, clearly set off from the vegetative portion. A feature that separates the Ophioglossales from other living ferns is that the fronds, instead of expanding from a closely coiled immature state (a

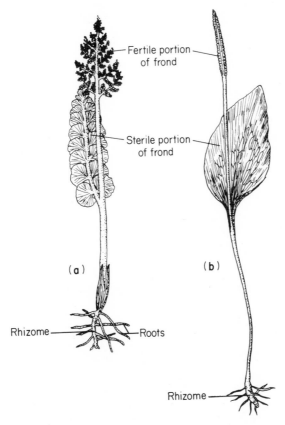

Fig. 6.5 **(a)** *Botrychium lunaria.* Habit. **(b)** *Ophioglossum vulgatum.* Habit. (Both after Lowe (1874–78). *Our Native Ferns.* London.)

condition known as 'circinate vernation'), grow marginally from a more or less flat primordium. That of *Botrychium lunaria* when young shows a distinct kind of folding, the upper margins of the pinnules being covered by the lower margins of the pinnules above.

Of the three genera of the Class, *Botrychium* (moonwort) and *Ophioglossum* (adder's tongue) are fairly widespread, the former mainly in the north temperate zone and the latter chiefly in the Tropics. Both genera include species native to the British Isles. The third genus, *Helminthostachys*, is restricted to the Polynesian Islands in the South Pacific and a few regions in the Asian Tropics, but is often locally abundant along roadsides.

In *Botrychium* (Fig. 6.5a), where the frond is annual, the vegetative and fertile parts are pinnately branched. *Helminthostachys* is basically similar, but the branches of the fertile part of the frond are very contracted. In *Ophioglossum* (Fig. 6.5b) the sterile part of the frond, which is reticulately veined, is elliptical and entire or, in a few epiphytic forms, dichotomously lobed. The fertile part is never anything other than a simple spike.

THE VASCULAR SYSTEM OF THE RHIZOME   The Ophioglossales are rhizomatous, growth taking place from a single apical cell. In *Botrychium* the rhizome of the young plant contains a medullated protostele (Fig. 6.6), but in the stele of older plants a parenchymatous area perforates the xylem anteriorly to the departing leaf trace. The endodermis remains wholly exterior. We thus arrive at a stele intermediate between a protostele and a solenostele, often referred to as a siphonostele (see Fig. 6.18). A rudimentary solenostele does in fact arise in some species of *Botrychium* as a result of an endodermis appearing on the inside of the xylem cylinder. The stele shows a number of points of anatomical interest. The metaxylem tracheids, for

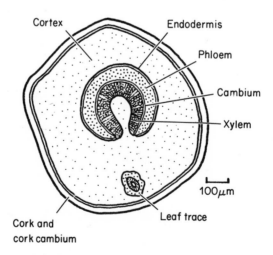

Fig. 6.6   *Botrychium lunaria*. Transverse section of rhizome.

example, bear bordered circular pits, found outside the Ophioglossales only in the gymnosperms and angiosperms. There is also cambial activity leading to secondary vascular tissue, otherwise unknown in living ferns. Apart from this feature, the anatomy of *Ophioglossum* and *Helminthostachys* resembles that of *Botrychium*.

The roots of all Ophioglossales tend to be fleshy. The central vascular strand is either diarch, or in the larger roots polyarch.

REPRODUCTION   The spherical sporangia of *Botrychium*, eusporangiate in origin, arise in two ranks on the ultimate branches of the fertile part of the frond. The spore mother cells are enclosed in a tapetum, several cells in thickness, in whose disintegration products the spores mature. Even at maturity the wall of the sporangium is massive, and stomata interrupt its outer layer. The spores, a few thousand in number, are released by transverse dehiscence. *Botrychium*, like the Ophioglossales as a whole, is homosporous.

The gametophyte of *Botrychium* is a flattened tuberous prothallus, subterranean and invested with an endophytic fungus, presumably in mycorrhizal relationship (Fig. 6.7). Gametophytes have also been raised in pure culture after attrition of the spores. Chlorophyll remains absent even in the light, and sugar is essential for successful growth. The antheridia are sunken, and each yields over a thousand multiflagellate spermatozoids. The archegonia, of quite normal construction, are partially immersed. The embryogeny of *Botrychium* is somewhat variable: in some species there is a suspensor and development is endoscopic, in others the suspensor is lacking and development is exoscopic. The first sporophytic organ to emerge is the root, infected from the first with the same endophytic fungus as the gametophyte. The

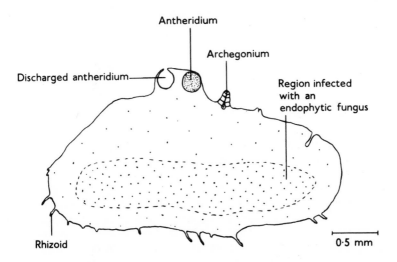

Fig. 6.7 *Botrychium virginianum*. Vertical section of gametophyte. (After Campbell (1905). *The Structure and Development of Mosses and Ferns*. Macmillan, New York.)

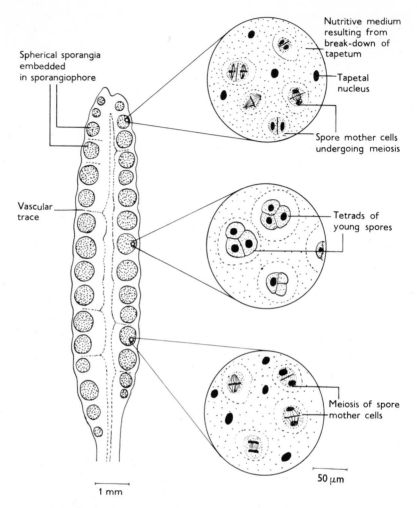

Nutritive medium
resulting from
break-down of
tapetum

Spherical sporangia
embedded
in sporangiophore

Tapetal
nucleus

Spore mother cells
undergoing meiosis

Vascular
trace

Tetrads of
young spores

Meiosis of spore
mother cells

50 μm

1 mm

Fig. 6.8 *Ophioglossum vulgatum*. Longitudinal section of young fertile spike.

young plant may remain subterranean in an immature condition for several years, and until the first leaf appears above ground the nutrition is presumably supplied principally by the mycorrhiza.

The reproduction of *Helminthostachys* is similar to that of *Botrychium*, but the dehiscence of the sporangia is longitudinal, and the embryogeny is regularly endoscopic. In *Ophioglossum* the sporangia, which occur as two rows partially embedded in the spike (Fig. 6.8), open by transverse clefts. Each contains numerous spores, in some species of the order of 15 000. The gametophyte of *Ophioglossum* is subterranean and cylindrical, sometimes approaching 5 cm in length. Both antheridia and archegonia are sunken. The embryogeny is exoscopic.

The phylogeny of the Ophioglossales

The Ophioglossales have no close relatives, and the evidence of distribution and comparative anatomy, particularly in relation to the massive sporangia and the stele, points to their being the relicts of an ancient lineage. The absence of a fossil history indicates that they were never very numerous.

Marattiales

The Marattiales are another small Order of ferns, in this instance wholly tropical. Although not conspicuous in contemporary vegetation, they have a remarkably rich fossil record, and the Class has been recognized with certainty as far back as the Carboniferous.

MORPHOLOGY AND ANATOMY   Of the living genera, the most common are *Angiopteris* and *Marattia*. Both have short upright trunks bearing large, pinnately branched and rather fleshy fronds (Fig. 6.9), sometimes reaching a length of 5 m, and showing circinate vernation. At the base of the petiole are two prominent stipules which persist after the leaf has fallen. *Christensenia*, a monotypic genus of the Indo-Malayan region, has a creeping rhizome with palmately divided fronds, which have the distinction of containing the largest stomata known in the Plant Kingdom. *Danaea*, a small genus confined to Tropical America, has one species with a simple, ovate frond, and another

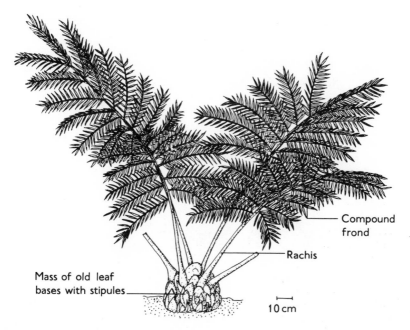

Fig. 6.9 *Angiopteris teysmanniana*. Habit of young plant. (After Bitter, in Engler and Prantl (1902). *Die naturlichen Pflanzenfamilien*, **1**, 4. Engelmann, Leipzig.)

with a small pinnate frond in which the lamina is pellucid and filmy. These forms, although revealing the diversity in the fronds of the Marattiales, are nevertheless unusual, and a massive angular construction is characteristic of the fronds of the Marattiales as a whole. The laminae normally show differentiation into palisade and mesophyll, with the stomata confined to the lower surface.

The stems of the upright forms grow not from a single apical cell, but from a more massive meristem, and in this they are unique amongst the living ferns. The leaves form an apical crown, and since each receives an extensive trace, consisting of several strands of vascular tissue, the form of the stele is highly complex. A transverse section of the stem shows a number of concentric cycles of partial steles (*meristeles*), and dissection reveals that the meristeles of each cycle anastomose freely, occasional anastomoses also occurring between adjacent cycles. Leaf traces originate from the outer cycle of meristeles. Root traces, which may arise at any depth, pass out obliquely into the cortex. An endodermis, although present in young plants, is usually absent from the stelar regions of the older.

The stems of the Marattiales also contain little if any sclerenchyma, but there is an abundance of mucilage canals and tannin sacs, as elsewhere in the plant (Fig. 6.10). These indicate a particular kind of carbon metabolism that seems to have been widespread amongst the ancient ferns.

REPRODUCTION　The fertile fronds resemble the sterile in most genera, and the sporangia, always eusporangiate in origin, are confined to the lower surface. In *Angiopteris* they arise in two ranks beneath veins towards the

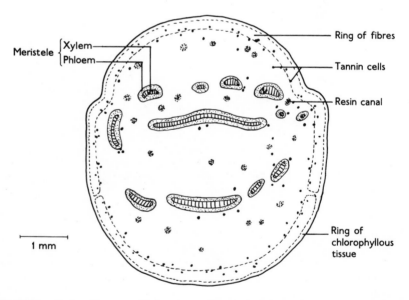

Fig. 6.10　*Angiopteris evecta.* Transverse section of secondary rachis.

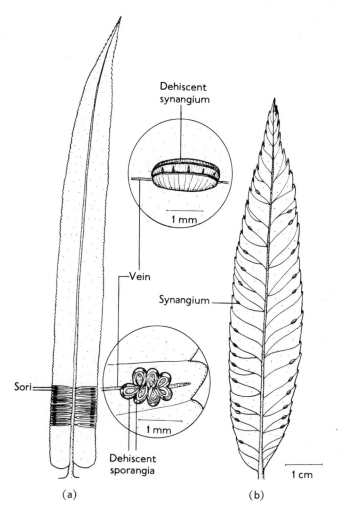

Fig. 6.11 Fertile pinnules of the Marattiales. **(a)** *Angiopteris.* **(b)** *Marattia.*

margins of the pinnules (Fig. 6.11a). The group is called a sorus and dehiscence of the sporangia, along a longitudinal stomium, is directed towards the mid-line of the sorus (Fig. 6.13b). In *Marattia* the fertile regions are similar, but the sporangia are congenitally fused into a synangium (Figs 6.11b and 6.12). As the synangium matures and dries, it splits longitudinally into two valves (Fig. 6.13a), and each compartment dehisces by a pore in the inner face. The number of spores in each sporangium (or synangial compartment) in the Marattiales reaches a thousand or more.

Given warmth, moisture and light, the spores germinate rapidly, and after passing through a brief filamentous phase, generate a green thalloid gameto-

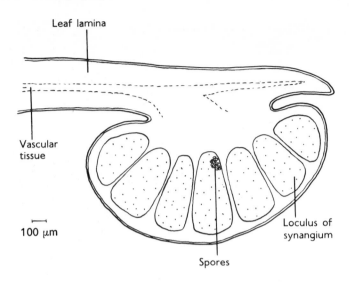

Leaf lamina

Vascular
tissue

100 μm

Loculus of
synangium

Spores

Fig. 6.12 *Marattia*. Vertical section of synangium.

phyte with apical growth, somewhat resembling the liverwort *Pellia*. Although autotrophic, the lower cells contain an endophytic, and presumably mycorrhizal, fungus. The gametophyte may be long-lived and old specimens reach a length of 3 cm or more. The antheridia are sunken, and occur on both surfaces, but the archegonia are confined to the median region of the ventral surface. The archegonia are also sunken, and the protruding neck cells form more of a cap over the egg than a neck (Fig. 6.14).

The first division of the zygote is by a wall transverse to the longitudinal axis of the archegonium. The subsequent embryogeny is endoscopic, and the embryo often emerges from the upper side of the gametophyte. A suspensor has been reported in *Danaea*, but is elsewhere lacking.

The fossil history of the Marattiales

The Marattiales are represented in the Carboniferous period by both vegetative and fertile material. *Psaronius*, for example, is the remains of a trunk surrounded by a mantle of descending roots (Fig. 6.15). The vascular tissue, which was wholly primary, formed a polycyclic array of anastomosing, band-like meristeles. Morphologically and anatomically *Psaronius* is so suggestive of an arborescent *Angiopteris* that there seems little doubt of its affinity. Fertile material is represented by *Scolecopteris* (Fig. 6.16) and *Eoangiopteris*, the sporangia of which were very similar to those of *Angiopteris*, although there were minor differences in the sorus. Fertile fronds of Marattiales, resembling those of various modern genera, are also found throughout the Mesozoic.

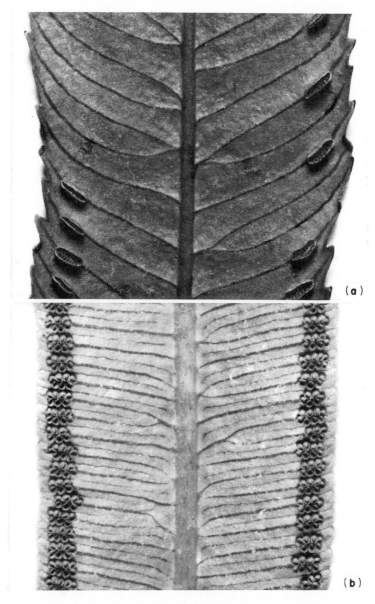

Fig. 6.13 *Marattia fraxinea*. The abaxial surface of a fertile pinnule showing the arrangement of the synangia. The synangia have dehisced. (× 5) **(b)** *Angiopteris evecta*. The abaxial surface of a fertile pinnule showing the grouping of the sporangia into sori, and the arrangement of the sori. The sporangia have dehisced. (× 6.5)

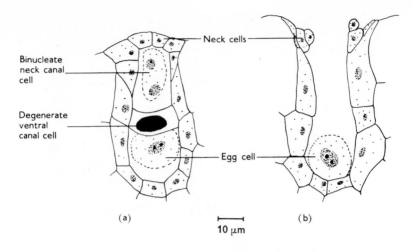

Fig. 6.14   *Angiopteris evecta*. Longitudinal section of archegonium. **(a)** Immature. **(b)** Prior to fertilization. (After Haupt, from Foster and Gifford (1959). *Comparative Morphology of Vascular Plants*. Freeman, San Francisco. Copyright © 1959.)

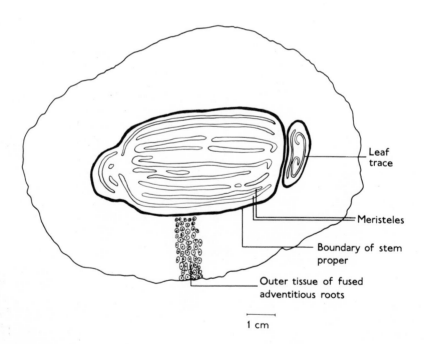

Fig. 6.15   *Psaronius conjugatus*. Transverse section of stem. From a specimen in the Oliver collection.

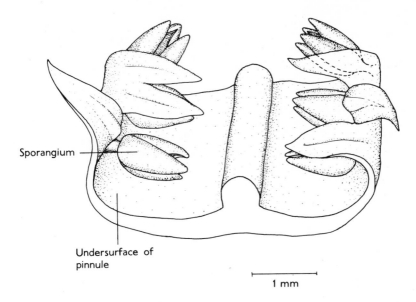

Sporangium

Undersurface of
pinnule

1 mm

Fig. 6.16 *Scolecopteris incisifolia*. Reconstruction of portion of fertile pinnule. (After Mamay, from Andrews (1961). *Studies in Paleobotany*. Wiley, New York.)

Filicales

The Filicales number some 300 genera, and they include all the common ferns of temperate regions. Taken as a whole they show such a great diversity of growth form, morphology and anatomy that there are only two features, and these are at a cellular level, which distinguish them collectively from the remainder of the ferns. These are, first, the origin of the sporangium, and, second, the plane of the first dividing wall of the zygote. In the Filicales the sporangium develops from a single initial cell, a condition termed *leptosporangiate*, in contradistinction to the eusporangiate condition of the Ophioglossales and Marattiales. Also in the Filicales, at least in those species which possess the typical heart-shaped gametophyte, the first dividing wall of the zygote is vertical or slightly oblique, parallel to the longitudinal axis of the archegonium, whereas in the Ophioglossales and Marattiales the first wall is perpendicular to the archegonial axis. The subsequent embryogeny of the Filicales is regularly endoscopic.

THE DISTRIBUTION AND ECONOMIC USES OF THE FILICALES   Although world-wide in distribution, most Filicales are tropical, and they flourish in constantly humid, warm-temperate situations, such as prevail within a certain range of altitude on tropical mountains. In these situations are found, besides the herbaceous upright and rhizomatous forms of temperate regions, numerous epiphytes and arborescent and scandent forms. There is also an epiphytic species of south-east Asia in which the rhizome is curiously inflated and specialized to house colonies of ants, an adaptation (termed *myrmecophily*) otherwise occurring only in flowering plants. A small number of wholly aquatic Filicales (Hydropterideae), all showing marked morphological specialization and heterospory, are widely distributed in fresh waters and swamps.

Few Filicales have any economic value. In Asia young croziers of *Pteridium* are cooked and eaten as human food (despite the presence of a carcinogen dangerous to cattle), and in North America *Matteuccia* is similarly used as a delicacy. In some areas the sporocarps of *Marsilea* serve as a source of starch. *Azolla* probably plays an important role in maintaining the fertility of rice padi by virtue of the nitrogen fixed by its endosymbiont *Anabaena*. The tenacity of the sclerenchyma of tree ferns is exploited in the less stable regions of the Andes where Indians use the trunks in preference to timber because of their resistance to shattering by earthquakes. Extracts and portions of a few ferns have minor medicinal uses. Several species are popular horticultural plants, particularly those mutants with striking modification of the form of the frond.

THE GROWTH OF THE STEM   The stems of the Filicales, with the occasional exception of some of the larger specimens of *Osmunda*, grow from a single, conspicuous, apical cell (Fig. 6.17), tetrahedral in shape. Daughter cells are cut off adjacent to its three posterior faces, although it is disputed whether the apical cell itself divides. Studies with polarized light have shown that the cellulose microfibrils in the tangential walls of the daughter cells lie in arrays transverse to the axis of the stem. Consequently tangential growth is constrained, and expansion is predominantly radial and longitudinal.

The meristematic activity in the apical cone diminishes towards its base. Below the apical cone, cell divisions are more generalized and variously directed. Leaf primordia arise in this region in a definite phyllotactic sequence. A leaf primordium, first visible as a slight protuberance, soon

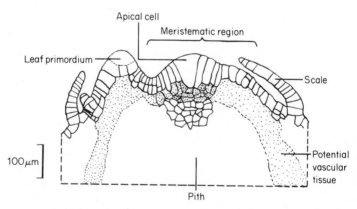

Fig. 6.17   *Dryopteris aristata.* Longitudinal section of stem apex. (After Wardlaw (1944). *Annals of Botany, New Series,* **8**, 173.)

develops its own apical cell. As the leaf primordia age and become separated by the expansion of the apex, bud primordia may be formed between them, but in some species buds do not appear at all so long as the apex is actively meristematic. Development of buds beyond the stage of primordia is rarely seen in the region of developing leaves.

A curious situation is found in *Pteridium* (bracken). The rhizomes of the mature plant are arranged in layers, the lowermost (up to 30 cm or more beneath the surface) consisting solely of 'long' shoots with extended internodes. Most of the fronds are borne on short stubby rhizomes near the soil surface. In the event of a 'front' of bracken invading a new area, the lowermost rhizomes head the advance.

Since the apices of many ferns are comparatively broad and accessible, they provide excellent material for experimental work on phyllotaxy. The results indicate that the young leaf primordia, each a centre of meristematic activity, suppress growth in their immediate vicinity. Thus, if the position in which a leaf primordium is expected to arise is isolated by radial incisions from the neighbouring recently initiated primordia, then the new primordium develops with unusual vigour and outgrows the others. Similar experiments also confirm the fundamental similarity of stems and leaves in the Filicales. For example, tangential incisions on the anterior side of very young primordia that would normally yield leaves result in the production of stem buds instead. Incisions on the posterior side are without any effect. Consequently the determination of the sub-apical primordia appears to depend upon their being initially traversed by gradients of metabolites originating in the apical meristem. If a primordium is isolated from these gradients by an anterior incision it yields the radially symmetrical structure of a stem instead of the dorsiventral symmetry of a leaf.

THE FORMATION AND MORPHOLOGY OF THE STELE   The cells which yield the vascular tissue first become recognizable as a distinctively staining ring shortly below the apical cell. The diameter of the ring increases in register with that of the apex as a whole, and beneath the leaf primordia its cells become confluent with crescentic strands of similar cells descending from them. Further down in the apex the cells of the ring are continuous with the procambium, and this in turn with the vascular tissue of the mature shoot. The position of the protoxylem is variable, but commonly mesarch. The large tracheids of the metaxylem have scalariform pitting, and in a few instances (e.g. in the rhizome of *Pteridium*) the oblique end walls have scalariform perforations. These vessel-like channels recall the situation in some species of *Selaginella* (see p. 157). The phloem consists of sieve cells with sieve areas confined to the oblique end walls. The vascular tissue is usually surrounded by a narrow zone of parenchyma, and then by an endodermis with a clear Casparian strip (see p. 295). The walls of the cortical cells adjacent to the endodermis are often thickened and made conspicuous by impregnation with phlobaphene. The endodermis and these thickened tangential walls probably together limit apoplastic transport between stele and cortex.

The form of the stele in the Filicales shows considerable variation

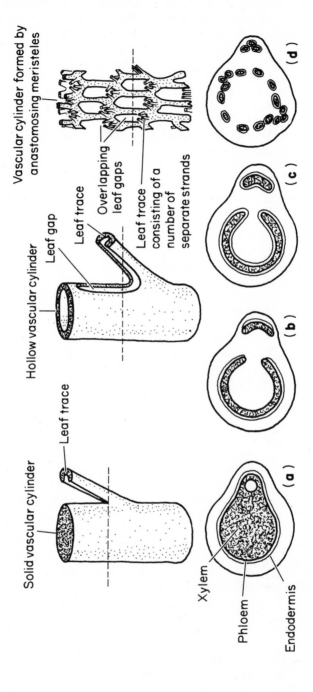

Fig. 6.18 Principal forms of fern steles. **(a)** Protostele. **(b)** Siphonostele. **(c)** Solenostele. **(d)** Dictyostele.

Labels within figure:

Solid vascular cylinder

Leaf trace

Hollow vascular cylinder

Leaf gap

Leaf trace

Vascular cylinder formed by anastomosing meristeles

Overlapping leaf gaps

Leaf trace consisting of a number of separate strands

Xylem

Phloem

Endodermis

(a)

(b)

(c)

(d)

(Fig. 6.18). In some species of *Gleichenia*, for example, the procambial tissue yields a solid core of tracheids from which the leaf traces depart without any break in the continuity of the xylem. Some other species of *Gleichenia* show a similar stele, but with the medullation of the tracheidal core leading to the production of a siphonostele (Fig. 6.18b). In *Osmunda* (Fig. 6.19) the stele is basically a siphonostele, and the phloem and endodermis remain wholly external. The continuity of the xylem, however, is broken at the departure of the leaf traces, leaving a so-called 'leaf gap' which closes again anteriorly. Since, when dissected, the xylem (but not the stele as a whole) has the appearance of a cylinder of netting, *Osmunda* is said to have a dictyoxylic siphonostele.

In some ferns in which the stele is cylindrical phloem and endodermis are present both inside and outside the xylem (Fig. 6.18c). This form of stele, found principally in rhizomatous ferns, is termed a **solenostele** (or amphiphloic siphonostele). Leaf gaps are regularly present and sometimes additional perforations unrelated to the departure of leaf traces. The internal and external phloem and endodermis are in continuity around the margins of the gaps in the xylem. If the leaf gap and other perforations are close together, as in *Dryopteris*, the stele in section appears as a ring of

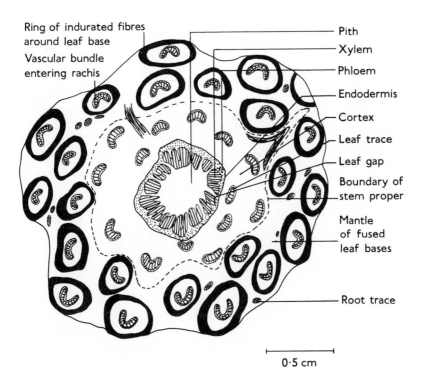

Ring of indurated fibres around leaf base

Vascular bundle entering rachis

Pith

Xylem

Phloem

Endodermis

Cortex

Leaf trace

Leaf gap

Boundary of stem proper

Mantle of fused leaf bases

Root trace

0·5 cm

Fig. 6.19 *Osmunda regalis*. Transverse section of rootstock.

anastomosing vascular bundles (Fig. 6.18d), each with internal xylem and concentric phloem. This type of stele, which is of widespread occurrence, is termed a ***dictyostele***. A complication, shown for example by the rhizome of *Pteridium* (Fig. 6.20) and the trunks of the tree fern *Cyathea*, is the presence of two or more concentric vascular systems, interconnected at intervals and usually all contributing to the leaf traces. These steles are said to be polycyclic. A point to be noted in passing is that steles are not always radially symmetrical. Dictyosteles in the ferns with creeping rhizomes, for example, are often markedly dorsiventral, the departure of the leaf traces being confined to the upper surface and flanks.

THE EXPERIMENTAL INVESTIGATION OF STELAR MORPHOLOGY   The form of the stele in the Filicales has also been the subject of experimental investigation. In *Dryopteris*, for example, if the apical region is isolated by vertical cuts, but left in contact below, it continues to grow and a solenostele differentiates behind it. As the apex expands and builds up a new crown, the stele gradually opens out to reform a dictyostele. In any one species, therefore, the size of the apex determines the form of the stele. This is also well shown in sporelings where a protostele is always present at the beginning. In protostelic species this merely increases in diameter as the plant develops, but in solenostelic and dictyostelic species the protostele of the sporeling becomes

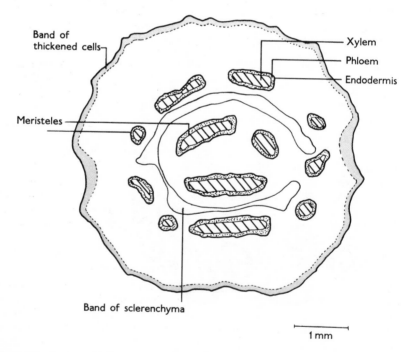

Fig. 6.20   *Pteridium aquilinum.* Transverse section of rhizome.

medullated, and phloem and endodermis appear within. The stele thus acquires its mature form in step with the increasing girth of the apex. This relationship between size and form is clearly the consequence of physiological equilibria, but they are undoubtedly complex and have yet to be resolved.

OTHER ANATOMICAL FEATURES OF THE STEM   In addition to the xylem, which is wholly primary, there are frequently bands or rods of sclerenchyma in the stem contributing to its rigidity. In the tree ferns, for example, some of which may reach a height of 10 m, mechanical stability is dependent almost entirely upon the extremely tough girdle of sclerenchyma in the outer cortex and in association with the leaf bases. In those Filicales which are believed, on the basis of fossil evidence, to be relics of very ancient groups (e.g. *Gleichenia*), the parenchymatous tissue of the stem often contains resin sacs and mucilage canals. Among the Filicales believed to be more recent in origin, however, these features are less evident.

ROOTS   After the primary root all subsequent roots are adventitious, in arborescent forms often being produced even in the aerial regions and providing a mantle of stubby outgrowths between the leaf bases. They show a distinct apical cell, but in some instances this may be quiescent, divisions being confined to the cells at its flanks. As in other Pteropsida a root cap is present and in many species root hairs. The xylem is commonly diarch, and in many epiphytic forms all but the protoxylem often remains unlignified.

The roots of the water ferns are usually delicate and the absence of soil makes them very easy to examine. In *Azolla*, for example, it is a simple matter to follow by serial sectioning the cell lineages derived from the three posterior faces of the apical cell, and the spatial interactions which lead to the generation of the axis can be readily understood.

THE MORPHOLOGY AND ANATOMY OF THE VEGETATIVE LEAVES   The leaves of the Filicales retain an apical cell during their development from the primordium. Some form of pinnate branching is usually present in the mature leaves. True dichotomous branching occurs very rarely (the frond of *Rhipidopteris peltata* provides one of the few examples), but cymose branching, superficially resembling dichotomy, is shown by the leaves of several species of *Gleichenia*. In a few Filicales meristematic areas are retained in the differentiated leaf, and these subsequently grow out to form either additional leaves (as in *Trichomanes proliferum* (Fig. 6.21)) or new plantlets (as in *Asplenium mannii* and *Camptosorus radicans*). These forms illustrate how in the living ferns, as in the extinct (see p. 185), the leaves sometimes display features suggestive of stems. All parts of the young leaf show circinate vernation, and the extension of the rachis and the unrolling of the pinnae clearly involve considerable co-ordination of growth in space and time. There is evidence that this is dependent upon the diffusion and varying relative concentrations of auxins in the expanding leaf, but these auxins are not necessarily identical with those in seed plants. The expanding leaves of some ferns,

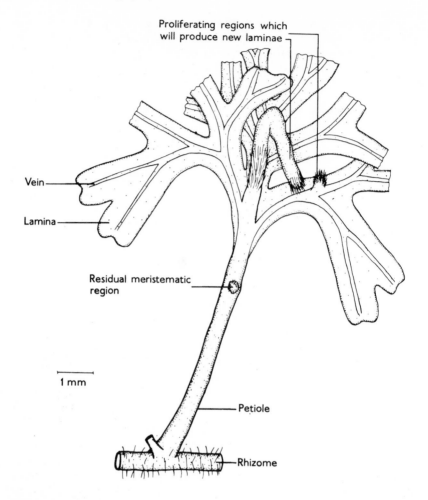

Fig. 6.21 *Trichomanes proliferum*. Frond, showing indefinite growth. (After Bell (1960). *New Phytologist*, **59**, 53.)

for example of the tropical *Dryopteris decussata*, are enveloped in mucilage, possibly with some protective function.

The lamina of the leaf is commonly differentiated into palisade and mesophyll, but the texture is very variable and in some species a thick cuticle on the upper surface gives the leaf a surprising harshness. 'Filminess', the possession of laminae only one cell in thickness, is found in *Leptopteris* and throughout the family Hymenophyllaceae. Filmy ferns are necessarily confined to situations of continuously high humidity. They are often able to thrive in irradiances far below those tolerated by flowering plants, and more akin to those of bryophyte communities, with which they are frequently intermixed. In some tropical epiphytes the sterile leaves are of two forms, one

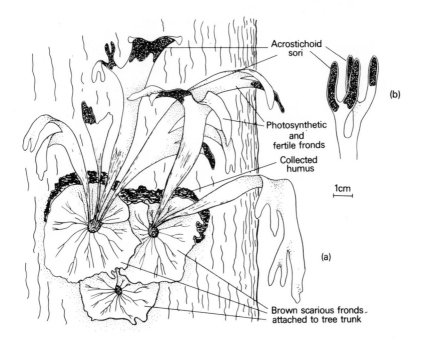

Fig. 6.22 *Platycerium.* **(a)** Habit, showing the two forms of leaves. **(b)** Lower surface of fertile portion of frond.

remaining photosynthetic and the other soon dying, but persisting in a rigid, scarious condition as a collector of humus and moisture. *Platycerium* (Fig. 6.22) provides a striking example of this kind of habit. In other epiphytes of similar situations the leaves, borne on a short upright rootstock, are stiff and tightly overlapping, so forming a funnel which traps rain and organic matter. *Asplenium nidus* provides a typical example of these 'nest ferns'. The material at the base of the funnel is freely penetrated by absorptive rootlets. The leaves of the Hydropterideae are wholly peculiar. In *Salvinia*, for example, the surface of the leaf is made unwettable by a covering of waxy hairs, and in *Azolla* the minute leaves contain cavities inhabited by the blue-green alga *Anabaena azollae*.

THE FERTILE LEAVES AND THE NATURE OF THE SPORANGIA   The fertile leaves of the Filicales are often quite similar to the sterile (as in *Dryopteris*), but dimorphy is not uncommon. In *Blechnum spicant*, for example, both sterile and fertile fronds are simply pinnate, but in the fertile the sterile part of the lamina is very reduced. The sporangia arise from single initial cells (except in *Osmunda* where a few additional cells are involved), either at the margin or on the lower surface of the leaf. The mature sporangium has a distinct stalk, the structure of which ranges from a broad multicellular stump to a delicate and relatively long column of cells. The wall of the capsule is only one cell

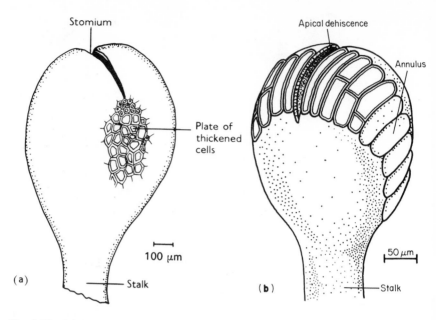

Fig. 6.23 **(a)** *Osmunda regalis*. Sporangium. **(b)** *Gleichenia*. Sporangium.

thick (again with the exception of *Osmunda*, where a thin inner layer may also be present), and it always contains a series of indurated cells and a well-defined stomium. In *Osmunda* the indurated cells are grouped laterally (Fig. 6.23a), and a linear stomium extends from them over the apex of the sporangium. In other Filicales the indurated cells are arranged in a single band (***annulus***) which encircles the sporangium, either transversely near the apex of the sporangium (as in *Aneimia* (Fig. 6.24) and other Schizaeaceae), or obliquely (as in *Gleichenia* (Fig. 6.23b) and the Hymenophyllaceae), or vertically (as in *Dryopteris* and most common temperate ferns). The annulus is interrupted by the stomium. Where the annulus is vertical, it is also interrupted by the stalk of the sporangium, the stomium then lying just in front of the stalk. The sporangium dehisces at the stomium as a consequence of tensions set up in drying. Although this process is of a general nature in *Osmunda*, it is more precise in those Filicales where the sporangia have annuli, particularly where the annulus is vertical. Here, as the cell sap in the annular cells diminishes by evaporation, asymmetrical thickening of the cell walls (Fig. 6.25) causes an increasing tangential tension which tends to reverse the curvature of the annulus. The stomium eventually breaks, and the upper part of the sporangium gradually turns back as if on a hinge (Fig. 6.25b). Tension in the cells of the annulus soon reaches a critical level; at this point the water remaining in the annular cells spontaneously becomes vapour. The tension is immediately released, and the upper part of the sporangium flies back to more or less its original position (Fig. 6.25c). These two movements effectively disperse the spores. It has sometimes been

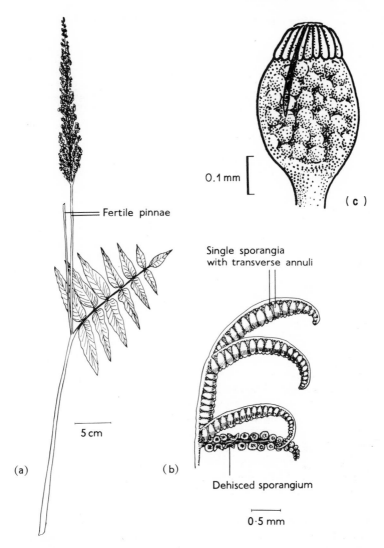

Fig. 6.24 *Aneimia phylliditis*. **(a)** Fertile frond. **(b)** Portion of fertile region. **(c)** Sporangium showing transverse annulus.

observed that the movements are repeated, a feature not so readily explained.

THE ARRANGEMENT OF THE SPORANGIA   In most Filicales the sporangia arise in distinct groups, called sori, usually beneath or near the extremities of veins, these two positions being called superficial and marginal respectively. Sometimes the sporangia are produced in a continuous line, referred to as a *coenosorus*, well shown, for example, by *Pteridium*. The sorus is often partly or wholly covered by an outgrowth of the lamina, called an *indusium*

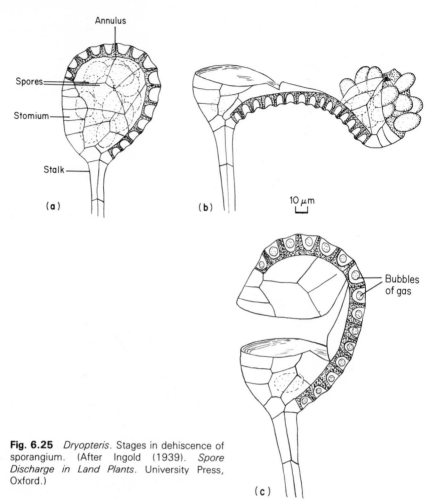

Fig. 6.25 *Dryopteris*. Stages in dehiscence of sporangium. (After Ingold (1939). *Spore Discharge in Land Plants*. University Press, Oxford.)

(Fig. 6.26), which adopts a characteristic form. In *Dryopteris*, for example, the indusium is reniform (Fig. 6.27), and in *Polystichum* peltate. In *Onoclea* the sori are protected by a rolling up of the fertile pinnules, a development carried to an extreme in the water ferns. In the family Marsileaceae, for example, the fertile segments of the frond are concrescent and form an indurated **sporocarp** at the base of the petiole. In some species of *Polypodium*, where an indusium is typically absent, the sori are immersed in the lower surface of the lamina. A few ferns show the so-called 'acrostichoid' condition in which the sporangia arise as a continuous felt on the lower surface of the fertile frond (e.g. *Platycerium*; see Fig. 6.22). In some forms (e.g. *Gleichenia*) the sporangia in a sorus are all of the same age (Fig. 6.28a), in others they are produced in spatial and temporal sequence on an elongated receptacle (as in the Hymenophyllaceae (Fig. 6.28d)), yielding a so-called

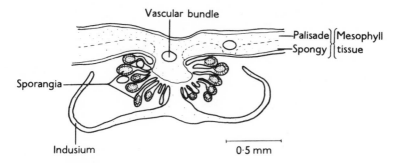

Fig. 6.26 *Dryopteris filix-mas.* Vertical section of sorus.

'gradate' sorus, and in yet others the sporangia are produced over a period but intermingled, leading to a so-called 'mixed' sorus (Fig. 6.28e and f) (as in *Dryopteris* and most common temperate Filicales).

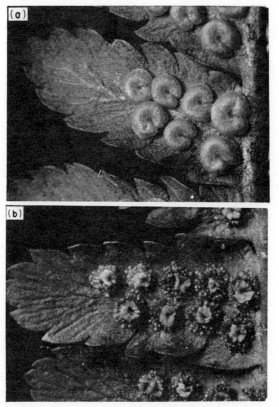

Fig. 6.27 The abaxial surface of fertile pinnules of *Dryopteris filix-mas.* **(a)** The sori in a young condition, before dehiscence of the sporangia. **(b)** Three weeks later, most of the sporangia having dehisced. (× 6)

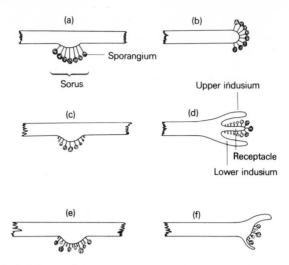

Fig. 6.28　Principal forms of sorus. **(a,b)** Simple, the sporangia all of the same age. **(a)** Superficial. **(b)** Marginal. **(c,d)** Gradate. **(c)** Superficial. **(d)** Marginal. **(e,f)** Mixed. **(e)** Superficial. **(f)** Marginal.

THE DEVELOPMENT OF THE SPORANGIA AND SPORES　In the development of the sporangium a cluster of spore mother cells becomes surrounded by a two-layered tapetum. In origin this is part of the wall tissue and not of the archesporium, as in eusporangiate sporangia. The cytological changes accompanying meiosis provide a model for sporogenesis in the land plants generally. As their nuclei enter prophase the spore mother cells become surrounded by a thickened wall, and cytoplasmic connections between them are extinguished. In the heterosporous ferns this wall contains callose, but its presence in the homosporous ferns is doubtful. During prophase the density of the cytoplasm diminishes and there is a loss in affinity for basic stains, largely accounted for by a fall in the frequency of ribosomes and hence in the concentration of ribonucleic acid. Spores are produced in tetrads. The first exine is secreted by the spore itself. At this stage the thickened wall of the mother cell, weakened by autolytic degradation, breaks open. The spores then separate and complete their development in the breakdown products of the tapetum.

Some 2^6 to 2^8 spores are produced in each sporangium, the higher numbers being characteristic of the families believed to be more primitive. The spores of the homosporous Filicales (those of the heterosporous are considered later) are usually of the order of 40 μm in diameter, but measurements of a representative sample will usually show a normal distribution with a range of about ± 10 μm about the mean. A few ferns, normally included with the homosporous, do in fact produce spores of different sizes. The most extreme example is seen in *Platyzoma*, a fern of very unusual appearance growing on deep sands in N.E. Australia. Although the mature sporangia are all about the same size, some produce 16 large spores and others 32 small ones. The difference in spore size seems to result from the greater availability of nutrients and space for spore growth when the number is lower.

The outer part of the spore wall (exine), the material of which is derived from the tapetum, is sometimes deposited in a characteristic pattern of bars and ridges. This is especially true of the Schizaeaceae, and it provides a feature that has been very useful in identifying fossil forms. In some ferns, principally the more recent, the spores are monolete instead of trilete. Monolete spores often have an additional translucent investment, called a perispore, formed from the remains of the tapetum.

THE GAMETOPHYTIC PHASE   In the Osmundaceae and sporadically in some other families the spores are green, but normally chlorophyll is absent. Non-green spores are often rich in lipid, amounting in some instances to 60 per cent of the dry weight. Green spores are short-lived, but non-green spores often remain viable for up to three years. Although moisture is essential for germination the requirement for light is variable. Some spores (e.g. *Pteridium*) germinate in the dark, but others require the stimulus of red light. The phytochrome system is probably involved, but the situation is not simple since in some instances the effect of red is reversed by both far-red and blue. After the first division the protruding daughter cell often yields a colourless rhizoid. Continued divisions give rise to an algal-like filament, the growth of which is predominantly apical, furnished with a few rhizoids. Germination and production of the first rhizoid appear to be resistant to inhibitors of transcription, so it is probable that these initial steps are dependent upon long-lived messenger RNA already present in the spore.

In some Filicales (the Hymenophyllaceae, for example, and some Schizaeaceae), the gametophyte remains filamentous and the sex organs are borne on lateral cushions of cells. In most, however, the cells at the apex of the filament soon begin to divide in a number of directions and so form a cordate (heart-shaped) gametophyte (Fig. 6.29), of which *Dryopteris* provides a familiar example. There is no general agreement about the cause of this change from one-dimensional to two-dimensional growth, but it seems likely that it is a consequence of a changing balance between carbohydrate and protein metabolism. When the protein metabolism is depressed in relation to the carbohydrate (as can be done experimentally by growing the cultures in red light) the gametophytes persist indefinitely as filaments with elongated cells. In blue light, which changes the balance of the metabolism in favour of protein, the cells divide more frequently and become progressively shorter, their width remaining little changed. Ultimately the apical cell becomes broader than long. In these conditions one-dimensional growth gives way almost immediately to two-dimensional. The change in the direction of division at the apex of the filament is in line with the concept that the plane of the new wall in a dividing cell is transverse to the principal stress in the plasmalemma. In cells which are longer than broad this stress, caused by the turgor of the cells, is longitudinal, and hence the new wall transverse. In cells broader than long the principal stress is transverse to the filament, and the new wall correspondingly longitudinal or almost so.

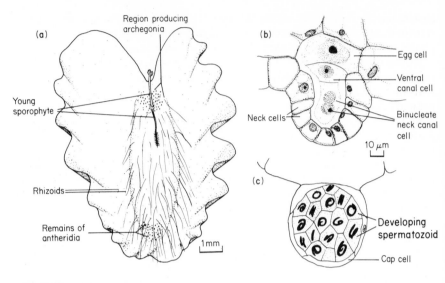

Fig. 6.29 **(a)** *Pteris ensiformis*. Lower surface of gametophyte, bearing young sporophyte. **(b)** *Pteridium aquilinum*. Median longitudinal section of young archegonium. **(c)** *Pteridium aquilinum*. Mature antheridium.

Despite their delicate structure, the gametophytes of some ferns have proved surprisingly resistant to desiccation and freezing. A few gameto-phytes have modifications which clearly promote survival. That of the Mediterranean *Anogramma leptophylla*, for example, has a tuberous portion resistant to drought. In a few tropical species of the Schizaeaceae the gameto-phyte is wholly subterranean and closely similar to that of *Psilotum* (see pp. 144, 223).

Many cordate gametophytes produce only antheridia in the first stages of growth, but subsequently the production of antheridia declines or even ceases, and archegonia then appear in sequence on the lower surface of the gametophyte behind the apical meristem. When gametophytes are grown in pure culture substances accumulate in the medium which stimulate the production of antheridia in other young gametophytes. The chemical nature of these substances, called antheridogens, varies with species but some are similar to gibberellins. As a gametophyte passes from the antheridial to the archegoniate phase it becomes insensitive to its antheridogen, but it is still secreted into the medium. The part played by antheridogens in Nature is not known, but it is conceivable that they regulate sexuality in clusters of gametophytes.

There is evidence that sexuality is also dependent on spore size, and that small spores give rise to wholly male gametophytes. Certainly when a sample of spores is sown, a proportion of the gametophytes never develops a clearly defined meristem. These *ameristic* gametophytes continue to produce antheridia indefinitely. A relationship between sexuality and spore size clearly seen in *Platyzoma*. The small spores give rise to gametophytes which remain depauperate and male, while the large spores yield complex gametophytes which are at first female and only later produce antheridia.

Antheridia usually contain 2^5 or 2^6 spermatocytes, each of which differentiates a spermatozoid. When the antheridia are mature flooding causes mucilage within to swell. The cap cells are forced off and the spermatocytes are extruded. The flagella of the spermatozoids then become active and each breaks free from its mucilaginous shell. The motile spermatozoid is about 5 μm long and takes the form of a helix with a left-handed screw when viewed from the anterior end. The anterior gyre is taken up with the multilayered structure and the associated mitochondrion, and the remaining gyres with the nucleus and microtubular ribbon. The flagella, confined to the anterior gyres, emerge tangentially and are directed posteriorly. They beat with a helical wave which drives the spermatozoid forward and at the same time causes it to rotate in the sense of the screw.

The necks of the archegonia, consisting of 4–6 tiers of cells, usually project conspicuously and are often recurved. The maturation of the egg, completed within about 24 hours, involves extensive interpenetration of nucleus and cytoplasm. The full significance of this remarkable cytology is not yet known, but it seems likely that it is concerned with the establishment of sporophytic growth. The mature egg is surrounded by a conspicuous lipoidal membrane. Flooding causes mucilage within the canal of the ripe archegonium to swell. The cap is forced off and the contents of the canal ejected. A clear passage, containing watery mucilage, now runs down to the surface of the egg cell.

In ferns with cordate gametophytes the production of archegonia continues only so long as growth remains ordered and symmetrical. This in turn is dependent upon the activity of the apical meristem, from which growth-regulating substances stream back into the gametophyte. Beyond a certain size the activity of the meristem declines. The gametophyte then begins to proliferate irregularly and a male phase is re-established. Some fern gametophytes appear to have lost the power to produce sex organs. Pale green thalli, resembling small liverworts and locally abundant in shaded rocky areas of parts of temperate N. America, are believed to be gametophytes of *Vittaria*, but sporophytes have never been observed. The filamentous gametophytes of some Hymenophyllaceae may have become similarly independent. Reproduction of these forms is solely by asexual gemmae.

THE HETEROSPOROUS REPRODUCTION OF THE HYDROPTERIDEAE   In the heterosporous Hydropterideae, where reproduction resembles that of *Selaginella* and *Isoetes*, some sporangia produce single megaspores and others up to 64 microspores. In *Pilularia* (Fig. 6.30), which is representative of the Marsileaceae, the spores are liberated by the eventual decay and rupture of the sporocarp. Germination begins at once; the megaspore rapidly gives rise to a single archegonium surrounded by a few somatic cells (which may develop chlorophyll), while the microspores each produce a single

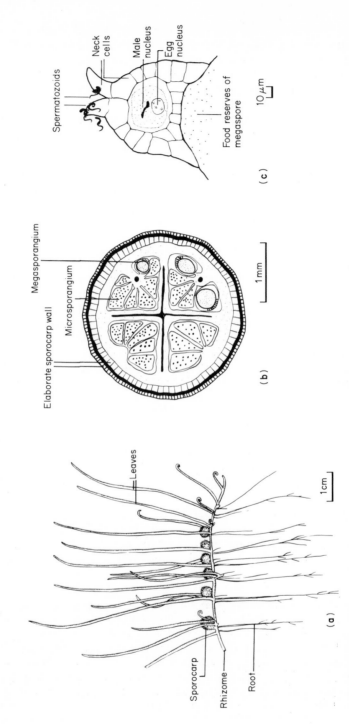

Fig. 6.30 *Pilularia globulifera.* **(a)** Habit. (After Hyde and Wade (1940). *Welsh Ferns.* National Museum, Cardiff.) **(b)** Transverse section of sporocarp showing the four sori. **(c)** Longitudinal section of inseminated archegonium.

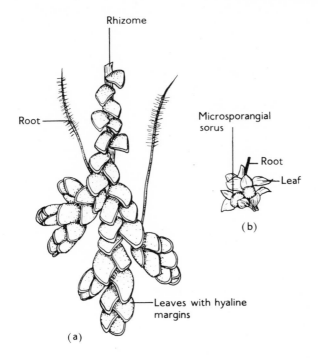

Fig. 6.31 *Azolla filiculoides.* **(a)** Habit. **(b)** Lower surface of shoot showing microsporangial sori. (After Campbell (1905). *The Structure and Development of Mosses and Ferns.* Macmillan, New York.)

antheridium containing 16 spermatocytes. In *Azolla* (Salviniaceae) (Fig. 6.31) the megaspore becomes surrounded by four frothy *massulae* (Fig. 6.32) formed from the tapetum, and these give the liberated megaspore buoyancy, but it is doubtful whether they keep it afloat indefinitely. A variable number of massulae are formed in the microsporangium (Fig. 6.33). Each includes a number of microspores at its periphery, and is furnished externally with peculiar anchor-like *glochidia*. These male massulae hook themselves to the female, and the complex then sinks. Germination of the spores, in the main similar to that of the spores of the Marsileaceae, then follows. The cutinized glochidia of *Azolla* can still be recognized in lake deposits of considerable antiquity.

FERTILIZATION AND EMBRYOGENY   Fertilization in the Filicales depends, as with most other archegoniate plants, upon the presence of water. The mucilage around the mouth of the archegonial canal, possibly because it contains traces of malic acid (known to possess chemotactic properties), attracts the spermatozoids and effectively confines them to the region of the archegonia. Several spermatozoids commonly enter an open archegonium, and occasionally some can be seen to swim out again. Where crosses are

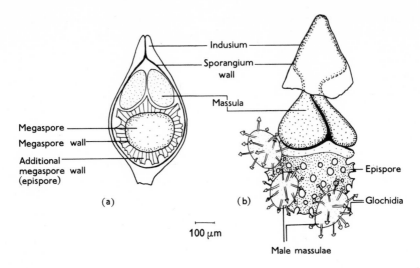

Fig. 6.32   *Azolla filiculoides.* **(a)** Longitudinal section of megasporangium. **(b)** Liberated megaspore with male massulae attached. (After Strasburger (1873). *Ueber Azolla.* Abel, Leipzig.)

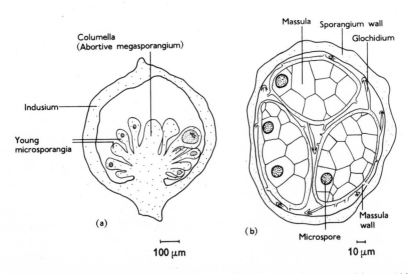

Fig. 6.33   *Azolla filiculoides.* **(a)** Longitudinal section of young microsporangial sorus (After Campbell (1905). *The Structure and Development of Mosses and Ferns.* Macmillan, New York.) **(b)** Transverse section of mature microsporangium. (After Smith (1955). *Cryptogamic Botany,* II. McGraw-Hill, New York.)

attempted between certain genera (e.g. *Athyrium* ♀ x *Dryopteris filix-mas* ♂) the mucilage immobilizes the foreign spermatozoids and no hybrids are produced. Barriers to fertilization in ferns are so far little studied. It is possible that in some species self-fertilization is prevented, or its chances lessened, by incompatibility mechanisms in the archegonial mucilage or at the surface of the egg, but the evidence so far is inconclusive. It appears however to be a general rule that only one spermatozoid enters the egg, although others may be seen pressed against its surface.

Division of the zygote follows about 48 hours after fertilization. The first vertical wall is succeeded by a horizontal so that in lateral aspect the zygote appears divided into quadrants. These quadrants indicate in a general way the course of the subsequent embryogeny. The upper anterior region, for example, goes to form the apex of the new sporophyte, the lower anterior the first leaf, while the posterior regions give rise to the first root and the foot (Fig. 6.34). There is no suspensor. Following fertilization, possibly a consequence of growth-regulating substances coming from the zygote, the growth of the gametophyte diminishes and the initiation of archegonia ceases. At the same time the cells of the archegonium immediately above the fertilized egg proliferate, forming a conspicuous cap (calyptra). Experiments have shown that this calyptra, probably by exerting mechanical pressure on the developing embryo, plays an important part in determining normal embryogenesis, recalling the situation in the mosses (p. 131). If it is removed from above a very young zygote, the zygote gives rise to a mass of parenchymatous tissue before producing differentiated growing points. With cordate gametophytes, an intact apical meristem, probably in consequence of the auxin it produces, is also essential for normal embryogenesis. If this meristem is destroyed, differentiation of the embryo is markedly slower and the emergence of the first root very much delayed.

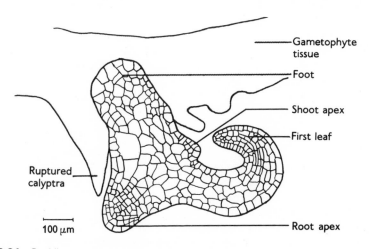

Fig. 6.34 *Pteridium aquilinum.* Vertical section of embryo.

Embryogenesis is much more rapid in the heterosporous ferns than in the homosporous. Fertilization occurs within about 12 hours of the germination of the spores, and an embryo emerges from the archegonium on the following day.

Aberrations of the fern life cycle are not uncommon. Apogamy, the production of a sporophyte without sexual fusion, occurs regularly in some ferns and can be induced experimentally in others. *Dyropteris borreri (D. affinis)* provides an example of a regular apogamous cycle. In this and similar species the final mitosis of the sporogenous cells is incomplete. Following division of the chromosomes the nucleus reforms. The cell remains undivided, although it has grown considerably as if a normal division were about to occur. Instead it becomes a spore mother cell. The restitution nucleus, which has twice the sporophytic number of chromosomes, enters meiosis. The four spores produced, reflecting the large size of the spore mother cell, have diameters of about 80 μm, and contain nuclei with the same number of chromosomes as the parent sporophyte. The spores germinate in the normal way and the gametophytes pass through a male phase. The spermatozoids are perfectly formed and functional, and are capable of fertilizing eggs of related sexual species. Subsequently, however, in place of archegonia, a sporophyte develops directly from the sub-apical region of the gametophyte. The cellular mechanisms underlying this kind of life cycle are not yet understood. The condition is however known to be genetically dominant since hybrids with sexual species have the same kind of apogamous cycle. Apogamy can sometimes be induced by withholding water from gametophytes and preventing sexual reproduction. Sporophytes have been raised in this way from gametophytes of *Dryopteris filix-mas* and *Dryopteris dilatata*. In *Pteridium* in pure culture apogamous outgrowths are promoted by a high level of sucrose. Such experimentally produced sporophytes are usually depauperate, and if they reach meiosis fail to produce viable spores. They naturally lack the pre-meiotic doubling of the chromosomes characteristic of the *D. borreri* kind of cycle.

The direct production of gametophytes from sporophytes, termed apospory, can be very readily induced in culture by placing fragments of leaves in sterile conditions on an agar medium. The gametophytes are usually sexually perfect and yield tetraploid sporophytes. This phenomenon may occur sporadically in the wild and lead to the production of natural autopolyploids.

The evolution of the Filicales

The evolutionary relationships of the living Filicales can be studied at two levels. First, by carrying out crosses between living species and examining the pairing behaviour of the chromosomes at meiosis we can obtain evidence of the extent of the genetic identity between them. Second, by comparing the morphology and anatomy of living species with the fossil we can obtain a general impression of the evolution of the contemporary fern flora.

Studies of chromosome homologies at meiosis, facilitated by the ease with which squash preparations can be made of developing sporangia, have shown that many familiar species are of hybrid origin. *Dryopteris filix-mas*, for example, is an allotetraploid probably derived from a hybrid between two diploid ancestors, one similar to *D. abbreviata (D. oreades)* and the other so far unknown. Particularly interesting is that some widespread species have been shown to be autotetraploids. Examples are *Asplenium ruta-muraria* and possibly the world-wide *Pteridium aquilinum* (bracken). The latter commonly has a chromosome number of n = 52, but a few populations are known with n = 26. Selfing experiments with the common form have revealed considerable genetic variation, presumably arising from recombination between duplicated unlinked loci (*homoeologous heterozygosity*). Hybridization and duplication of chromosome number undoubtedly accounts for much of the diversity in living ferns, but provides for little profound anatomical or morphological change.

The study of the fossil Filicales has given valuable information about the evolutionary status of the present-day families. It is clear, for example, that the Osmundaceae have a very long history. A transverse section of the stem of the late Permian *Thamnopteris* is very similar, even to the extent of the arrangement of the sclerenchyma and the packing of the leaf bases, to a section of *Osmunda*. The record of this family, which includes preserved fronds and sporangia, continues through the Mesozoic to the present. The Schizaeaceae also have a well-established fossil history extending back to the Palaeozoic, and that of the Gleicheniaceae is similar. Other families first appear in the Mesozoic. Nevertheless, despite these evidences of antiquity, most of the living Filicales either have no fossil record, or no record extending back further than the Tertiary. This is particularly true of the large family Polypodiaceae, and we must suppose that these ferns are comparatively recent, probably having evolved towards the end of the Cretaceous period and subsequently.

Comparison of the living Filicales with the fossil, quite apart from tracing particular lineages, also reveals those features which can be regarded in a general way as primitive. Protostelic and solenostelic vascular systems, the simultaneous production of the sporangia in the sori, short thick sporangial stalks, the indurated cells of the sporangial wall aggregated laterally or arranged in a transverse annulus, and a large number of spores in each sporangium are all features of the early Filicales. Conversely, dictyosteles, the production of mixed sori, long and delicate sporangial stalks, vertical annuli, and low spore numbers are all features of Filicales with little or no fossil record.

On the basis of these criteria it is possible to assign the families of living ferns to three grades according to their evolutionary advancement. It is also possible to arrange them in two series, according to whether the sporangia are marginal or superficial in origin, but recent research has thrown considerable doubt upon the significance of this feature. Nevertheless, it is tentatively retained here, and the position of some of the more important families and genera in this double classification is shown in Table 6.1. The Osmundaceae are considered as contributing to both series, since the sporangia are marginal

Table 6.1

	Features of evolutionary significance	Position of origin of sporangia	
		Marginal	Superficial
	Steles, commonly dictyosteles. Sori mixed. Sporangia with vertical annuli interrupted at stalk	*Pteridium* *Davallia*	*Dryopteris* *Polystichum* *Asplenium* *Athyrium* *Polypodium*
	Steles, dictyosteles or solenosteles. Sori gradate. Sporangia with oblique annuli	Hymenophyllaceae	Cyatheaceae
	Steles, protosteles or solenosteles. Sori simultaneous. Indurated cells of sporangial wall aggregated, or in transverse or oblique annuli	Schizaeaceae *Osmunda* (Osmundaceae)	Gleicheniaceae *Todea* (Osmundaceae)

(Left margin vertical labels: "Numbers of spores/sporangium" ranging from 2^{5-6} (top) to 2^{7-8} (bottom); "Reduction in number of cells forming sporangial stalk" with upward arrow.)

in *Osmunda*, but superficial in *Todea*. This classification does not of course imply that living families and genera have evolved from each other; it merely illustrates relative primitiveness. The arrangement is substantiated by the fossil record which is of greater duration in respect of the Filicales at the bottom of the table than of those at the top.

The origin of the Filicinae and the morphological nature of the megaphyll

Since the Cladoxylales, Coenopteridales, Marattiales, Filicales, and possibly the Ophioglossales, were all in existence together towards the close of the Palaeozoic era, it seems clear that these Orders represent parallel lines of evolution within the Filicinae, only three of which have survived. There are sufficient similarities between the Orders, however, to suggest a common ancestor in some earlier period. A significant feature of many Filicinae is the close resemblance, both in appearance and behaviour, between leaves and branch systems. This has given rise to the view that the origin of the Filicinae should be sought in some early *Psilophyton*-like form. It is significant that some species of *Psilophyton* from the Lower Devonian of both Norway and North America did bear sprays of lateral branches very similar to the fronds of some of the early ferns.

　It is difficult to distinguish between frond and axis in some living ferns. In *Stromatopteris*, for example, a rare fern of New Caledonia with a *Psilotum*-like gametophyte, and a fern which on the basis of the criteria discussed earlier (p. 221) would be considered primitive,.fronds and branches arise from the creeping rhizome in an extremely similar manner. Moreover the frond, which consists of a rachis with a row of pinnules on each side, is morphologically similar to the shoot of *Tmesipteris* (p. 144). In most ferns of course the distinction between shoot and frond is much clearer, but, as we have seen (pp. 185,205), indications of the shoot-like nature of the megaphyll are encountered throughout the Filicinae.

The evidence at present available is, therefore, fully in accord with the idea that the frond of Filicinae originated in a lateral branch system, and that the Filicinae evolved from some early group of psilophytes in which the plant body became differentiated into main and lateral axes.

7

The Tracheophyta, III
(Pteropsida: Gymnospermae)

[handwritten margin notes: Cycads, Conifers, Ginkgo, Gnetales]

The gymnosperms,[42] although a diverse Class with possibly more than one origin from the earliest land plants, are with few exceptions archegoniate and almost entirely arborescent. Their fossil record rivals that of the ferns in richness and variety.

Gymnospermae

Sporophyte usually arborescent; branching and leaves various. Secondary vascular tissue always present, consisting of tracheids (in a few forms also of vessels) and sieve cells. Sporangia borne on specialized structures, probably of axial origin. Heterospory general. The megasporangium enclosed within a specialized tissue (nucellus) of the sporophyte, this in turn surrounded by a distinctive sheath (integument), perforated at the apex by a narrow channel (micropyle), the whole termed the ovule. Neither male nor female gametophytes autotrophic. Fertilization by multiflagellate spermatozoids, or by male cells with no specialized means of locomotion, occurring within the ovule, either before or after its being shed. Embryogeny endoscopic, the embryo remaining contained within the seed developed from the ovule.

All the early seed plants, with the exception of the specialized *Lepidocarpon* (see p. 168) and *Miadesmia* of the Lycopsida, are referable to the Gymnospermae. Gymnospermy, a term which implies the bearing of naked seeds, unenclosed in any carpellary structure, can thus be legitimately regarded as the most primitive form of the seed habit.

The first plants recognized as gymnosperms are found in the late Devonian period. These are probably early members of the two Orders Pteridospermales and Cordaitales, both now extinct. Other Orders of the Gymnospermae treated here are the Coniferales, Ginkgoales and Gnetales, all with living representatives, and the extinct Bennettitales. The Caytoniales, also extinct, are considered in relation to the possible origin of angiospermy (p. 328).

Pteridospermales (Cycadofilicales)

Investigators of the fern-like fronds found in Carboniferous rocks soon became aware that not all these were in fact referable to ferns. Some were undoubtedly associated with seeds, and others with stems in which there were secondary thickening and other anatomical features not found in the Filicales. Nevertheless, the habit of these plants was probably something like that of the Marattiales. The leaves were megaphyllous, compound and pinnately branched, and borne on stems of varying height.

VEGETATIVE FEATURES   One of the best-known pteridosperm stems of Carboniferous age is *Lyginopteris*. This ranged in diameter from about 0.5 to 4.0 cm, and occasionally branched. There was a central pith, which contained nodules of thick-walled cells (similar to the groups of stone cells in the flesh of a pear), surrounded by a ring of primary xylem strands (Fig. 7.1). Exterior to these was a relatively large amount of secondary xylem. The tracheids of the secondary xylem, like those of the metaxylem, were furnished with bordered pits, but in the somewhat smaller tracheids of the secondary xylem they were absent from the tangential walls. A girdle of phloem, rarely well preserved, lay outside the xylem. A characteristic feature of *Lyginopteris* was the anatomy of the outer cortex. This contained radially elongated bands of fibres which anastomosed freely and clearly gave

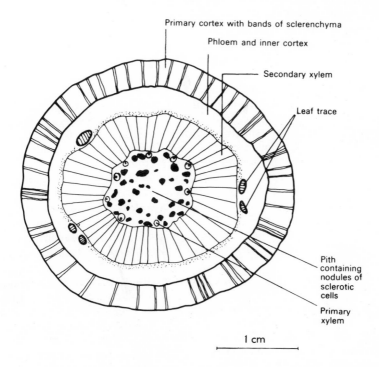

Primary cortex with bands of sclerenchyma

Phloem and inner cortex

Secondary xylem

Leaf trace

Pith containing nodules of sclerotic cells

Primary xylem

1 cm

Fig. 7.1 *Lyginopteris oldhamia*. Transverse section of stem. The glandular epidermis has been lost. The leaf trace divides into two strands in the cortex, but these reunite to form a V-shaped trace at the base of the petiole.

considerable mechanical support to the stem. The outer surface of the stem was furnished with peculiar multicellular glands.

The leaves of *Lyginopteris* (originally described as *Sphenopteris*) were borne in a 2/5 phyllotaxy and when young showed circinate vernation. In mature leaves, which sometimes reached a length of 50 cm, the rachis dichotomized at about half its length, but the remainder of the branching was pinnate and the ultimate segments were narrow pinnules. All surfaces of the leaf bore glands similar to those of the stem.

THE REPRODUCTIVE ORGANS The seeds (originally described as *Lagenostoma*) terminated axes which were probably branches of otherwise normal leaves. Each seed (or ovule) was partially enclosed in a cup formed by a number of glandular and basally fused bracts (Fig. 7.2a). This structure, called a cupule, in some forms contained more than one seed. The seed itself was an upright, radially symmetrical structure, about 0.5 cm long and a little less broad. The central part (Fig. 7.2b), the nucellus, possibly a specialized archesporial tissue, was surrounded by an integument of two layers, the outer of which contained a sclerenchymatous sheath. A single vascular bundle entered the base of the seed and divided symmetrically into nine which ascended the inner fleshy part of the integument. The upper part of the integument around the micropyle was shallowly lobed, the lobing corresponding to the intervals between the ascending veins. A female gametophyte, surrounded by a distinct membrane and bearing archegonia on its upper surface, lay within the nucellus. This gametophyte probably developed from a megaspore, formed as one of a tetrad within the nucellus, the remaining megaspores degenerating.

A peculiar feature of pteridosperm seeds was the form of the upper part of the nucellus. The central part of the nucellus was prolonged as a column, and this was surrounded by a sheath of similar tissue, a cylindrical space being left around the column. Pollen grains have been found in this chamber, and it may have been here that the microspores germinated.

The male reproductive organ of *Lyginopteris* is not yet known with certainty. It is, however, very probable that it consisted of a small ovate plate, about 2 mm in length, terminating a branchlet and bearing about six bilocular sporangia (Fig. 7.3a). Each sporangium was about 3 mm long and 1.5 mm wide. The microsporangiophores were borne on a branch system (the whole being known as *Crossotheca*) which may have formed part of a *Lyginopteris* leaf.

The pollen grains (Fig. 7.3b), which bore triradiate scars, were presumably distributed by wind. By analogy with the pollination of living gymnospermous ovules, it is thought the grains were trapped by a drop of sugary fluid which protruded from the micropyle, and that subsequent absorption of this drop drew the grains down into the nucellar chamber. Germination of the grains is thought to have been proximal (i.e. at the site of the triradiate scar as with the spores of ferns and bryophytes). This contrasts with the regular distal germination of conifer and angiosperm pollen, and pteridosperm pollen in consequence is often referred to as 'prepollen'. No seeds have

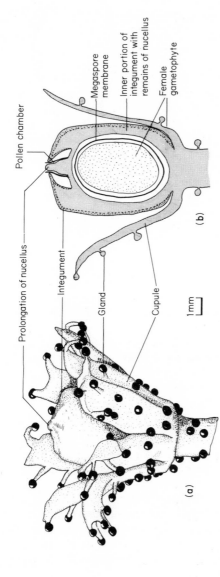

Fig. 7.2 *Lagenostoma lomaxi.* **(a)** Reconstruction of seed. (After a drawing by Oliver.) **(b)** Longitudinal section of seed. (After Walton (1940). *An Introduction to the Study of Fossil Plants.* A. and C. Black, London.)

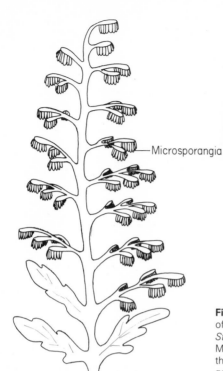

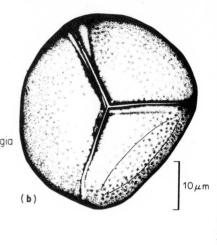

Microsporangia

10 µm

(b)

Fig. 7.3 *Crossotheca.* **(a)** Reconstruction of fertile shoot. (After Andrews (1961). *Studies in Paleobotany.* Wiley, New York.) **(b)** Microspore. The deep cleavage at the site of the triradiate scar indicates proximal germination. (From a preparation by R. Kidstun, photographed by W.G. Chaloner.)

(a)   1 cm

been found which indicate whether fertilization was brought about by spermatozoids or by unspecialized male gametes. It is also curious that no seeds are known containing embryos. Fertilization may therefore have occurred after the seed was shed.

The later pteridosperms

Lyginopteris is representative of a wide range of Carboniferous pteridosperms, but towards the end of the Palaeozoic other, more complex, forms became prominent. They were more like tree-ferns in habit, and the leaves had pinnules conspicuously larger than those of the earlier pteridosperms. The seeds, about six times the size of those of *Lyginopteris*, were more clearly in association with foliar organs, and the microspores, which were monolete, were produced in large cup- or spindle-shaped synangia. The stems had a complicated stelar structure. Vascular tissue differentiated from a number of successive and often discontinuous cambia, so that it became interspersed with much parenchyma.

The origin of the pteridosperms

The pteridosperms probably had their origin in the same group of axial plants

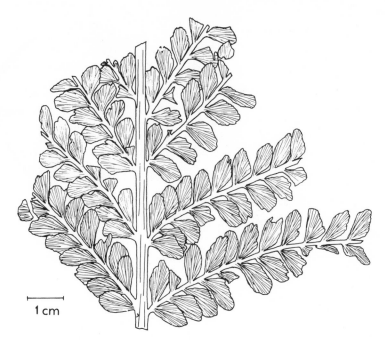

Fig. 7.4   *Archaeopteris* sp. Compression of a portion of a vegetative shoot. (After Andrews (1947). *Ancient Plants and the World They Lived In*. Comstock, New York. Used by permission of Cornell University Press.)

as that which yielded the ferns, but were the consequence of a trend in which the evolution of megaphylls was accompanied by increasingly pronounced heterospory, and by the formation of secondary vascular tissue within the stem. An intermediate stage in the progression envisaged is shown by the pinnately branching *Archaeopteris* of the late Devonian (Fig. 7.4), a well-known example of an early, fern-like, heterosporous plant. At least one species was borne on a stem with pronounced secondary thickening, and the rachis had a stem-like anatomy.

The radial symmetry of many pteridosperm seeds (on account of which the ovules are termed radiospermic) suggests that they may have been derived from a tassel of megasporangia (resembling the male organ *Crossotheca*) in which a central megasporangium was closely surrounded by a ring of similar megasporangia. Sterilization of the outer ring, but not of the centre, would then have resulted in an integumented megasporangium. Although no intermediate stages in such a transformation have yet come to light, it is significant that in one of the earliest pteridosperm seeds from the Lower Carboniferous (*Salpingostoma*) the integument consists of a ring of finger-like processes, each containing a vein, fused only in the basal region. Progressive fusion of components of this kind might then have led to the entire integuments of the pteridosperm seeds of the Upper Carboniferous, the shallow lobing often seen in the micropylar region of these integuments being the only remaining indication of their compound origin.

Another seed which can be interpreted in this way is *Archaeosperma* (Fig. 7.5). This is of additional interest, not only for its great age (it comes from the Upper Devonian of Pennsylvania), but also because of the undoubted occurrence of three aborted spores (Fig. 7.5b) above the functional megaspore of the seed. The arrangement and markings of these spores clearly indicate that the symmetry of the tetrad was tetrahedral, and the situation in the nucellus of *Archaeosperma* thus bears a striking resemblance to that in the megasporangium of a heterosporous fern, such as *Pilularia* (p. 215).

The descendants of the pteridosperms

It seems beyond doubt that some pteridosperms persisted into the early Mesozoic, although the remains become much less frequent after the close of the Palaeozoic era and of an unfamiliar form. There is also evidence that the

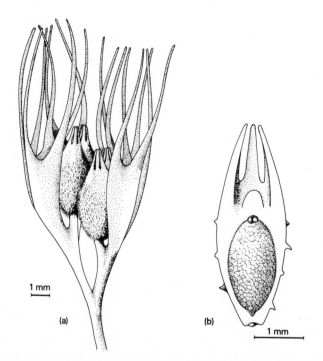

1 mm

(a)   (b)   1 mm

Fig. 7.5 *Archeosperma arnoldii*. **(a)** The two-seeded cupules were borne in pairs, the two pairs of seeds facing each other. **(b)** Reconstruction of an individual seed showing the three abortive megaspores at the top of the one which is functional. The integument is deeply lobed above. (After Pettitt and Beck (1968). *Contributions to the Museum of Paleontology, University of Michigan*, **22**, 139.)

evolution of the pteridosperms in the northern and southern hemispheres diverged at this time, but the two floras remained in contact in certain regions of Africa. The early Mesozoic rocks of these regions have yielded a number of puzzling plants, almost certainly derived from the pteridosperms and probably of very great importance in the evolution of the later seed plants. Some of them had leaves with reticulate venation and an appearance strikingly like that of the leaves of some modern flowering plants.

Cycadales

The cycads are seed plants with upright, naked ovules, resembling in general features and size those of the later pteridosperms. The fossil evidence indicates that the cycads came into prominence at the beginning of the Mesozoic era.

Although in Mesozoic times the cycads were distributed as far north as Siberia and Greenland, they are today confined to tropical and sub-tropical regions in both the Old and New Worlds. There are nine genera in all, the commonest being *Zamia, Macrozamia, Cycas, Encephalartos, Dioon* and *Ceratozomia*, but only *Cycas*, extending eastwards from the Malagasi Republic into Polynesia, has anything approaching a wide distribution. *Encephalartos woodii* of Africa is represented only by male plants. The female is presumably extinct.

HABIT   The sporophyte is upright and has either a short, stocky stem, often with a large portion below ground (Fig. 7.6), or is much taller, reaching heights of up to 15 m. Below ground is a massive tap root which bears, together with normal roots, others which are negatively geotropic and which break up at the soil surface into coralloid masses. These contain endophytic fungi and blue-green algae.

Apart from stature, all cycads have a similar growth form. The thick stem, usually unbranched, bears an apical rosette of large pinnate leaves. The rate of growth is very slow and, although there are no recognizable annual rings in the stem, the age of any specimen can be calculated approximately from the rate of leaf production and the number of leaf bases. A specimen of *Dioon* only 2 m high was estimated in this way to be about 1000 years old. *Bowenia*, confined to N. Queensland, is anomalous in possessing a tuberous rootstock bearing only one or two leaves at a time.

THE VEGETATIVE ANATOMY   The stem grows from a massive apex in which there is generalized meristematic activity, and considerable centrifugal expansion, as well as growth in length. Behind the apical initials a core of central tissue, which soon becomes distinguishable from the peripheral, differentiates into the vascular tissue and pith. The peripheral zone becomes cortex. The primary vascular tissue consists of a ring of bundles with endarch protoxylem, and these surround an extensive pith. The secondary xylem is traversed by wide parenchymatous rays and the radial walls of its tracheids are furnished with several series of circular bordered pits (except in *Zamia*

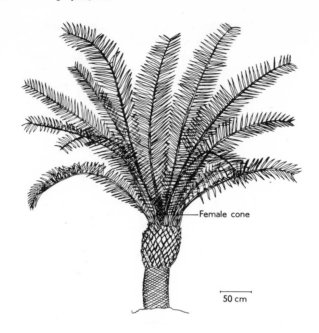

Female cone

50 cm

Fig. 7.6   *Encephalartos hildebrandtii.* Habit of plant bearing female cone. (After Eichler, from Eichler in Engler and Prantl (1889). *Die natürlichen Pflanzenfamilien*, II, 1. Engelmann, Leipzig.)

and *Stangeria* where the pits are of the narrower kind characteristic in ferns). The first cambium is of limited activity, and it is followed by others which arise successively outside the vascular cylinder. These cambia are of diminishing activity, the last producing merely a few concentric bundles lying out in the cortex. The stele is thus highly parenchymatous, and the main mechanical support of the stem comes from its armour of sclerenchymatous leaf bases. A stem with soft spongy xylem of this kind is termed *manoxylic*. Mucilage canals, tannin cells, and cells containing crystals of calcium oxalate occur in the pith and cortex of mature stems.

The leaves are of two kinds, foliage leaves (or fronds) and scale leaves, sequences of each following each other in a definite phyllotactic arrangement. The scale leaves cover the apex and the upper part of the stem (Fig. 7.7), often disintegrating to form a fibrous sheath below. The foliage leaves in some genera show circinate vernation (as in *Cycas*), but in others a vernation similar to that seen in the fern *Botrychium*. In all cycads the leaves are pinnately branched, but are twice pinnate only in *Bowenia*. The leaf trace, which is horse-shoe shaped in section in the petiole, has a complex origin in the stem. Some strands arise opposite the insertion of the leaf and girdle the stem obliquely upwards into the leaf base. In an individual bundle of the trace much of the metaxylem is adaxial to the protoxylem, but characteristically a few tracheids, often separated by parenchyma, lie on the abaxial side adjacent to the phloem ('centrifugal xylem').

Fig. 7.7 *Zamia* sp. Young plant with female cone. Although produced terminally, the female cone becomes pushed to one side by the sympodial growth and appears to be lateral. (Photograph by Frank White.) (Approx. × 0.5)

The venation of the pinnae is various, but any branching is dichotomous and open. In section small patches of ***transfusion tissue*** (anatomically intermediate between parenchyma and tracheids) are often present on each side of the xylem. In *Cycas*, where each pinna has only a midrib, a sheet of similar cells extends from the midrib to the margin ('accessory transfusion tissue').

The pinnae have a leathery texture. A conspicuous cuticle is usually present, and an epidermis (often accompanied by a hypodermis), palisade and mesophyll are well differentiated. The cell walls of the lower epidermis are straight or slightly sinuose, and the stomata, although usually sunken, are surrounded by a simple ring of subsidiary cells (***haplocheilic*** stomata). These epidermal features, which remain clearly evident in fossil material, are of great value in distinguishing extinct Mesozoic forms from superficially similar contemporary plants (see Fig. 7.14).

THE REPRODUCTIVE STRUCTURES   The mega- and microsporangiophores of the cycads are aggregated into separate strobili borne on different plants. The female cone either terminates the main axis (in which case subsequent growth is sympodial (Fig. 7.7)) or it is lateral, according to the genus. The situation in *Cycas* is exceptional for here the main axis, having given rise to a sequence of megasporangiophores, continues to be active and reverts to the production of normal vegetative leaves.

The female cones vary in compactness and in the number of ovules borne on each megasporangiophore. At one extreme stands *Cycas*, in which the female cone consists of a loose aggregate of megasporangiophores, the distal, sterile portions of which are more extensive than in any other cycad. Several pairs of ovules are attached in the proximal region (Fig. 7.8a), the micropyles of the ovules being directed obliquely outwards. At the other extreme are *Zamia* and *Encephalartos*, in which small, peltate sporangiophores (Fig. 7.8b) are tightly packed in a distinct ovoid cone. Each sporangiophore bears two ovules, the micropyles of which are directed towards the centre of the cone. Both the cones and ovules in the cycads generally are of extraordinary size. In *Encephalartos* female cones have been recorded weighing as much as 45 kg, and in *Macrozamia* the ovules reach a length of 6 cm. The whole of the female reproductive system is thus on a much larger scale than in any other living plants.

The male cones of the cycads (Fig. 7.9) are also either terminal or lateral. Where terminal, subsequent growth is always (including that of *Cycas*) sympodial. Where lateral, growth is monopodial and the cones may be present in considerable numbers. In *Macrozamia*, for example, 20–40 cones may be produced in rapid sequence around the lower part of the apex. There is much more uniformity in the structure of the male cones than in that of the female, although again considerable variation in size. In some species of *Encephalartos* the male cones reach a length of 50 cm, but in *Zamia* only 5 cm. The microsporangiophores are in the form of scales, closely adpressed during growth and when mature covered on their lower surfaces with several

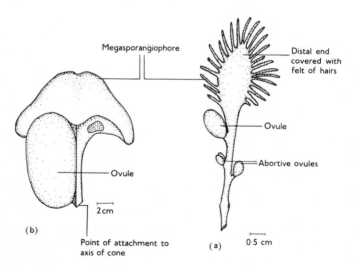

Fig. 7.8 (a) *Encephalartos hildebrandtii*, megasporangiophore. **(b)** *Cycas revoluta*, megasporangiophore.

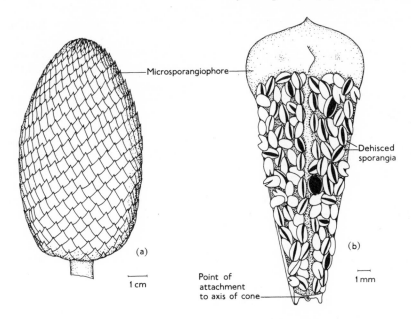

Dehisced
sporangia

(a)

(b)

1 cm

Point of
attachment
to axis of cone

1 mm

Fig. 7.9   *Cycas* sp. **(a)** Male cone. **(b)** Microsporangia, viewed from below.

hundred microsporangia (Fig. 7.9b). The sporangia, about 1 mm in length and structurally resembling those of the fern *Angiopteris*, are grouped in sori, each consisting of 3–4 sporangia. Their origin is eusporangiate and their formation almost simultaneous. In *Encephalartos* temperatures some 15°C above the ambient have been recorded in male cones at the time of meiosis, a consequence of the intense respiratory activity throughout the cone at this stage of development. The microspores of the cycads are uniformly trilete, and in most species each sporangium produces some hundreds of spores. Germination begins in the sporangium, and each spore when shed already contains three cells, namely a single prothallial cell, a generative cell and a tube nucleus (Fig. 7.11a).

THE DEVELOPMENT OF THE GAMETOPHYTES AND FERTILIZATION   The ovules of the cycads are very similar to those of the later pteridosperms. They are distinctly radiospermic and the integument is differentiated into sclerenchymatous and fleshy layers. In the immature ovule the megaspore mother cell, which is deeply embedded in the nucellus, undergoes meiosis and gives rise to four megaspores in a linear tetrad.[21] The three outer megaspores degenerate, but the inner (*chalazal*) megaspore remains viable and develops a layered wall of surprising complexity. It germinates *in situ* and begins to form the female gametophyte, the initial development of which consists of a sequence of free nuclear divisions (Fig. 7.10a). The gametophyte enlarges with the expanding ovule and eventually a vacuole forms at its centre. Wall formation then begins at the periphery of the gametophyte and continues

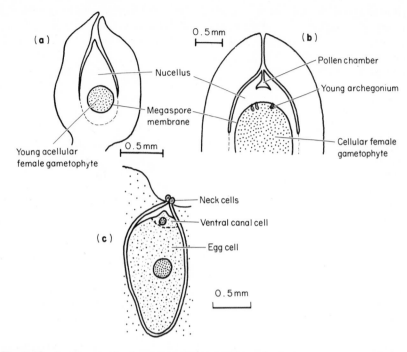

Fig. 7.10   Female reproductive system of cycads. **(a)** Longitudinal section of young ovule. **(b)** Completion of development of female gametophyte. **(c)** Archegonium containing newly formed egg. (All after McLuckie and McKee (1958). *Australian and New Zealand Botany*. Horwitz, Sydney.)

towards the centre. The archegonia (Fig. 7.10b) appear at the micropylar end of the gametophyte where the cells are comparatively small. A distinct and thickened boundary, known as a 'megaspore membrane', persists between the haploid gametophyte and the diploid nucellus.

Although at first sight unfamiliar, the archegonia of the cycads can be seen from their development to be quite similar to those of the lower archegoniate plants. A single initial cell divides into an outer primary neck cell (which subsequently gives rise to one tier of neck cells) and an inner central cell. The latter rapidly expands, and then divides to form the egg and a small superficial ventral canal cell (Fig. 7.10c), which degenerates as the egg becomes ready for fertilization. Maturation of the egg involves considerable cytological activity. The nucleus enlarges and becomes weakly staining, and in *Zamia* small bodies, which appear in the light microscope as refractive droplets, are seen to stream away from its surface into the cytoplasm. The cycads possess the largest egg cells known amongst land plants. The diameter may reach or exceed 3 mm or more, and even that of the nucleus may be as much as 0.5 mm.

Pollination, which occurs during the closing stages of the growth of the female gametophyte, is brought about by the microspores, which are distributed by wind (and in some species possibly by insects), being caught in a sugary 'pollination drop' at the orifice of the micropyle. This fluid, probably secreted by the cells at the tip of the nucellus, is now withdrawn, carrying the microspores with it. These now become lodged in a shallow pollen chamber formed by autolysis at the tip of the nucellus (Fig. 7.10b). Here each grain puts out a tube from the distal part of the grain (i.e. on the side away from the centre of the original tetrad) laterally into the nucellus.[29] This tube, which is of limited growth, has a purely haustorial function, and only the tube nucleus enters it. In the presence of the pollen, and possibly accelerated by enzymes secreted by it, the upper part of the nucellus continues to break down, until all that remains above the mature archegonia is a small pool of fluid. The development of the male gametophyte meanwhile continues. The generative cell, the only cell to show further activity, divides, giving rise to a body cell and a stalk cell (Fig. 7.11b). The proximal part of the male gametophyte now bends over the archegonia, and the nucleus of the body cell ultimately divides to form two coiled, multiflagellate spermatozoids (Fig. 7.11c). These are constructed on the same principles as those of the lower archegoniates, but are much larger and in some species may reach 300 μm in diameter. These are finally released from the proximal part of the tube close to the ruptured exine directly into the fluid above the archegonia. One or more eggs become fertilized. The penetrating spermatozoid sheds its cytoplasm (including the flagella) in the cytoplasm of the egg, and its nucleus enters and disperses in the large female nucleus.

This account of the development of the male gametophyte and fertilization is based on the events in *Macrozamia, Dioon, Zamia* and *Cycas*. The distal germination of the microspores in these genera coupled with the proximal

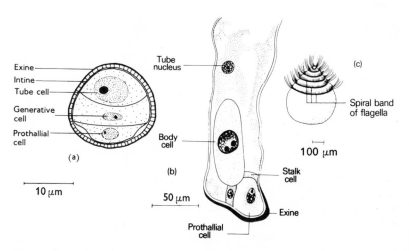

Fig. 7.11 *Cycas* sp. Development of the male gametophyte. **(a)** Vertical section of microspore at liberation. **(b)** Longitudinal section of pollen tube. **(c)** Spermatozoid. (**(a, b)** After Swamy (1948). *American Journal of Botany,* **35**, 77; **(c)** modified after Swamy (1948). *American Journal of Botany,* **35**, 77.)

release of the gametes suggests a position intermediate between the prepollen of most of the pteridosperms and the pollen of the conifers and angiosperms. Many features of sexual reproduction in the cycads however need confirmation and re-investigation by modern techniques.

EMBRYOGENESIS   Although the development of the male and female gametophytes and the interval between pollination and fertilization is prolonged and may extend over months, the formation of the proembryo follows immediately after fertilization.[40] After a period of free nuclear division, in which as many as 256 (2^8) nuclei may be formed, the proembryo becomes cellular. Further growth takes place at the chalazal end, and the embryogeny is evidently endoscopic. At the extreme base of the proembryo is a group of small meristematic cells which develop into the embryo proper, in some species protected on the outside by a layer of cap cells which later degenerate. Above the embryonic cells are a number of elongating cells which form a conspicuous suspensor. The mature suspensor may reach several centimetres in length, but the resistance it meets in driving the young embryo into the nutritive tissue of the female gametophyte causes it to be highly twisted and coiled (Fig. 7.12).

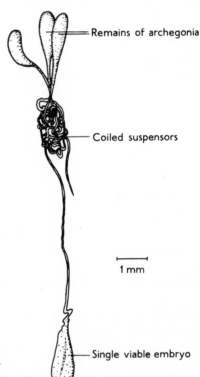

Remains of archegonia

Coiled suspensors

1 mm

Single viable embryo

Fig. 7.12 *Cycas* sp. Embryo and suspensor. (After Swamy, (1948). *American Journal of Botany*, **35**, 77.)

The embryo grows and differentiates at the expense of the female gameto-phyte, occasionally termed – by analogy with the angiosperms (see p. 311) – a 'pseudo-endosperm'. In the mature seed it has two or several cotyledons (depending upon the species), directed away from the micropyle and enclosing the stem apex (plumule). Although a short axis is present below the cotyledons (hypocotyl), a root is still lacking at this stage. The fully formed embryo is surrounded by the exhausted remains of the female gametophyte and nucellus, and externally by the integument. Germination occurs as soon as conditions are favourable and the seed has imbibed suffi-cient water. The hypocotyl pushes its way through the micropylar end of the seed and then begins to develop a strong tap root which persists throughout the life of the plant. The cotyledons remain partially enclosed in the seed, but the plumule emerges and gives rise to mature leaves, the first of which have only a few pinnae.

The fossil history and possible origin of the cycads

The first cycads are found in the late Palaeozoic, both petrified male cones (resembling those of *Stangeria*) and megasporangiophores of the *Cycas* kind having now been discovered in North American rocks. A cycad of the Mesozoic era is *Bjuvia simplex* from the Rhaetic of Sweden, the remains being sufficient to permit a reconstruction of the plant (Fig. 7.13). There was a general resemblance to *Cycas*, but the leaves were entire instead of pinnate. The megasporangiophores, which showed little development of the distal sterile region, formed a loose terminal cone. *Beania* is a female cycad cone from the Jurassic, resembling that of *Zamia*, but less compact. The same beds yield what are almost certainly male cycad cones and also compressions of leaves, the epidermal features of which are quite similar to those of living cycads.

In their radiospermic seeds and in the complexity of their stelar structure the cycads so closely resemble the later pteridosperms that it seems beyond doubt that they had their origin in some common stock. If so, both leaves and sporangiophores would have been derived from lateral branch systems. If we compare the megasporangiophores of *Cycas revoluta*, with its pinnate distal portion, with that of *Zamia*, where the sterile distal portion is lacking, we may see the process by which an ovuliferous megaphyll lost its sterile region and became wholly reproductive. This specialization seems to have been accomplished earlier in the male inflorescences, since in all cycads, even those of Palaeozoic age, the microsporangiophores have little sterile tissue.

Bennettitales (Cycadeoidales)

The Bennettitales are wholly fossil, their record extending from the Triassic to the Cretaceous periods. The frequency of their remains is such that they were probably a more conspicuous element of the Mesozoic floras than the Cycadales. In habit there was a general resemblance to the cycads. Some Bennettitales were upright, sparingly branched plants, while others were squat, bearing a crown of leaves near the soil surface.

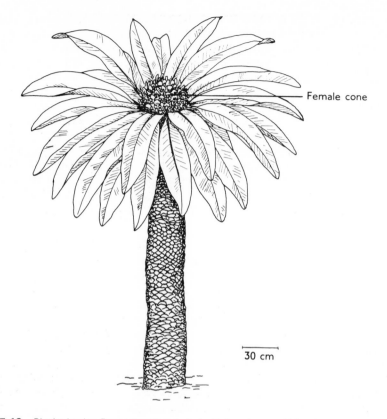

Female cone

30 cm

Fig. 7.13   *Bjuvia simplex*. Reconstruction of plant with female cone. (After Florin, from Arnold (1947). *An Introduction to Paleobotany*. McGraw-Hill, New York.)

MORPHOLOGY AND ANATOMY   The leaves of the Bennettitales were entire or pinnate, and very similar to those of the cycads. They were not in fact easily distinguishable from these until it was discovered that the epidermal features were quite different (Fig. 7.14). In the Bennettitales the walls of the epidermal cells were highly sinuose, and the guard cells and subsidiary cells appear to have had their origin from the same mother cell, giving rise to so-called *syndetocheilic* stomata of characteristic form.

The stem structure, so far as it is known, was similar to that of the cycads.

REPRODUCTION   The female cone of the Bennettitales consisted of an axis bearing upright ovules interspersed with sterile scales. The male inflorescence was a whorl of microsporangiophores which produced either microsporangia or large complex synangia. The pollen grains were monocolpate, similar to those of the later pteridosperms. In most forms the inflorescence was

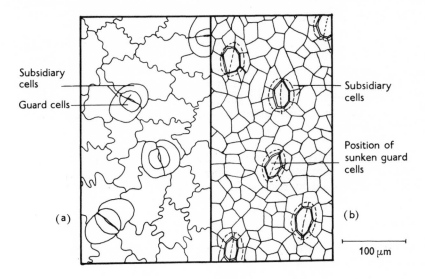

Fig. 7.14 Fossil cuticles of: **(a)** the Bennettitalean type; **(b)** the Cycadalean type. (From a preparation by W. G. Chaloner.)

bisexual, the male portion being below the female (Fig. 7.15). In *Cycadeoidea* the ovules are often well preserved, and this genus provides the oldest known example of megaspores produced in a linear tetrad. The seeds were similar in structure and symmetry to those of cycads and pteridosperms. In some forms embryos were preserved. A massive hypocotyl was directed towards the micropyle, resembling the situation in the cycads.

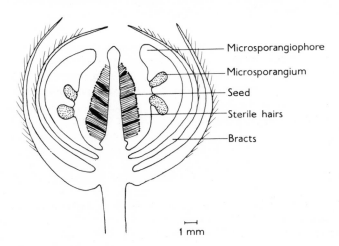

Fig. 7.15 *Williamsoniella coronata.* Longitudinal section of fertile shoot. (After Harris, from Andrews (1961). *Studies in Paleobotany.* Wiley, New York.)

The relationships and origins of the Bennettitales

The general anatomical and reproductive features of the Bennettitales indicate that they also had their origin in some pteridospermous stock, and in some respects the resemblance is closer than with the cycads. The microsporangiophores, for example, can readily be envisaged as derived from condensed microsporangiate pteridosperm fronds. The fossil record indicates that the Bennettitales evolved parallel to the cycads. There is no obvious explanation of why they should have become extinct, whereas the Cycadales survived. It is noteworthy, however, that the microsporangiophores of the Bennettitalean cone commonly ascended close to the ovules, and self-pollination may have been the rule. This would have limited genetic recombination and consequent adaptability.

Caytoniales

The Caytoniales, Jurassic in age and also with radially symmetrical seeds, are considered later (p. 328).

Cordaitales

The Cordaitales, which vegetatively resembled some modern conifers, must have been amongst the most impressive of the seed-bearing plants of the later Palaeozoic. So far as is known, they were all arborescent with columnar trunks, many probably reaching heights of 30 m and diameters of 1 m. The leaves, confined to the upper branches, were spirally arranged and strap-shaped (Fig. 7.16a). In some forms they were as much as 1 m in length and 15 cm in width. There was regular parallel venation interspersed with longitudinal bands of hypodermal fibres, a structure not dissimilar to that of the leaves of the modern conifer *Araucaria araucana*.

VEGETATIVE FEATURES   In general the vascular tissue of the cordaitalean trunks consisted of a large amount of secondary xylem, traversed by narrow parenchymatous rays, surrounding a medullated primary stele. Stems with dense wood of this kind are termed *pycnoxylic*. The primary xylem tended to diminish in later forms, leaving a ring of mesarch bundles surrounding an extensive pith, often broken up into lenticular diaphragms. The secondary tracheids showed several series of circular bordered pits on their radial walls, and were closely similar to those of living *Araucaria*. The leaf traces, which passed outwards from the primary xylem, were simple in origin and commonly consisted of two parallel strands. The roots of the Cordaitales are quite well known, since they often became petrified while they were penetrating decaying remains of other plants. They show a triarch stelar structure, and a distinct root cap at the growing tip.

THE REPRODUCTIVE STRUCTURES   The reproductive organs of the Cordaitales (known as *Cordaianthus*) were borne on slender branches.

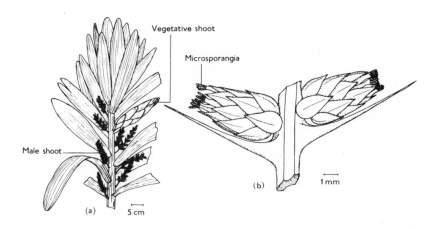

Fig. 7.16 **(a)** Reconstruction of a Cordaitalean shoot. (After Grand'Eury, from Andrews (1961). *Studies in Paleobotany.* Wiley, New York.) **(b)** *Cordaianthus concinnus.* Reconstruction of two male shoots (After Delevoryas (1953). *American Journal of Botany,* **40**, 144.)

Although male and female were separate, they possibly occurred on the same tree. Each reproductive region was bascially an axis, from 10 to 30 cm in length (Fig. 7.16a), bearing two rows of bracts in a complanate distichous arrangement. The male and female shoots occurred singly in the axils of the bracts.

The individual male shoots (Fig. 7.16b) were about 1 cm long. Each consisted of a short, stout axis bearing a large number of linear-lanceolate scales, each with a single vein, in a close spiral. The lower scales were sterile, and were acute or obtuse at their apices, but the upper were emarginate and terminated in several (usually six) cylindrical microsporangia. Since both sterile and fertile scales lay in one spiral, they appear to have been of similar morphological nature. The pollen grains were surrounded by air bladders formed by the separation of the layers of the wall, the two layers remaining in contact however in one region, possibly the site of liberation of the gametes. This thin area was opposite the small triradiate scar at the proximal pole of the grain. It thus appears that the Cordaitales had true pollen with distal germination.

The female shoots were of similar organization, but the fertile scales terminated in ovules instead of microsporangia. In earlier forms (e.g. *Cordaianthus pseudofluitans* (Fig. 7.17a)) the fertile scales (megasporangiophores) projected conspicuously from the shoot, branched, and carried more than one seed. In the later, however, the megasporangiophores were shorter and unbranched, terminating in only one seed concealed amongst the sterile scales.

The seeds of the Cordaitales were not radially symmetrical, but bilateral, the margin of the seed often being extended as a wing. Because of their characteristic flattened appearance, these seeds are termed platyspermic, and

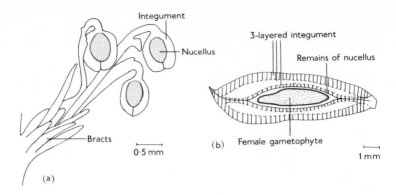

Fig. 7.17 **(a)** *Cordaianthus pseudofluitans*. Reconstruction of female shoot. (After Florin, from Andrews (1961). *Studies in Paleobotany*. Wiley, New York.) **(b)** *Kamaraspermum leeanum*. Transverse section of seed. (After Kern and Andrews, from Andrews (1961). *Studies in Paleobotany*. Wiley, New York.)

they are readily distinguishable from the radiospermic seeds of the pteridosperms. The integument of a cordaitalean seed (Fig. 7.17b) has the appearance of having been formed by two valves, each containing a vein, coming together and enclosing a megasporangium. This view is supported by the occasional occurrence in female shoots of what are interpreted as abortive megasporangia subtended by two unfused lobes. Each seed would thus have three components, the two outer which form the integument possibly having been derived from sterilized sporangia. It is noteworthy in this connection that the microsporangia were commonly produced in multiples of three.

Platyspermic seeds, similar to those seen in cordaitalean inflorescences, are frequently found detached in Carboniferous deposits and their structure is now well known (Fig. 7.17b). They are about 1 cm in height and only a little less in their major transverse diameter; the minor is of the order of 0.5 cm. The integument was differentiated into one or more layers, at least one of which was sclerenchymatous. The nucellus, except for the basal region, appears to have been separate from the integument. A female gametophyte, surrounded by a distinct 'megaspore membrane', developed within the nucellus, and archegonia were produced on its upper surface, the nucellus above this region becoming differentiated as a pollen chamber. Reticulate markings on the pollen grains lying in this chamber were once regarded as indicating endosporic germination of the grain, but this interpretation is now questioned. The pattern is more probably a relic of the sculpturing of the wall than of an internal cellular structure. Gametes were possibly liberated into fluid above the archegonia. As with the pteridosperms, no seeds have been found containing embryos.

The origin and fossil history of the Cordaitales

Little is known of the origin of the Cordaitales, but it possibly lay well before the Carboniferous period. Remains of substantial woody plants, the xylem of which showed araucarian pitting (see p. 247), have been found as early as the Middle Devonian. In the leaves of some of the Cordaitales of the Lower Carboniferous the nerves branched as they approached the tip, possibly indicating an origin in a fan-shaped structure. Axes bearing leaves of the kind envisaged are in fact known from the Middle Devonian, but the relationship of these fossils (placed in the genus *Barrandeina*) to the Cordaitales is quite unproven. Nevertheless, the impression is that the Cordaitales were derived from axial heterosporous forms in much the same way as the pteridosperms, but that in the Cordaitales the megaphyll condensed into the characteristic strap-shaped leaf and the integument of the seed evolved in a slightly different way. The earliest platyspermic seeds come from the later Devonian.[7]

The Cordaitales probably persisted into the beginning of the Mesozoic, but then became extinct.

Coniferales

The conifers are the most widespread of all the groups of gymnosperms, and they form the climax vegetation at high altitudes and in the colder regions of the temperate zones, particularly the north. They are much less common in the Tropics, and here they are usually confined to mountains and are often mixed with angiospermous trees. Of all the vascular plants discussed so far the conifers are the first of significant economic importance. They are almost all arborescent and the wood is used extensively as timber and as a source of pulp for paper-making and related industries.

GROWTH FORMS The growth form of a conifer is frequently pyramidal, the conspicuous main axis being the principal source of the valuable timber. A few conifers of this form attain remarkable sizes and ages. Specimens of *Sequoia*, for example, in California frequently exceed 100 m in height, their trunks reaching diameters of several metres and showing over 2000 growth rings. *Pseudotsuga* in the forest of the Olympic Peninsula of the Pacific north-west may attain even greater heights (but not girth). The oldest living conifers are probably specimens of *Pinus aristata* at high altitudes on the arid White Mountains of the California-Nevada border. Modern techniques of dating show that some of these are almost 5000 years old.

Some conifers (such as the junipers) are bushy, and a few (confined to Australasia) are dwarf, heather-like shrubs of boggy alpine situations. Occasionally the growth form is markedly influenced by the habitat. *Pinus montana*, for example, is a pyramidal tree when growing in acid situations on lower hills, but a straggling shrub with no evident main axis (*Krummholz*) when on limestone at higher altitudes. Most conifers tend to be surface rooted, and many species produce stubby rootlets in the humus layer which are associated with mycorrhizal fungi. *Taxodium distichum* (swamp cypress) which grows in swamps in the warmer parts of eastern North America, is outstanding amongst conifers in producing negatively geotropic aerophores which rise above the surface of the water. These specialized roots are, however, rarely produced by specimens planted outside the native habitat.

LEAVES   The leaves of the conifers take a variety of forms (Fig. 7.18), but they are nearly always small and simple in shape ranging from needle-like structures several centimetres in length (*Pinus*, Fig. 7.18a) to closely adpressed scales reaching only a few millimetres (as in Cupressaceae, Fig. 7.18c). *Araucaria araucana* is unusual in having broadly lanceolate leaves 5 cm or more in length (Fig. 7.18b). In the Cupressaceae the plant frequently passes through a juvenile phase in which it produces needle-like leaves. Cuttings or grafts of the juvenile phase sometimes go on producing needle-like leaves indefinitely, and these so-called *Retinospora* forms are common in gardens. The venation of conifer leaves is never reticulate. There are either a number of parallel veins (as in some species of the Araucariaceae, Fig. 7.18b), or a single median vein, often showing a double structure (as in *Pinus*). In some conifers (e.g. *Pinus*) the leaves are borne wholly or principally on short shoots. The leaves of most conifers persist for several seasons; in only a few genera (e.g. *Larix*) are the leaves truly deciduous. In some Taxodiaceae (e.g. *Taxodium, Metasequoia*) the leaves are confined to the ultimate branchlets, and the branchlets (phyllomorphs) are shed at the end of the growing season.

THE ANATOMY OF THE STEM AND LEAVES   The stems of conifers grow from a group of meristematic cells. In some genera, notably *Araucaria*, the apex is

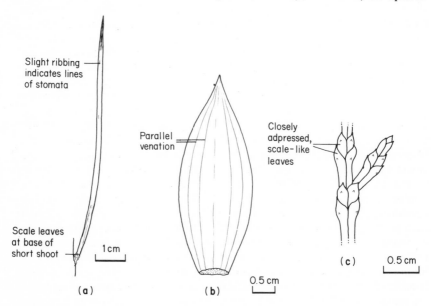

Fig. 7.18 Types of conifer leaf. **(a)** *Pinus monophylla.* **(b)** *Araucaria araucana.* **(c)** *Chamaecyparis obtusa.*

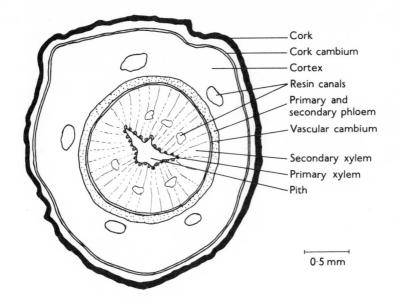

Cork
Cork cambium
Cortex
Resin canals
Primary and
secondary phloem
Vascular cambium

Secondary xylem
Primary xylem
Pith

0·5 mm

Fig. 7.19 *Pinus sylvestris.* Transverse section of young stem with only one season's secondary vascular tissue.

organized into a distinct tunica, in which divisions are principally anticlinal, and a central corpus, where divisions are in several planes. The latter gives rise to the pith and primary vascular tissue. The mature stems of conifers are mostly secondary wood, the pith and primary xylem being relatively inconspicuous (Fig. 7.19). At the outside, the phloem, cortex, and periderm form a comparatively narrow band. The dense secondary xylem consists of radial files of tracheids, traversed by narrow parenchymatous rays. Wood parenchyma is not conspicuous. In *Pinus* it is confined to the epithelium of the resin canals, and it is entirely lacking in *Taxus*. The tracheids are usually differentiated in distinct annual rings, those formed towards the end of a season's growth being narrower than those at its beginning (Fig. 7.20). In *Pinus* the tracheids, which rarely exceed 4 mm in length, bear bordered pits, usually in a single row, on the radial walls. The central part of the pit membrane is thickened and forms the torus (Fig. 7.20b); thickenings are often present along the margins of the pits, the 'Rims of Sanio'. In *Araucaria* the pits are similar, but in 3–4 rows, the pits of adjacent rows alternating. The tracheids of *Araucaria*, but not of *Pinus*, occasionally have small trabeculae (initially of cellulose, but subsequently lignified) extending across the lumen. These so-called 'Bars of Sanio' also occur in a number of other genera.

Considerable differentiation is sometimes present in the parenchymatous rays of conifer woods. In *Pinus*, for example, the cells of the upper and lower margins in the xylem portion of the ray may form radially orientated

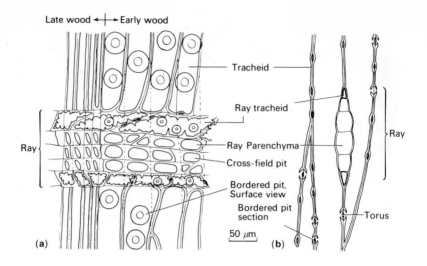

Fig. 7.20   *Pinus sylvestris.* **(a)** Radial longitudinal section of secondary xylem at junction of early and late wood. **(b)** Tangential longitudinal section of a ray similar to that shown in **(a)**.

tracheids (Fig. 7.20), and cells in a similar position in the phloem closely apply themselves to the sieve cells and beome conspicuously rich in cytoplasm. The rays provide an important means of transporting materials laterally in the growing stem.

The resin canals of the conifers, which are schizogenous in origin and interconnected, run longitudinally in the leaves (Fig. 7.21) and xylem, and also transversely in some of the larger rays. The resin itself (a complex acidic substance containing oxidized phenols and terpenes) is synthesized in the epithelium of the canals, probably mostly in the younger tissues, but the actual site of synthesis in the cells is not yet exactly known. The resin system can be tapped by driving a gutter-shaped steel wedge into the xylem near the base of the tree (Fig. 7.22), and from some species considerable quantities of commercially valuable resin can be collected. Pine resin, for example, is the source of turpentine and colophony, both widely used in the paint and varnish industry. The male bark beetle of *Pinus ponderosa* transforms a component of the resin into the sex pheromone of the species.

Many features of anatomical and physiological interest are presented by conifer leaves (Fig. 7.21). The cuticles, for example, are often furnished, especially in the region of the stomata, with distinctive patterns of tubercles and ridges. Palisade and spongy mesophyll are commonly present, and in *Pinus* the walls of the mesophyll cells have ridges projecting into the cell (Fig. 7.21). A well-defined hypodermis, the cells of which may be lignified, is present in many leaves. The vascular bundles are often surrounded by transfusion tissue. Resin canals are frequent, and in some leaves (as of *Thuja*) a prominent gland on the back of the leaf contains a fragrant oil. The leaves of conifers at high altitudes, and of arctic regions, are able to withstand

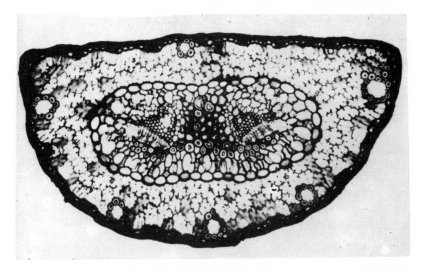

Fig. 7.21 *Pinus sylvestris* A photomicrograph of a transverse section of a leaf, the flat adaxial side uppermost, mid-way along its length. The epidermal cells are lignified and the stomata obscured, but the cells of the hypodermis are visible and adjacent to them the resin canals. Two vascular bundles, separated from the mesophyll by a common endodermis, lie at the centre of the leaf. Thickened and lignified cells are found between these vascular bundles and below the phloem, but the cells elsewhere within the endodermis are largely transfusion tissue. (Approx. × 100)

extreme cold, are remarkably resistant to frost damage, and are able to carry out photosynthesis at unusually low temperatures.

The roots of conifers have a simple primary structure, similar to that found in the ferns. The apical meristem is protected by a root cap, and root hairs are produced from a zone immediately behind it. Unlike the ferns, however, secondary vascular tissue begins to be formed at a very early stage, often before the primary tissues are fully differentiated (Fig. 7.23). Resin canals are abundant in the secondary xylem, rays, and cortex.

THE REPRODUCTIVE STRUCTURES   As the name implies, the male and female reproductive organs of the conifers are commonly borne in cones (Figs 7.24 and 7.25). Most conifers are monoecious but diclinous, the male and female cones being produced in different regions. In *Pinus*, for example, the female cones are produced near the apex of the tree and occupy the positions of main lateral buds, while the male cones are produced on the lower branches, usually in groups, each cone occupying the position of a short shoot. A few conifers (e.g. *Taxus* and *Juniperus communis*) are dioecious. The reproductive cones are usually compact, but in the Podocarpaceae the female cones are either lax or reduced, and in the Taxaceae the female reproductive region is not cone-like at all. Nevertheless, the general affinities of the Taxaceae are clearly with the conifers.

The male cones are fairly uniform in structure, although they range widely

Fig. 7.22 *Pinus   pinaster* (maritime pine). The trunk of this tree (photographed in Portugal) has been tapped for resin. This species of pine is the principal source of natural resin in Europe.

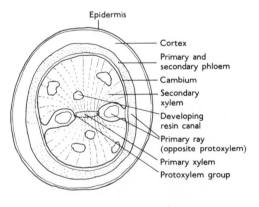

Fig. 7.23 *Pinus sylvestris*. Transverse section of young root.

Fig. 7.24 *Larix decidua* (larch). Portion of a shoot showing the short shoots and the female cone in its second year. Note that the female cone is negatively geotropic and that it terminates a lateral axis having a position equivalent to that of a short shoot. (× 4/5)

in size. Those of *Taxus*, and of the Podocarpaceae and Cupressaceae are globose, hardly reaching 0.5 cm in diameter, but those of other conifers are commonly elongated, and in *Araucaria* they may exceed 20 cm in length and 3 cm in width. All, however, consist of a central axis bearing regularly arranged microsporangiophores (Fig. 7.25). These take the form of scales, somewhat peltate in shape, a variable number of pollen sacs being attached to the head and lying parallel to the stalk. The pollen grains of many species are winged and readily identifiable. The grains of *Pinus* (Fig. 7.26), for example, have two asymmetrically placed air bladders (formed by local separation of the layers of the exine) between which the pollen tube emerges. Other grains have characteristic ornamentation; those of *Cryptomeria*, for example, possess a peculiar cuticular hook on one side. The pollen grains often begin to develop internally before being shed. In *Pinus* (Fig. 7.31d) the pollen grain when liberated contains two degenerating prothallial cells, a tube cell and a generative cell (or nucleus). The grains of the Taxaceae, Taxodiaceae and Cupressaceae, however, lack prothallial cells and are uninucleate when shed, whereas the mature grains of the Araucariaceae contain up to 15 prothallial cells. The pollen grains of all conifers germinate distally, i.e. away from the centre of the original tetrad.

Of the female reproductive regions, that of *Taxus* is best considered first as it facilitates an understanding of the more complex situation in *Pinus* and other conifers. In *Taxus* the ovule terminates a short shoot bearing three pairs of decussate bracts. The ovule itself is upright and bilaterally symmetrical

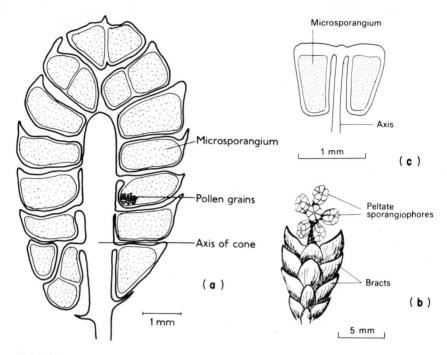

Fig. 7.25 **(a)** *Pinus sylvestris*. Longitudinal section of male cone. Each microsporangiophore bears two pollen sacs. **(b,c)** *Taxus baccata*. Mature male cone, each microsporangiophore bearing 6–8 pollen sacs, and longitudinal section of microsporangiophore.

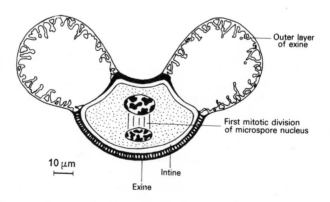

Fig. 7.26 *Pinus banksiana*. Median section of pollen grain showing the first division of the microspore nucleus and the nature of the bladders.

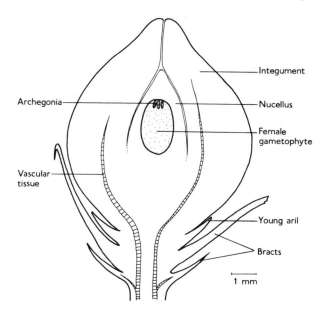

Archegonia

Vascular tissue

Integument

Nucellus

Female gametophyte

Young aril

Bracts

1 mm

Fig. 7.27 *Taxus baccata*. Longitudinal section of young ovule. When fully mature, the nucellus bears the pollination drop at its tip.

(Fig. 7.27). The single integument contains a sclerenchymatous layer, and two vascular bundles, diametrically opposed, ascend in the fleshy portion adjacent to the nucellus. A female gametophyte arises in the nucellus as in other gymnosperm ovules, and when mature it bears immersed archegonia in its micropylar surface. The minute short shoot terminating in the ovule is axillary to another short shoot furnished with spirally arranged scale leaves. This whole complex is itself borne in the axil of a normal foliage leaf. Both short shoots of the female reproductive system in *Taxus* are highly condensed and can be seen only by a careful dissection.

In the female cone of *Pinus* (Fig. 7.28a) we are again concerned with an axis bearing spirally arranged scales in the axils of which are ovuliferous structures (Fig. 7.28b). In *Pinus*, however, the ovuliferous structure is also scale-like, and it is largely fused with and ultimately projects beyond the bract scale in whose axil it arises (Fig. 7.28c). This, however, is not always the situation, even in the Pinaceae. In *Abies*, for example, the bract and ovuliferous scales remain separate, and in some species (e.g. *A. venusta*) the bract scale projects far beyond the ovuliferous. In the Pinaceae the ovuliferous scale bears two inverted ovules near its base (Fig. 7.28b), but in other families the number of ovules and their orientation vary. In *Araucaria* (Fig. 7.29) the ovuliferous scale produces and ultimately entirely surrounds a single inverted ovule. A specialized ovuliferous scale of this kind, also found in some Podocarpaceae, is termed an **epimatium**. Ovules throughout the conifers are regularly bilaterally symmetrical and the seeds are often winged.

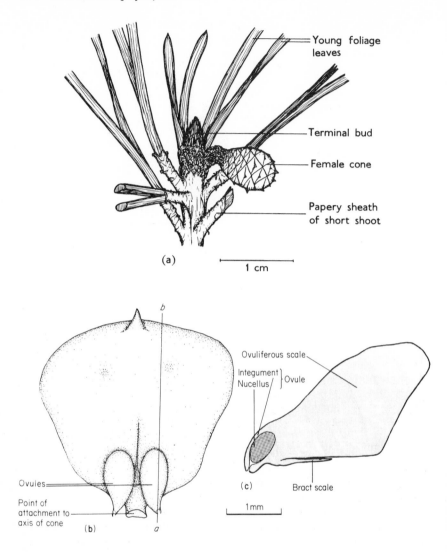

Fig. 7.28 *Pinus sylvestris*. **(a)** Female cone in the early summer of its first year. **(b)** Single scale from cone in **(a)**, viewed from above. **(c)** Longitudinal section along line *a–b*.

The morphological and anatomical evidence, now supported by the palaeobotanical (see p. 260), points to the ovuliferous scale being a highly modified shoot. The vascular supply to the bract scale, for example, consists of vascular bundles of which the xylem is adaxial, the orientation normal for a leaf trace. The bundles passing to the ovuliferous scale, however, are not only similar in position to those entering an axillary shoot, but the xylem of each is also abaxial, an orientation often seen at the base of a shoot trace. The

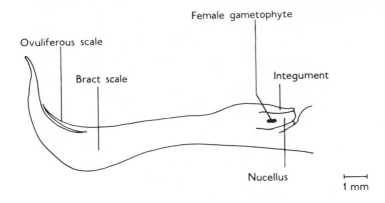

Fig. 7.29 *Araucaria araucana.* Longitudinal section of young ovule. (After Hirmer (1936). *Bibliotheca Botanica*, **28**, 27.)

female region of *Taxus* can be readily related to the cone of *Pinus* if the primary axis with its spirally arranged bracts is regarded as a cone, only one scale of which is fertile. The secondary axis with its decussate bracts and terminal ovule is then equivalent to an ovuliferous scale (Fig. 7.30).

POLLINATION AND FERTILIZATION Pollination, which in temperate climates occurs in the spring, involves in most species a pollination-drop mechanism of the usual kind. The development of the female cone is so co-ordinated with that of the male that at the time of release of the pollen the axis of the female cone undergoes general elongation, thus opening the scales and

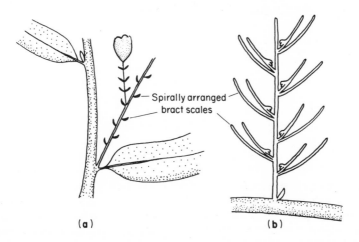

Fig. 7.30 Diagrammatic representations of female reproductive regions of **(a)** *Taxus* and **(b)** Pinaceae, showing how *Taxus* can be regarded as bearing a cone only one scale of which is fertile.

allowing penetration of the pollen. Following pollination, rapid growth of the scales causes them to be tightly packed once again.

In *Pinus sylvestris* the germination and development of the male gameto-phyte within the female cone are very slow and extend over a whole season (Fig. 7.31), coinciding with meiosis in the nucellus and the initiation of the female gametophyte. A pollen tube emerges from the grain in this first season's growth (Fig. 7.31e and h), and the generative cell divides into a stalk cell and a body cell. Little further occurs in the winter, but in the following spring development is resumed. After a period of free nuclear division the female gametophyte becomes cellular (Fig. 7.31i) and 1–6 archegonia are formed in its upper surface (Fig. 7.31j), each surrounded by conspicuous jacket cells. The pollen tubes grow towards the archegonia, and when a tube has come to within a short distance of an archegonium the body cell, which has moved into its tip, divides into two male ('sperm') nuclei of unequal size. The tube eventually penetrates the archegonium and the two sperm nuclei are liberated into the egg cytoplasm. The larger sperm nucleus passes into the egg nucleus, whilst the smaller sperm nucleus, the stalk cell and the tube nucleus all degenerate. Several archegonia in one ovule may be penetrated by pollen tubes, and this can result in the formation of several zygotes and subsequent polyembryony.

Reproduction in *Pinus* is representative of that of the conifers generally. Amongst the principal variations is the 'dehiscent' or 'explosive' pollen of the Cupressaceae. Following hydration the grains swell and the exine is shed. In some conifers (e.g. *Larix*, larch) the pollen is trapped by a stigmatic flap of the integument before being transferred to the micropylar canal. In *Araucaria* the pollination drop mechanism is entirely absent. The pollen germinates between the scales of the female cone forming a freely branching, multinucleate, mycelium-like weft, many of the branches penetrating the nucellus. In the Taxodiaceae the female gametophyte often produces many archegonia, up to 60 being present in *Sequioa*. Pollen tubes may discharge gametes, which in this family are similar in size, above adjacent archegonia, each then being fertilized.

EMBRYOGENESIS   A curious feature of the zygote in many conifers is that the male gamete in moving to the egg nucleus takes its cytoplasm with it. Follow-ing karyogamy the male cytoplasm flows around the zygotic nucleus, and a new cytoplasm ('neocytoplasm') is generated from it. This in turn becomes the cytoplasm of the embryo, and the original egg cytoplasm is largely displaced. Germination of the zygote frequently involves free nuclear division, but in *Pinus* this is not extensive, only four nuclei being so formed (Fig. 7.32a). These move to the bottom of the archegonium and form a plate, walls then being laid down between them. These cells divide longitudinally, the cells of each column behaving synchronously. This leads (Fig. 7.32) to the formation of a suspensor, tetragonal in section, terminating below in four groups of embryonic initials, each capable of yielding an embryo. This so-called cleavage polyembryony may be further complicated by additional embryos budding off from the basal suspensor cells. Usually only one of

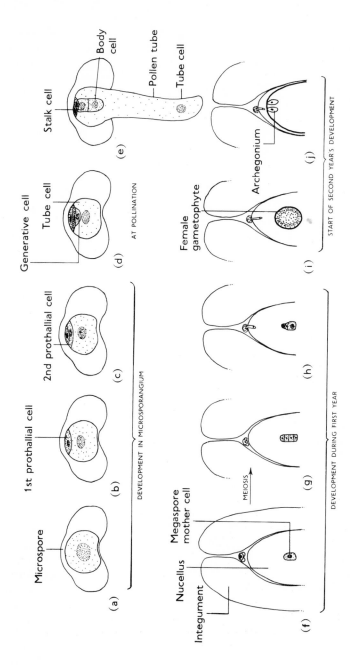

Microspore

1st prothallial cell

2nd prothallial cell

Generative cell

Tube cell

Stalk cell

Body cell

Pollen tube

Tube cell

(a)

(b)

(c)

(d)

(e)

DEVELOPMENT IN MICROSPORANGIUM

AT POLLINATION

Integument

Nucellus

Megaspore mother cell

MEIOSIS

Female gametophyte

Archegonium

(f)

(g)

(h)

(i)

(j)

DEVELOPMENT DURING FIRST YEAR

START OF SECOND YEAR'S DEVELOPMENT

Fig. 7.31 *Pinus* sp. Development of the pollen and ovule. Diagrammatic, not to scale.

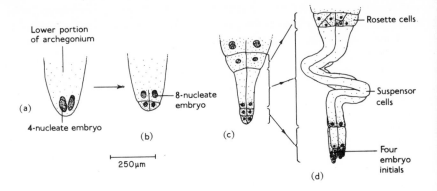

Fig. 7.32   *Pinus* sp. Stages in the development of the embryo. (After Bucholz, from Foster and Gifford (1959). *Comparative Morphology of Vascular Plants*. Freeman, San Francisco. Copyright © 1959.)

these many potential embryos reaches maturity.

The development of the zygote in other conifers differs only in detail from that seen in *Pinus*. In *Sequoia*, for example, there is no initial free nuclear division, and in the Podocarpaceae the cells of the proembryo pass through a binucleate stage, a feature believed peculiar to this family. In other conifers polyembryony seems less common than in *Pinus*. The embryos of many conifers have several cotyledons; as many as 12 may be present in *Pinus*.

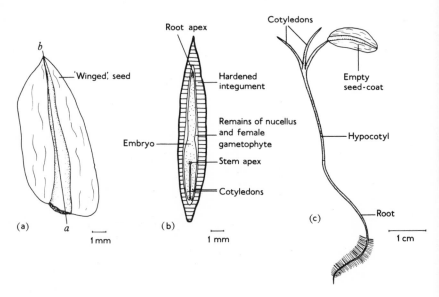

Fig. 7.33   *Sequoia gigantea (Sequoiadendron giganteum)*. **(a)** Seed. **(b)** Longitudinal section (at *a–b*) of well-soaked seed. **(c)** Young seedling.

THE FORMATION AND LIBERATION OF THE SEEDS   The mature embryo lies in the remains of the female gametophyte and nucellus, and is surrounded by a hard seed coat formed from the integument (Fig. 7.33). In some conifers (e.g. *Pinus*) this is expanded as a conspicuous wing assisting the distribution of the seeds by wind (Fig. 7.33a). The female cone often becomes dry and woody during the formation of the seeds and sometimes does not open until a long period after the seeds are mature. In *P. sylvestris* the cone opens and releases the seeds in the second year after pollination (the whole process of reproduction thus extends over three years), but in the 'closed cone' pines of the Pacific coast of North America the cones remain closed indefinitely and the seeds are released by decay of the scales or as a consequence of the singeing of the cones by a forest fire. The cones of some pines are extraordinarily large; those of *P. coulteri*, for example, may reach 40 cm in length and 2 kg in weight. In some Podocarpaceae and in *Juniperus* the ovuliferous scales become fleshy in fruit, the 'berries' of *J. communis* being used to flavour gin. In some other podocarps and in *Taxus* the seed becomes surrounded by a succulent aril which grows up from the base (Fig. 7.34). The bright red aril of *Taxus* is sought after by birds and is probably an aid to dispersal.

GERMINATION   In most conifers germination is initiated by the root pole of the embryo elongating and breaking through the seed-coat. The vigorous primary root soon anchors the seedling, and the elongating hypocotyl raises the remains of the seed, from which several cotyledons are rapidly withdrawn (Fig. 7.33c). All conifer seedlings, so far as is known, become green in the

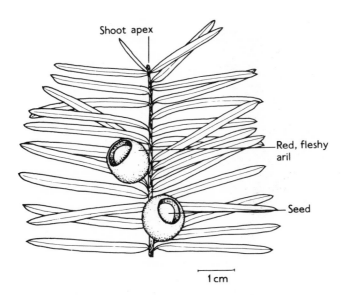

Shoot apex

Red, fleshy aril

Seed

1 cm

Fig. 7.34   *Taxus baccata.* Shoot bearing seeds surrounded by mature arils.

dark, a remarkable property that distinguishes them from the seedlings of most angiosperms.

The seeds of *Araucaria* often germinate in the cone before it falls apart. In some species the hypocotyl swells to form a tuber, and the seedling is capable of 'resting' in this condition for several months. It was this curious feature that facilitated the transmission of the first specimens of *A. araucana* from Chile to Europe in the eighteenth century.

THE EVOLUTION AND ORIGIN OF THE CONIFERS   The current geographical distribution of the conifers presents a number of features of evolutionary significance. Floristically, for example, the conifers of the northern hemisphere are strikingly different from those of the southern, and some families (notably the Pinaceae in the north and the Araucariaceae in the south) hardly cross the equator. Fossils of Quaternary and Tertiary age give no indication that this is a recent segregation, but they do reveal that the distribution of some families was formerly much more extensive. *Sequoia*, for example, now confined to the west coast of North America, was once widespread in the northern hemisphere, and in the Tertiary period *Taxodium* swamps occurred in Spitzbergen. The distribution of the conifers has clearly contracted with the rise of the angiospermous forests.

The fossil record of the conifers extends back well into the Carboniferous, and the early forms show a close relationship, both vegetative and reproductive, with the Cordaitales. In *Lebachia* (Fig. 7.35), representative of

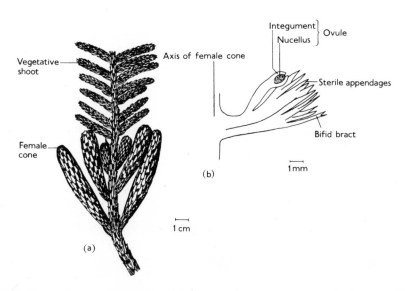

Fig. 7.35   *Lebachia* sp. **(a)** Reconstruction of a shoot bearing upright female cones. **(b)** Longitudinal section of ovuliferous structure. (Both after Florin, from Delevoryas (1962). *Morphology and Evolution of Fossil Plants*. Holt, Rinehart and Winston, New York.)

Upper Carboniferous forms, the female cone consisted of an axis, about 8 cm long, bearing spirally arranged, bifid bracts. In the axil of each was a radially symmetrical, but flattened, dwarf shoot terminating in a single erect ovule (Fig. 7.35b). This fertile short shoot corresponds on the one hand with the female flower of the Cordaitales and on the other with the ovuliferous scale of a modern conifer. There is in fact a series of fossils of late Palaeozoic and early Mesozoic age in which the vegetative and fertile parts of the female short shoot become progressively less distinct, leading ultimately to a structure extremely similar to an ovuliferous scale. Some families became distinct earlier than others. The curious lax cone of some Podocarpaceae, for example, is recognizable as far back as the Lower Triassic, and the characteristic female shoot of *Taxus* seems to have evolved by the end of the same period.

The male cones of the early conifers closely resembled those of the modern, and thus differed sharply from those of the Cordaitales, although the pollen grains of the Cordaitales and early conifers were quite similar. Nothing is known of fertilization in the early conifers, and spermatozoids may have been produced. The arrangement seen in modern conifers, which disposes of the necessity for free fluid at the time of fertilization and possibly reduces the hazards of copulation, may on the other hand have evolved quite early. It was perhaps its advantages which saved the Coniferales from the extinction that befell the Cordaitales.

There are sufficient similarities between the Coniferales and Cordaitales to warrant the assumption of a common origin, the two Orders possibly not diverging until the early Carboniferous.

Ginkgoales

This Order is represented today by a single genus and species, *Ginkgo biloba* (maidenhair tree). This remarkable tree, with a striking pagoda-like arrangement of the main branches, was unknown to the Western world until the seventeenth century. It was first discovered in Japan and subsequently in China, but always in cultivation. Suggestions that wild stands of *Ginkgo* may occur in remote parts of China, although not improbable, have never been confirmed. *Ginkgo* is now common in cultivation in all parts of the world.

THE VEGETATIVE FEATURES Fully grown specimens of *Ginkgo* are tall, deciduous trees reaching a height of 30 m or more. The lateral branches bear both long and short shoots (Fig. 7.36), and leaves occur on each. Damage to a long shoot will cause one or more adjacent short shoots to behave as long shoots, indicating that their manner of growth is not irreversible, and that the maintenance of the dwarf condition probably depends upon the presence of growth-regulating substances produced by the meristem of the long shoot. Anatomically, the apices of the long and short shoots are similar and show well-defined zonation, although no distinct tunica and corpus are present. Growth takes place from a superficial group of apical initial cells. A large proportion of a mature stem consists of secondary xylem, penetrated by

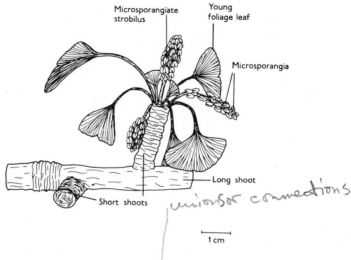

Fig. 7.36   *Ginkgo biloba*. Short shoot bearing male strobili. The leaves have not reached mature size.

narrow parenchymatous rays. The tracheids have bordered pits, usually in a single row, on their radial walls.

The leaves of *Ginkgo* are fan-shaped, usually with a distal notch (hence the specific name). Two vascular bundles ascend the petiole and dichotomize in the lamina, with occasional anastomoses. Short resin ducts may lie between the veins. The distal margin of the leaf is usually irregular, a feature much more marked in juvenile leaves where the distal part of the leaf may even be segmented.

REPRODUCTION   *Ginkgo* is dioecious, and sex determination appears to be chromosomal since the male possesses a heteromorphic pair of chromosomes. The male reproductive structures (Fig. 7.36) consist of small strobili, resembling catkins, which arise in the axils of scale leaves of the short shoot. The axis of the strobilus bears a number of microsporangiophores arranged in a loose and irregular spiral. Each microsporangiophore is slightly peltate and the sporangia, usually two, are attached beneath the head. The pollen grains, which have a characteristic furrow in the wall, contain four nuclei when shed, two of the nuclei being associated with rudimentary prothallial cells, and the others identified as the generative and tube nuclei.

The ovules are usually borne in pairs, two sessile ovules being symmetrically attached at the end of a stalk-like sporangiophore. Not infrequently, however, the sporangiophore branches irregularly and bears more than two ovules. The sporangiophore itself arises, as in the male, in the axil of a scale or a leaf on a short shoot. The ovules (Fig. 7.37) are about 0.5 cm long and about as broad, and are surrounded at the base by a cushion-like swelling of the broad end of the sporangiophore. They possess a single integument into

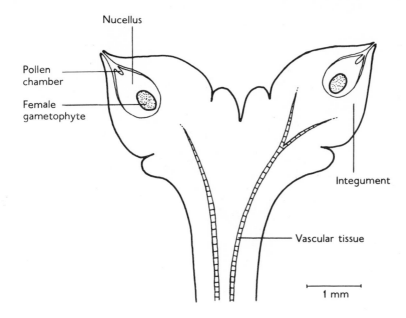

Fig. 7.37 *Ginkgo biloba.* Longitudinal section of female shoot with young ovules.

which two diametrically opposed bundles ascend. This bilateral symmetry is reflected in the micropyle which is slightly two-lipped at its tip. A megaspore is formed within the nucellus and this yields a female gametophyte bounded by a conspicuous membrane. Two archegonia arise at the micropylar end, and the upper part of the nucellus develops a pollen chamber.

Pollination is assisted by a 'pollination drop' at the micropyle, and the pollen chamber, having received the pollen, then becomes closed above. The cavity of the nucellus progressively deepens, carrying the pollen with it, until it reaches the female gametophyte, the centre of which is extended upwards to form a so-called 'tentpole' (a feature seen also in many fossil seeds). The germinating pollen forms a tube, but, as in *Cycas*, this has a haustorial function. The tube grows backwards into the nucellus, the swollen part of the expanded grain hanging in the chamber above the archegonia. As the archegonia mature the generative cells of the male gametophyte divide. Each yields a stalk and a body cell, and the latter divides again to form two multi-flagellate spermatozoids whose major diameters are of the order of 100 μm. These spermatozoids are released into the fluid above the archegonia and bring about fertilization.

The zygote, which may not be formed until after the ovule has been shed, begins its development by free nuclear division, leading to a proembryo containing about 256 nuclei. Walls then differentiate and a flask-shaped proembryo is formed, the lower part of which becomes the embryo proper. A clearly defined suspensor is thus absent. The mature embryo has two

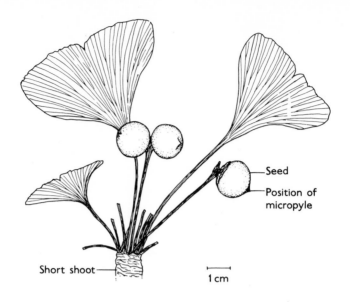

Fig. 7.38   *Ginkgo biloba*. Tip of short shoot bearing ripe seeds.

cotyledons. Usually only one of the paired ovules on the sporangiophore develops into a seed. In the mature seed (Fig. 7.38), which reaches a diameter of about 2 cm, the outer layer of the integument becomes fleshy and resinous, and the inner hard. The formation of the seed is completed in one season.

The fossil history and possible origin of the Ginkgoales

There is fossil evidence of the Ginkgoales having existed in the Mesozoic and at that time having been much more widely distributed than today. The leaves of the extinct Ginkgoales were usually much more like the juvenile than the mature leaves of *G. biloba*. The origin of the Ginkgoales is obscure, but it seems likely that they were derived as an offshoot of some cordaitalean stock. The bilateral symmetry of the seed and the dense secondary xylem are indications against affinities with the cycads and pteridosperms.

Gnetales

The Gnetales consist of only three genera, *Ephedra*, *Gnetum* and *Welwitschia*. Not only are the Gnetales very different from other gymnosperms, but the genera also differ so markedly amongst themselves that many have considered each to be worthy of independent classificatory rank. They have been studied extensively by morphologists because of certain features which make them appear intermediate between gymnosperms and angiosperms. However, although the Gnetales indicate how certain characteristics

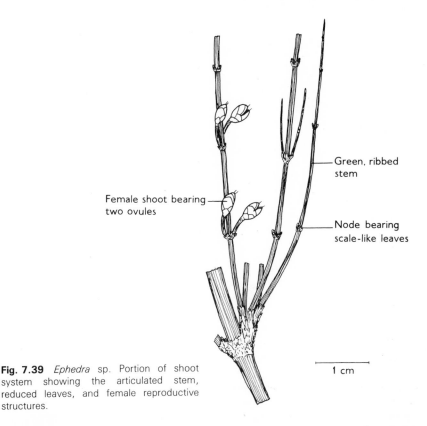

Green, ribbed stem

Female shoot bearing two ovules

Node bearing scale-like leaves

1 cm

Fig. 7.39 *Ephedra* sp. Portion of shoot system showing the articulated stem, reduced leaves, and female reproductive structures.

of angiosperms may have arisen, they themselves appear to be specialized offshoots from the main evolutionary trends. Unfortunately the fossil record of the Gnetales is very fragmentary and does not extend further back than the Tertiary. For the most part it merely indicates changes in geographical distribution.

Ephedra

Ephedra is widely but discontinuously distributed. Some 35 species occur in the Mediterranean region, Asia and in the Americas. They are typical 'switch plants', consisting of densely branched axes, the younger of which are green and photosynthetic (Fig. 7.39). The leaves consist of whorls of small scales which soon become scarious. Many species grow in extremely arid situations, such as sand dunes and scree slopes, and these not unexpectedly have an extensive root system. The young twigs of some species have medicinal uses, and the genus is the source of the alkaloid ephedrine.

THE VEGETATIVE FEATURES   The stem of *Ephedra* grows from a group of meristematic cells, and a distinct tunica and corpus are recognizable in the

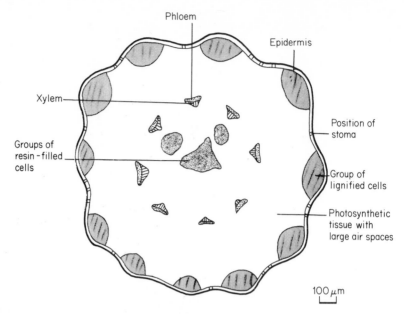

Fig. 7.40 *Ephedra* sp. Transverse section of young stem.

apex. The primary vascular system consists of a number of bundles symmetrically placed around a central pith (Fig. 7.40), the bundles being linked at the nodes by a transverse vascular ring, as in *Equisetum*. The primary xylem becomes surrounded by secondary, traversed by broad parenchymatous rays. The tracheids have bordered pits on their radial walls, well developed tori also being present. Many of the tracheids are arranged in columns, and the end walls are so extensively perforated that they can be legitimately regarded as vessel segments with foraminate perforation plates. The phloem consists of sieve cells and parenchyma, the sieve cells, like those of the conifers, having highly inclined end walls.

REPRODUCTION   *Ephedra* is dioecious. The male reproductive regions are cone-like terminations of short shoots which arise in the axils of the scale leaves. The short shoot bears a number of bracts in decussate pairs, and in the axil of each bract is a male flower (Fig. 7.41). This consists of a microsporangiophore, bearing at its summit 2–8 microsporangia and enclosed in the basal region by 2 medianly placed bracteoles. The pollen grains are ellipsoidal and furnished with prominent longitudinal ridges. Nuclear divisions occur immediately after the formation of the grains and when mature they contain 4 or 5 nuclei. The first 2 daughter nuclei, which do not again divide, are regarded as prothallial.

The female reproductive organ is similar in structure to the male, but only the uppermost pair of bracts is fertile. Each subtends an upright ovule which is surrounded by a sheath, probably homologous with the two bracteoles of

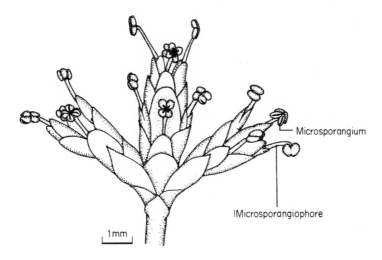

Microsporangium

IMicrosporangiophore

1mm

Fig. 7.41 *Ephedra altissima.* Microsporangiate shoot.

the male flower. The radially symmetrical ovule (Fig. 7.42) is bounded by a papery integument, the apex of which is prolonged into a micropylar tube, highly cutinized at maturity. Megasporogenesis is initiated in the nucellus and in the usual way only one megaspore of the tetrad persists. The female gametophyte passes through a period of free nuclear division before becoming cellular. Two, rarely more, archegonia are differentiated at the micropylar end, and they are unusual in being quite deeply sunk into the somatic tissue of the gametophyte. As the eggs mature, the upper part of the nucellus breaks down to form a pollen chamber, and complex cytological phenomena, among them the amoeboid migration of nuclei and endomitosis, occur in the upper cells of the gametophyte.

The pollen is distributed by wind and possibly also by insects, and a pollination-drop mechanism is present in most species. Germination of the pollen occurs directly on the surface of the female gametophyte and a pollen tube pushes its way into an archegonium. Two sperm nuclei, produced by division of the body cell, enter the egg cell. One fuses with the nucleus of the egg cell and the other, in some species at least, fuses with the ventral canal cell. This 'double fertilization' does not, however, appear to have any special significance, since only the zygote undergoes any further development. The interval between pollination and fertilization, in contrast to the prolonged period in the conifers, may amount to no more than 24 hours in *Ephedra*.

The embryology of *Ephedra* is not well known, but it appears established that the zygote first undergoes free nuclear division, the eight nuclei so formed being the initial cells of proembryos. There is thus potential poly-embryony, but only one embryo reaches maturity. A suspensor, of complex, compound origin, drives the embryo into the central region of the female gametophyte, rich in food reserves. The mature embryo has two cotyledons

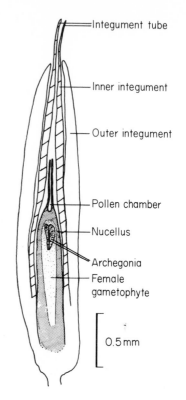

Integument tube

Inner integument

Outer integument

Pollen chamber

Nucellus

Archegonia
Female
gametophyte

0.5 mm

Fig. 7.42 *Ephedra* sp. Longitudinal section of ovule with archegonia. (From a photograph by W.R. Ivimey-Cook in McLean and Ivimey-Cook (1951). *Textbook of Theoretical Botany*. Longmans, London.)

and lies surrounded by the membranous remains of the ovular tissues and the hardened integument. In many species the bracts below the ovules become hard and wing-like in fruits, but in the alpine *E. helvetica* they become fleshy and brightly pigmented.

Gnetum

Gnetum is a tropical genus, occurring in Asia, Africa and South America. Many species are lianes, but others are small trees. In contrast to *Ephedra* the leaves are well developed (Fig. 7.43), and possess broad, oval laminae with reticulate venation, some of the veins ending blindly in areolae. Vegetatively, therefore, *Gnetum* has very much the appearance of an angiosperm.

THE VEGETATIVE FEATURES   The stem of *Gnetum* usually has a small pith, surrounded by a little primary xylem. Most of the xylem is secondary and is interspersed with broad parenchymatous rays. In the climbing forms the stem is eccentric, and successive cambia give rise to a polycyclic stelar structure, and asymmetry in any particular region depending upon its spatial orientation. In general features, therefore, the stem is closer to that of the cycads and pteridosperms than to that of the conifers. A striking difference, however, is

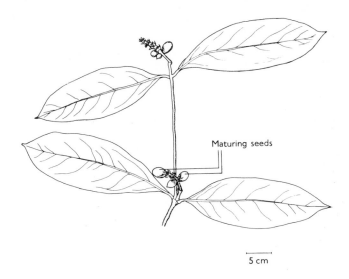

Maturing seeds

5 cm

Fig. 7.43 *Gnetum gnemon*. Portion of shoot bearing female strobili. (After Madhulata, from Maheshwari and Vasil (1961). *Gnetum*. CSIR, Delhi.)

the even closer approach in *Gnetum* than in *Ephedra* to the differentiation of authentic vessels in the secondary xylem. Another peculiarity is that in *Gnetum* parenchymatous cells are closely associated with the sieve cells, recalling the companion cells of angiosperms. The cortex adjacent to the phloem is rich in fibres, occasionally used as cord. In some species the stems contain laticifers, one of the very few instances of these tissue elements occurring outside the flowering plants.

REPRODUCTION   The retention of *Gnetum* in the gymnosperms is justified by the nature of the reproduction. Both the male and female reproductive regions are again strobili, usually terminating lateral axes. In the male strobilus (Fig. 7.44a) the axis bears a succession of gallery-like sheaths, usually about eight in number, probably formed from coalesced bracts. In the axil of each sheath are whorls of male flowers, in some species surmounted by whorls of abortive ovules. The male flower (Fig. 7.44b) consists of a single microsporangiophore, terminating in two microsporangia, surrounded at its base by a delicate membranous sheath. The pollen grains are trinucleate when shed, one nucleus possibly being prothallial.

In the female strobilus (Fig. 7.44c) each sheath encloses a whorl of female flowers. Each flower consists of a single, radially symmetrical ovule (Fig. 7.45a) surrounded by three integuments, the outer of which is possibly homologous with the basal sheath of the male flower. The inner integument is extended into a cutinized micropyle, and the nucellus beneath becomes transformed into a pollen chamber. One or more of the tetrad of megaspores formed in the nucellus enters into the formation of the acellular female gametophyte.

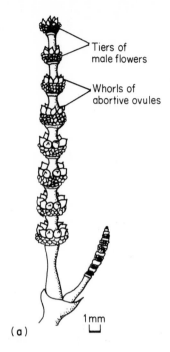

Tiers of
male flowers

Whorls of
abortive ovules

1mm

(a)

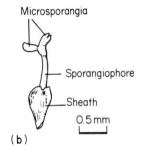

Microsporangia

Sporangiophore

Sheath

0.5mm

(b)

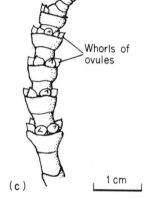

Whorls of
ovules

(c)

1 cm

Fig. 7.44 *Gnetum gnemon.* **(a)** Young microspor-angiate strobilus with whorls of abortive ovules. **(b)** Dehiscent microsporangia. **(c)** Megasporangiate strobilus. (All from Maheshwari and Vasil (1961). *Gnetum.* CSIR, Delhi.)

Pollination, again involving a pollination drop mechanism, initiates renewed growth of the ovule. As the pollen grains pass down into the pollen chamber, the cells lining the micropyle proliferate and occlude the tube. Germination of the enclosed grains stimulates rapid expansion of the ovule and growth of the female gametophyte. The gametophyte, or, as it is more usually called in *Gnetum*, embryo sac, is shaped like an inverted flask (Fig. 7.45b). Much of the cytoplasm, in which the nuclei are irregularly scattered, lies at the base of the sac, but the remainder is distributed as a thick layer around the periphery, a large vacuole occupying the centre. While the embryo sac is completing its development, the pollen tubes, having pene-trated the nucellus, approach the sac. One of the nuclei in the male game-tophyte moves to the tip of the tube and divides into two sperm nuclei. Mean-while one or more nuclei in the upper part of the embryo sac in the region adjacent to the closest pollen tube become conspicuously large. The pollen tube, having by now made contact with the sac, discharges the two sperm nuclei into it. They immediately migrate to the nearby large nuclei of the sac, which can thus be identified as egg nuclei.

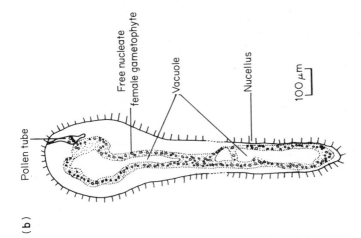

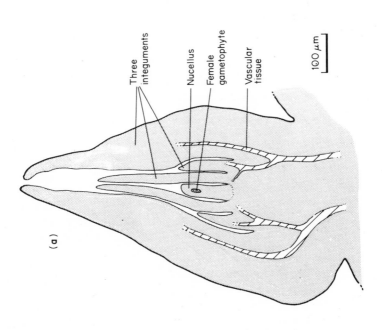

Fig. 7.45 *Gnetum ula*. **(a)** Longitudinal section of very young ovule. **(b)** Longitudinal section of mature female gametophyte (embryo sac) showing entry of pollen tube. (Both after Vasil, from Maheshwari and Vasil (1961). *Gnetum*. CSIR, Delhi.)

The entry of the male nuclei stimulates general division of the somatic nuclei within the embryo sac. The contents of the sac become cellular, the cells often containing several nuclei, which subsequently fuse. The male nucleus and egg nucleus meanwhile coalesce and form a zygote lying within what can now be regarded as a cellular endosperm. Development of the zygote proceeds at once, but does not involve any free nuclear division. A complicated suspensor, possibly partly haustorial in function, is formed before the embryo proper. Although, since more than one egg nucleus may be present in the embryo sac and several male nuclei may be discharged into it, there is potential polyembryony, only one embryo usually comes to maturity. The embryo has two cotyledons.

Although the inflorescence is quite different, the seed of *Gnetum* recalls that of some Bennettitales. When ripe, the outer integuments of the *Gnetum* seed become fleshy, and in some species are edible.

Welwitschia

Welwitschia (Fig. 7.46), in respect of habit, is one of the most peculiar plants in existence. The genus is monotypic, and the single species is confined to desert regions of south-west Africa. The stem is short and upright, and mostly below soil level. At the upper end it bears two strap-shaped leaves with indefinite basal growth. Developmentally these are the first pair of leaves after the cotyledons, and growth soon becomes confined to them. A further pair of decussate leaf primordia is formed in the young plant, but these differentiate into horn-like protuberances. Below, the stem passes into a long tap root which gives rise to an extensive root system. The stem contains much secondary tissue, and it shows anatomical peculiarities similar to those of *Gnetum* and *Ephedra*.

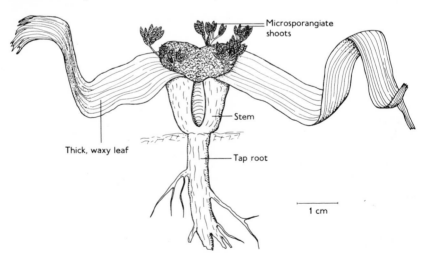

Fig. 7.46   *Welwitschia mirabilis.* Habit of the male plant. (After Hooker (1863). *Transactions of the Linnean Society, London,* **24**, 1.)

REPRODUCTION   *Welwitschia* is dioecious, the cone-like inflorescences terminating small branch systems arising in the axils of the leaves. The cones consist of a series of scales, arranged in decussate pairs, in the axils of which are the individual flowers. The male flowers consist of a short axis bearing first two decussate pairs of small scales, and then a ring of trilocular synangia, the filaments supporting them being fused together at the base into a membranous cylinder. In the centre of the flower, and terminating its short axis, is an abortive ovule. The pollen grains are binucleate when shed, but neither nucleus can be regarded as belonging to a prothallial cell.

The female flower consists solely of an ovule with two integuments, terminating an axis with occasionally two minute lateral outgrowths. The outer integument, which is broadly winged tangentially to the cone and is traversed by several vascular bundles, may be homologous with the upper pair of bracteoles of the male flower. The membranous inner integument is extended at its apex into a cutinized micropylar tube. The basic symmetry of the ovule seems to be radial rather than bilateral.

Pollination and the initiation of the embryo sac take place much as in *Gnetum*, but in the later stages, especially at the time of pollination, there are features peculiar to *Welwitschia*. After the initial free nuclear division in the sac, walls are laid down, many of the cells so formed being multinucleate. As the pollen tubes penetrate the nucellus, some of the multinucleate cells in the upper part of the embryo sac give rise to tubular processes. These grow up towards the descending pollen tubes and potential egg nuclei move to their tips. When a pollen tube and process make contact the separating walls dissolve and the sperm and egg nuclei fuse. The zygote then becomes ensheathed in cytoplasm and a cell membrane forms. There is no free nuclear division in the development of the zygote. A suspensor is formed and below it the embryo proper. Only one zygote yields a mature embryo. There are two cotyledons. In fruit the outer integument of the seed persists as a broad wing which assists aerial dispersal.

The possible origin of the Gnetales

The origins of the Gnetales remain conjectural. Anatomically, especially in the features of the secondary xylem, they recall the pteridosperms rather than the Cordaitales. This view is strengthened by the symmetry of the seeds, which is primarily radial. Nevertheless, the Gnetales are clearly far from the pteridosperms of the Carboniferous and probably the consequence of quite prolonged evolutionary change. Features of the distribution suggest that *Ephedra* originated in Eurasia and *Gnetum* in Gondwana, a vast southern continent which broke up in the Mesozoic era.

The morphological significance of gymnospermy

The diversity of the gymnosperms taken as a whole, in both vegetative and reproductive features, indicates that gymnospermy must be regarded as a grade of evolutionary advance. It was probably first attained in the second half of the Devonian period. Increasingly the plant remains of this age are yielding evidence of gymnospermous anatomy associated with heterosporous reproduction, and seeds are now known even older than *Archaeosperma*.[18] This has given rise to the concept of 'progymnospermy'. Although the limits of the progymnosperms, which range from near-psilophytes at one extreme to the immediate antecedents of the earliest seed plants at the other, are inevitably ill-defined, it is clear that this intermediate level of evolution was of great significance in the phylogeny of vascular plants. The evolutionary tendencies seen in the progymnosperms affected not only the reproductive organs, but also anatomy. Some Devonian representatives, for example, showed fibres in the cortex and secondary phloem, features well represented in the early gymnosperms.

The seed habit may of course have arisen independently in more than one group of heterosporous plants. The affinities of the cycads, for example, appear to be pteridospermous, those of *Ginkgo* cordaitalean. Here then are two groups of plants whose evolution has probably been independent for many millions of years, but which are at the same level of advancement in respect of the reproductive process. In both, despite their arborescent form, fertilization is still brought about by flagellate spermatozoids, liberated from very similar male gametophytes. The amount of fluid required in the fertilization process is of course very much reduced. In the conifers and *Ephedra*, despite the retention of archegonia, spermatozoids are eliminated. The pollen tube, at first possibly always a haustorium, as in the cycads and *Ginkgo*, has now taken on the function of delivering the male gametes to the egg.

The highest grade of gymnospermy is clearly shown by the Gnetales. The peculiarities of the female gametophyte in *Gnetum* and *Welwitschia* indicate the kind of developments which, in some early transitional forms, may have led to the angiospermous embryo sac. It is significant that these striking features of the reproduction in the Gnetales should be accompanied by anatomical developments which also foreshadow the angiosperms. The relationship between Gnetales and the angiosperms is thus analogous to that between the progymnosperms and the gymnosperms.

8
The Tracheophyta, IV
(Pteropsida: Angiospermae)

The angiosperms are the most abundant and widely distributed tracheophytes. They are of outstanding economic importance, being the source of many durable hardwoods, most of our vegetable foodstuffs, and about one quarter (in monetary value) of commercially marketed drugs. They number some 200 000 species and show remarkable diversity in growth form, morphology and physiology.[43]

Angiospermae

Sporophyte herbaceous or arborescent; branching usually axillary. Leaves various, but regarded as megaphyllous in origin. Secondary vascular tissue commonly present. Vascular system usually consisting of vessels and tracheids, and sieve tubes with distinctive companion cells. Heterospory as in the Gymnospermae, but the ovules borne within a characteristic structure (carpel), usually closed, the pollen germinating on a specialized region of the exterior (stigma). Female gametophyte always an embryo sac, lacking archegonia. Fertilization by unspecialized male cells, characteristically double, yielding in each embryo sac a zygote and a mostly triploid endosperm nucleus. Embryogeny endoscopic. Various forms of asexual reproduction not uncommon.

The growth forms of angiosperms

Examination of an angiosperm flora will usually reveal that it consists of a number of distinct growth forms, ranging from large woody trees to minute herbs. The extent of the representation of these different morphologies in a given vegetation presents intricate ecological problems, the discussion of which is facilitated by Raunkiaer's concept of life forms.[32] A life form is defined by the length of life of the shoots and the position and protection of the resting buds (Fig. 8.1). Those plants in which the shoots are persistent and the buds are carried well above the soil surface are termed *phanerophytes*, those with resting buds closer to the surface *chamaephytes*, and those with resting buds at the surface *hemicryptophytes*. Familiar examples of these three classes are, respectively, the larger woody plants,

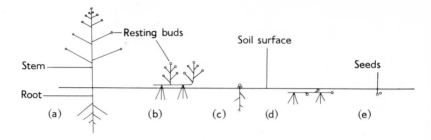

Fig. 8.1 Life forms of angiosperms. The circles indicate the positions of the resting buds. **(a)** Phanerophyte. **(b)** Chamaephyte. **(c)** Hemicryptophyte. **(d)** Cryptophyte. **(e)** Therophyte.

small bushes such as *Calluna*, and rosette plants such as *Taraxacum*. The classification, except for certain small specialized categories, is completed by the *cryptophytes* (geophytes) where the resting buds are below the soil surface, and the *therophytes*. The latter are those annuals and ephemerals which tide over unfavourable periods as embryos enclosed in seeds. By use of this classification it is possible to show in a precise statistical manner that, for example, the vegetation of the humid tropics consists predominantly of phanerophytes, and that in the northern hemisphere the percentage of hemicryptophytes in general increases with latitude (Table 8.1).

Table 8.1 The relationship between life form and latitude. (Data from Raunkiaer.[32] Succulents, water plants and specialized epiphytes are omitted.)

Flora	Approx. latitude	Percentage representation Ph	Ch	H	Cr	Th
Amazonian rain forest	0°	95	1	3	1	0
St Thomas & St Jan (West Indies)	18°N	58	12	9	3	14
Denmark	57°N	7	3	50	22	18
Iceland	65°N	2	13	54	20	11
Spitzbergen	77°N	1	22	60	15	1

Ph: phanerophyte   Ch: chamaephyte   H: hemicryptophyte
Cr: cryptophyte   Th: therophyte

For convenience of description we can regard the plant body of an angiosperm as consisting of three morphological categories, stem, leaf and root, although, as we shall see later, there are good reasons for regarding the leaf as a megaphyll and hence of axial origin. The main stem is usually upright, displaying negative geotropism and positive phototropism. Branching occurs in the axils of leaves. The final orientation of these laterals is probably determined by a combination of complex geotropic (termed plagiogeotropic) and phototropic responses. Although such branch systems are usually aerial, they may be subterranean (when, of course, light will no longer affect morphogenesis). The genus *Parinarium*, for example, is represented by

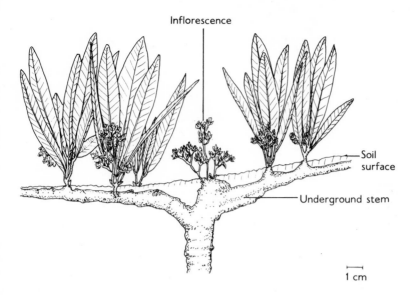

normal trees in tropical Africa, but in *P. capense* (Fig. 8.2), a species of colder South Africa, the stem, although of similar woodiness and ramification, is below soil level, only small shoots appearing at the surface. This so-called *suffrutescent* habit is also encountered, although less strikingly, in alpine willows.

In the development of the branch system the main axis may originate in two ways. Either the apical bud continues its vegetative growth indefinitely, or it

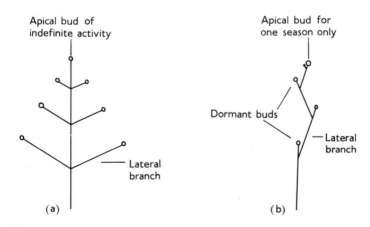

Fig. 8.3 The basic types of branching. **(a)** Monopodial. **(b)** Sympodial.

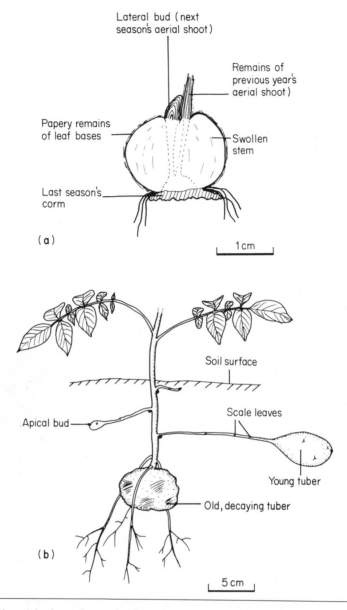

Lateral bud (next season's aerial shoot)

Remains of previous year's aerial shoot)

Papery remains of leaf bases

Swollen stem

Last season's corm

(a)

1 cm

Soil surface

Scale leaves

Apical bud

Young tuber

Old, decaying tuber

(b)

5 cm

is extinguished at the end of each season, when either it gives rise to a reproductive system of limited growth, or it aborts. Where the apical bud remains active (Fig. 8.3a), the main axis is of simple origin, and growth is said to be **monopodial**. Where it is extinguished (Fig. 8.3b), growth is continued in the following season by the uppermost lateral, and subsequent positional readjustments are often such that season discontinuities are hardly

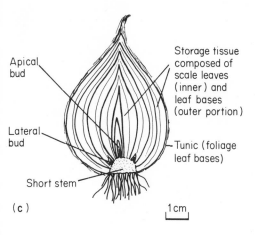

Apical bud

Lateral bud

Short stem

Storage tissue composed of scale leaves (inner) and leaf bases (outer portion)

Tunic (foliage leaf bases)

(c)

1 cm

Fig. 8.4 Stems modified for storage. **(a)** Longitudinal section of the corm of *Crocus* (winter condition). **(b)** Stem tuber formation in *Solanum tuberosum*. **(c)** Longitudinal section of the bulb of *Allium cepa*.

visible in the mature axis. The axis is, nevertheless, of compound origin, and growth is said to be *sympodial*. These two kinds of growth are represented amongst plants of all life forms. Amongst phanerophytes, for example, the growth of the ash (*Fraxinus*) is monopodial and that of the lime (*Tilia*) is sympodial. The factors which cause abortion of terminal buds in plants of sympodial growth are complex, but in trees day length is often of paramount importance. The shortening days of later summer and autumn stimulate the synthesis of a growth inhibitor in the leaves. This is in turn transmitted to the terminal buds, causing either dormancy or in some species abortion. In long days the inhibitor is lacking so that in experimental conditions plants of this kind can be made to assume a quite different manner of growth.

Aerial stems may sometimes twine in a regular fashion. Familiar examples are the hop (*Humulus lupulus*), where the stem generates a helix with a left-handed screw (clockwise when viewed from above), and bindweed (*Convolvulus*), where the screw is right-handed. In some flowering plants, as in the Filicales, the main axis takes the form of a horizontal rhizome. In *Aegopodium podagraria* (goutweed), and probably in other species, the rhizome is able, by adjusting the orientation of its growing region, to maintain itself at an almost constant depth beneath the surface of the soil. The physiological mechanism underlying this remarkable behaviour is still not wholly known. Other highly modified stems are seen in corms, bulbs and some tubers, in which the stem, which is adapted for the storage of food, either in itself (corms (Fig. 8.4a), stem tubers (Fig. 8.4b)) or in the associated swollen leaf bases or scale leaves (bulbs (Fig. 8.4c)), shows very little elongation. A very peculiar stem, resembling a thallose liverwort, is found in the Podostemaceae (Fig. 8.5), a family of small plants growing on rocks by tropical streams.

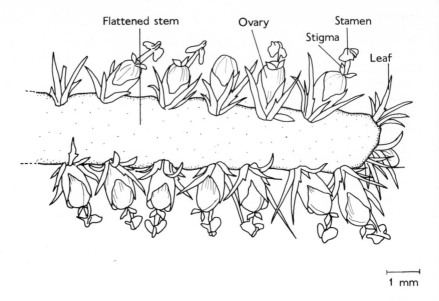

Fig. 8.5   *Butumia marginalis* of the Podostemaceae. Flowers are produced at the margin of the flattened stem. (After Taylor (1953). *Bulletin of the British Museum (Natural History), Botany*, **1**, 53.)

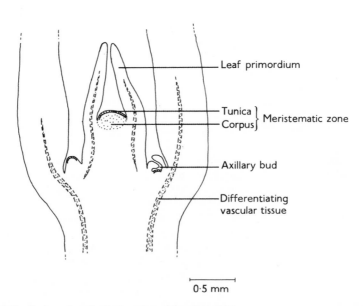

Fig. 8.6   *Syringa*. Longitudinal section of the stem apex.

Internal features of the stem

GROWTH AND ANATOMY   The stem of the angiosperm grows from a group of meristematic cells lying near the summit of the apex.[9] The apex itself (Fig. 8.6) is usually organized into two distinct zones, recognizable by their geometry and the directions of the divisions of the cells. In the centre is the **corpus**, where divisions are in various directions, thus adding to the width of the apex as well as to its length. The corpus is covered by the **tunica**, a layer of cells in which divisions are principally anticlinal, thus increasing the surface of the apex. The tunica is usually two cells thick, but in some plants the thickness may amount to only a single cell and in others to four or five. That these two zones are to some extent independent is shown by the existence of chimaeras in which the cells of the tunica acquire (as a consequence, for example, of treatment with colchicine) aneuploid or polyploid nuclei and retain them through all subsequent divisions, while the nuclei of the cells of the corpus remain diploid. The tunica yields the leaf primordia and ultimately the cortex of the mature stem, and the corpus the vascular tissue and associated parenchyma. Although this kind of apical organization is foreshadowed in the conifers (particularly in *Araucaria*, see p. 247), it is nowhere so distinct as in the angiosperms.

The angiosperms fall into two major groups, referred to as the monocotyledons and dicotyledons, in which the embryos are commonly furnished with one and two cotyledons respectively. These groups also differ

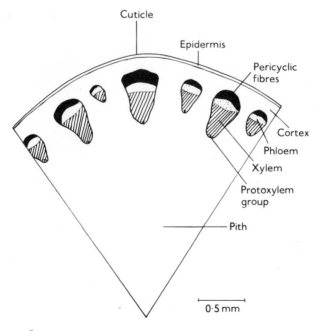

Fig. 8.7 *Dahlia*. Transverse section of a segment of a young stem.

in many other features, among them the form of the primary vascular tissue. In the dicotyledons, which are the more numerous, the primary vascular tissue usually consists of a ring of collateral bundles in which the differentiation of the xylem is centrifugal and of the phloem centripetal (Fig. 8.7). The xylem and the phloem usually remain separated by a thin layer of undifferentiated cells, later recognizable as the *intrafascicular* cambium. In the monocotyledons, however, the bundles, although often collateral and mostly orientated with the xylem adaxial, are not in one ring but are irregularly scattered in a parenchymatous matrix, referred to as the ground parenchyma (Fig. 8.8). The bundles of the monocotyledons also lack potentially meristematic tissue between the xylem and the phloem, and are consequently said to be closed.

Usually secondary vascular tissue soon appears in a dicotyledonous stem. The undifferentiated layer in the primary bundle becomes meristematic and continuous with a similar layer between the bundles (the *interfascicular* cambium), and subsequent activity of the cambial ring resembles that already

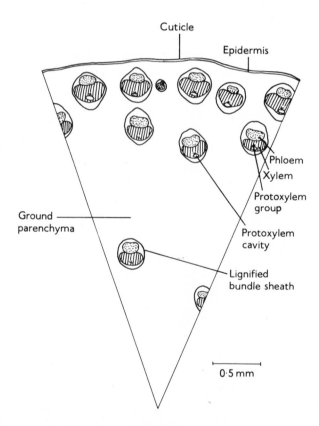

Fig. 8.8 *Zea mays*. Transverse section of a segment of stem.

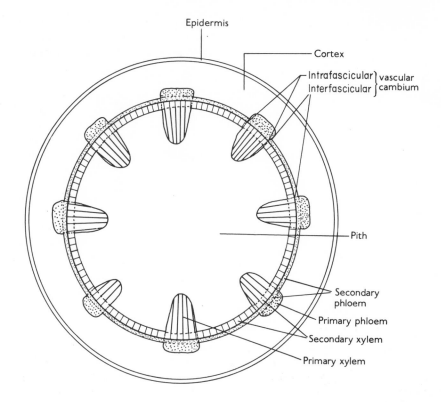

Fig. 8.9 Diagram showing the method of production of secondary vascular tissue in a dicotyledonous stem.

seen in the gymnosperms (Fig. 8.9). The first cambium does not always continue to function indefinitely; other cambia may arise outside the vascular cylinder, recalling the situation in *Cycas* (p. 232). In a few arborescent mono-cotyledons a meristematic zone at the periphery of the stem gives rise to additional collateral vascular bundles as the girth of the stem increases. this activity is, however, limited and in most arborescent monocotyledons, such as the palms, the plant remains for several years as a widening bud bearing a rosette of leaves only a little above soil level. When the bud approaches its mature diameter elongation of the stem takes place very rapidly. For this reason, coconut palms, for example, are rarely seen with their trunks half extended. The manner of growth of the monocotyledonous trees is thus quite different from that of dicotyledonous forms.

The vascular tissue of almost all angiosperms is distinguished from that of all but a few living gymnosperms by the presence of vessels in the xylem and of companion cells in the phloem. Vessels are long tubes, commonly about 10 cm long, but in some plants, such as lianes, reaching or exceeding a length of 5 m. These tubes are composed of segments (Fig. 8.10), each of which is

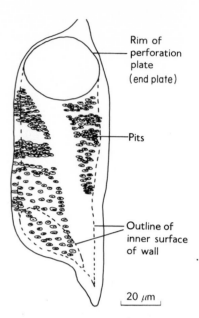

Rim of
perforation
plate
(end plate)

Pits

Outline of
inner surface
of wall

20 µm

Fig. 8.10   *Fraxinus.* A single vessel segment
from a macerate of the secondary wood.

derived from a single cell and is equivalent to a single tracheid. During
differentiation of a vessel, however, the end walls of the segments are
resorbed, leaving mere rims marking their position, or highly perforated
diaphragms, referred to as end plates. Vessels are often accompanied by
tracheids, and there are a few angiosperms in which the conducting elements
of the xylem remain wholly tracheidal.

The sieve cells of the angiosperms are arranged in longitudinal rows, the
whole column being referred to as a sieve tube (Fig. 8.11). The end walls of
the sieve cells (or 'sieve tube elements') are usually inclined and bear one or
more sieve plates. The contents of the sieve cells remain organized, although
the cytology is peculiar and the organelles, including the nucleus, are partly
degenerate. Deposits of an amorphous polysaccharide, callose, often appear
on the sieve plates at the end of the growing season. The companion cell or
cells, closely applied to the sieve cell, may be essential for its function. The
companion cell has a dense cytoplasm and a prominent nucleus; it arises from
the same mother cell as the adjacent sieve cell.

ECONOMIC PRODUCTS   The stems of angiosperms frequently contain
abundant fibres, and the formation of periderm in the outer cortex or even
closer to the vascular tissue is often extensive. Some trees generate cork very
freely, that of *Quercus suber* (cork oak) in the Mediterranean region and
Portugal being periodically harvested and finding many uses in commerce. A
large part of the economic value of angiosperms in fact lies in the stems. Quite
apart from the often very valuable timber, stems yield materials as diverse as
starch (e.g. sago from the palm *Metroxylon* and arrowroot from the rhizome

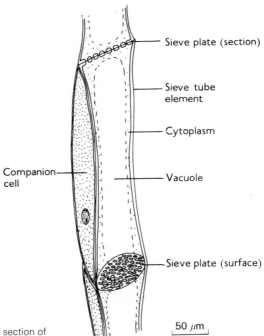

Sieve plate (section)

Sieve tube element

Cytoplasm

Companion cell

Vacuole

Sieve plate (surface)

50 μm

Fig. 8.11 *Cucumis*. Longitudinal section of a sieve tube and companion cell.

of South American *Maranta*), sugar (from sugar cane and in the form of syrup from the stem of *Acer saccharinum*, the sugar maple), fibres, spices (e.g. cinnamon, the bark of *Cinnamomum zeylanicum*), rubber (formed from the latex of *Hevea*, and less importantly from that of certain other species) and drugs (e.g. quinine from the bark of *Cinchona*).

Leaves

THE FORMS OF LEAVES   The leaves of angiosperms show a wide range of size and shape, and all forms of phyllotaxy from distichy to decussate, whorled and spiral arrangements are found. The factors determining these different arrangements have not yet been identified. There are however theoretical grounds for believing that spiral arrangements (which, with upright stems, result in the minimum amount of shading of a leaf by others above it) are the consequence of a periodicity set up by interacting morphogens at the apex. In most mature leaves a petiole and lamina are usually distinguishable, and two small lateral outgrowths at the base of the petiole, termed stipules, are often present. In many species the leaves are pinnately branched, sometimes even as far as the third order, causing the whole to resemble a lateral branch system. This resemblance is occasionally enhanced (e.g. in *Thalictrum aquilegifolium* (Fig.8.12)) by the presence of small outgrowths, recalling stipules, at the points of branching. At the other extreme leaves may be a little

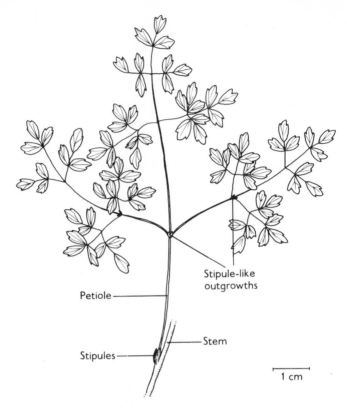

Fig. 8.12   *Thalictrum aquilegifolium.* Compound leaf structure.

more than scales, photosynthesis being carried out principally if not entirely in the stem. In *Casuarina,* a tree of the Tropics and sub-Tropics of the Old World, with leaves of this kind, the leaves are arranged in whorls so that the whole shoot comes to bear a striking external likeness to that of *Equisetum.* In a few plants (e.g. *Ruscus aculeatus,* butcher's broom) the reduction of the leaves to scarious scales has been accompanied by the transformation of lateral shoots of limited growth into flattened, leaf-like structures called *phylloclades* or *cladodes* (Fig. 8.13).

An intermediate condition is where photosynthesis is carried on in what morphologically are broadened and flattened petioles, the laminae being absent or rudimentary. These structures, called *phyllodes,* are found in many species of *Acacia.*

A classification has been developed for leaves, depending upon the area of the mature lamina, and clear relationships have emerged between climate and leaf size. In the tropical rain forest, for example, the leaves of many species tend to be large and of similar area, but in the dry, scrubby vegetation of the Mediterranean region the leaves by contrast show a much wider range of con-

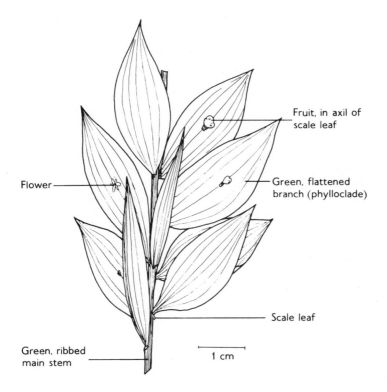

Fruit, in axil of
scale leaf

Green, flattened
branch (phylloclade)

Flower

Scale leaf

Green, ribbed
main stem

1 cm

Fig. 8.13 *Ruscus aculeatus*. Habit of shoot.

siderably smaller areas. Amongst other relationships, mostly of unknown
significance, is the rarity of leaves with toothed margins in tropical vegetation, and their abundance in temperate vegetation. Leathery leaves, with
their apices extended into 'drip tips', are also characteristic of vegetation in
regions of high rainfall,[35] and species growing by streams often tend to have
long narrow leaves (stenophylls).

Although the leaves of most species are differentiated into a lamina and a
petiole, in some, the grasses providing familiar examples, a petiole is absent
and the leaves are consequently termed sessile. Most leaves are orientated
with the plane of the lamina more or less horizontal, and the insertion into the
stem transverse. There are, however, some plants (all with distichous
phyllotaxy) in which the plane of the leaf is vertical and the base of the leaf
clasps the stem as a rider a horse, an insertion consequently termed equitant.
In some species motor cells, usually aggregated into a distinct *pulvinus*, are
able to alter the orientation of the leaf in a striking fashion. In the tropical
rain tree (*Pithecolobium saman*), for example, the leaves appear to collapse
with the diminishing light of the afternoon (a so-called 'sleep movement'),
and in *Mimosa pudica* (sensitive plant) similar movements take place if the
leaves are mechanically disturbed. The leaves of the 'compass plants', of

which *Lactuca scariola* (a wild lettuce) is a notable example, move in relation to the sun so that only one edge is fully insolated at any one time. The motive effects of pulvini depend upon rapid changes in turgor and the ability of the cytoplasm in certain conditions to move water actively in and out of vacuoles.

In many monocotyledons the leaf bases become swollen and form a subterranean bulb which overwinters. The cells of the bulb are filled with food reserves, and are often also rich in secondary metabolites. The lachrymatory thiopropanol S-oxide of the onion bulb is particularly well known.

THE ANATOMY AND DEVELOPMENT OF ANGIOSPERM LEAVES   Although the leaves of angiosperms are structurally similar to those of gymnosperms, there are a number of new features. The palisade tissue, for example, is frequently sharply differentiated from the rest of the mesophyll and in some plants, especially in those that are shade tolerant (e.g. *Impatiens parviflora*, balsam) the form of the palisade is markedly influenced by irradiance (Fig. 8.14). The change in cell shape in the shade exposes a greater proportion of the chloroplasts to the incident light with the result that over a wide range there is

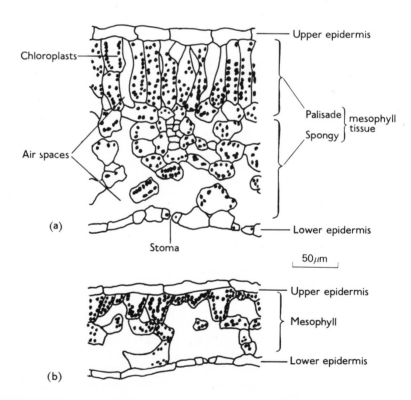

Fig. 8.14 *Impatiens parviflora*. Transverse section of leaf: **(a)** From plant growing in full sunlight. **(b)** From plant growing in 7% sunlight. (Both after Hughes (1959). *Journal of the Linnean Society, Botany*, **56**, 161.)

little variation in net assimilation. Equitant leaves are usually bifacial, stomata and palisade being symmetrically placed on each side. Amongst other structures found in leaves are oil glands, often giving the leaf a fragrance when crushed, and, usually at the margins or on the petiole, nectaries (as, for example, in cherry). In some species of tropical plants stipular nectaries are associated with symbiotic bacteria ('leaf nodules'). The venation of the lamina is commonly reticulate, patterns of extreme intricacy often being generated by the minor veins, many of which end blindly in the areolae (Fig. 8.15). The symmetry of the vascular supply ascending the petiole is usually clearly dorsiventral, but it becomes almost perfectly radial in the petioles of peltate leaves (e.g. *Tropaeolum majus*, the garden nasturtium).

The leaf primordium grows from initial cells at its margin, and at first cell division predominates over expansion. The pattern of the main veins soon, however, becomes established, and also in dicotyledons that of any branching of the leaf. This is brought about by the marginal meristem becoming discontinuous and its activity confined to definite areas of the periphery. Segmentation in the leaves of monocotyledons (e.g. palms) is a more complicated process and involves the degeneration of tracts of tissue between areas which will subsequently become pinnae.

As growth of a leaf primordium proceeds, cell expansion comes to predominate over division, one of the last products of cell division being the guard cells of the stomata. It is evident that the surface growth of the lamina is closely co-ordinated with the extent of the vascular framework. In some plants it is possible to disturb this co-ordination by allowing the primordium to form in one day length, but to expand in another, leading to deformed leaves. These experiments reveal that the metabolic factors influencing the growth of the vascular skeleton are different from those influencing the

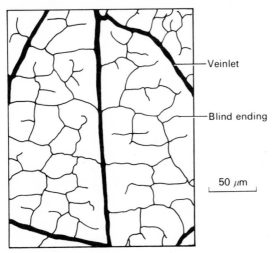

Fig. 8.15 *Impatiens parviflora*. The pattern of leaf venation. (After Hughes (1959). *Journal of the Linnean Society, Botany*, **56**, 161.)

expansion of the lamina. Standing apart from this general scheme of development are again the leaves of some monocotyledons. The leaf of a grass, for example, grows from a meristem at the base of the lamina, above which there is continuous basipetal differentiation until the leaf reaches its mature length.

In some species leaves are deciduous, being severed from the axis by a distinct abscission layer. Partial digestion and disarticulation of cell walls at this site are brought about by hydrolases, including cellulase. In temperate regions abscission usually occurs during the shortening days of autumn, but some tropical trees also regularly shed their leaves, the periodicity having no evident relation to season.[35]

Leaves of angiosperms show a wider range of surface coverings than those of any other land plants. These may take the form of plates or crystals of wax (as on the upper surface of the leaf of *Zea mays*) or of hairs or scales. Markedly pubescent leaves are often found in desert plants, and such investments may reduce the absorbance of photosynthetically active light by as much as 60 per cent. There is evidence that this is an adaptive response, the degree of hairiness being such that net assimilation is matched to the water available.

The vascular bundles in leaves are frequently surrounded by a chlorophyllous sheath. In some plants this sheath consists of a single layer of large cells containing conspicuous chloroplasts, often with prominent starch. This so-called 'Kranz' anatomy is associated with a particular kind of photosynthesis in which carbon dioxide is taken first into phosphoenolpyruvate instead of directly into the ribulose bisphosphate of the Calvin cycle. These two kinds of assimilation are termed C_4 and C_3 respectively. C_4 assimilation has less need of water than C_3 and it is able to continue at lower concentrations of carbon dioxide, a feature of significance in conditions of light saturation. It is therefore not surprising that C_4 assimilation is frequently present in plants of dry and sunny habitats, such as deserts, and in plants of saline soils (halophytes).

JUVENILE AND MATURE FORMS OF LEAVES   As in the conifers, the leaves of young plants sometimes differ from those of the mature in both form and arrangement. *Eucalyptus globulus* (Fig. 8.16) provides a striking example of this phenomenon. The juvenile leaves are ovate, decussately inserted, and bifacial in structure, whereas the mature are falcate, spirally inserted and possess normal dorsiventral structure. A young tree has a cone of juvenile foliage within a crown of mature foliage, showing that the change takes place more or less simultaneously at all the apices when a tree reaches a certain maturity. Sometimes as in *Hedera helix* (ivy), mature foliage does not appear until the approach of reproduction. Once the production of mature foliage has begun, the system is remarkably stable. Reproductive branches of *Hedera*, for example, can be struck as cuttings, and these yield small bushes quite different in appearance and growth form from the scandent vegetative plant. It is very difficult, even by experimental means, to cause the apices of mature plants to revert to juvenile growth. It appears likely that maturity is a consequence of complex and co-ordinated changes in the ribonucleic acids

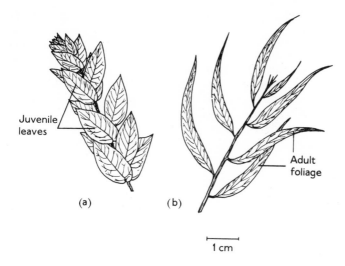

Fig. 8.16 *Eucalyptus globulus*. **(a)** Shoot with juvenile leaves. **(b)** Shoot with adult leaves. (Both after Niedenzu in Engler and Prantl (1898). *Die natürlichen Pflanzenfamilien*, III, 7. Engelmann, Leipzig.)

and proteins in the meristematic cells, and that the original system is only readily re-created during sexual reproduction.

THE ECONOMIC AND ECOLOGICAL IMPORTANCE OF ANGIOSPERM LEAVES   Angiospermous leaves are a rich source of foodstuffs and raw materials. Fodder crops often consist largely of leaves, and the leaves of many species are prized as pot-herbs because of the aromatic oils they contain. The leaves of a wide range of species find medicinal uses, and they are the commercial source of a number of important drugs, among them atropine and hyoscyamine (from *Atropa* and other members of the Solanaceae) and cocaine (from *Erythroxylon*). The hallucinogens of *Cannabis* accumulate within the resinous secretions of the leaves, particularly the bracts of the female inflorescence.[39] The amount produced depends upon temperature and in cool temperate regions is trifling. The leaves of many monocotyledonous species yield valuable fibres, and in tropical regions palm leaves provide a ready and efficient material for thatching. Apart from such direct utilization, angiospermous leaves play a large part in maintaining the fertility of the soil. Since they contain a higher proportion of nitrogenous substances and fewer antiseptic materials, such as tannins and phlobaphene, the leaves of angiosperms decay more rapidly than those of gymnosperms. Angiospermous litter is thus quickly reduced to humus, the organic matter that is the basis of soil fertility. The productivity of much natural vegetation in tropical regions depends upon the steady deposition of litter. Thoughtless removal of this natural cover and destruction of the ecosystem can result in areas of persistent and intractable barrenness.

Roots

The roots of the angiosperms are in general organized similarly to those of the gymnosperms. Those of some species are able to develop chlorophyll when illuminated. In *Taeniophyllum* (Fig. 8.17), a peculiar Malaysian epiphytic orchid, photosynthesis is in fact confined to band-like aerial roots, the stem remaining little more than a bud until the production of the inflorescence. A number of angiosperms, especially those of temperate regions that are biennial in habit, develop a swollen tap root which overwinters. These are often used as vegetables, carrots and parsnips being familiar examples. The

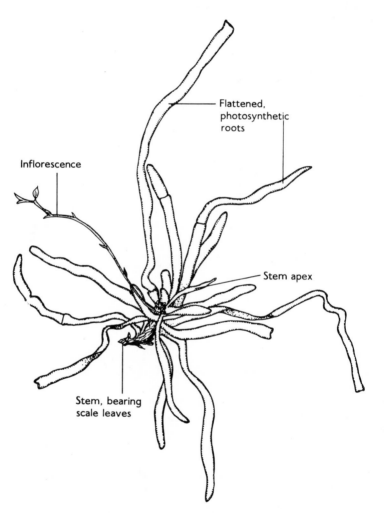

Flattened, photosynthetic roots

Inflorescence

Stem apex

Stem, bearing scale leaves

Fig. 8.17 *Taeniophyllum* sp. Habit showing flattened, photosynthetic roots. (After Goebel (1933). *Organographie der Pflanzen*. Fischer, Jena.)

swollen part of the radish is the transitional region between root and stem. Some swamp plants, notably the mangroves (which colonize brackish estuaries in the Tropics), produce negatively geotropic aerophores from the root system, a development foreshadowed in the conifer, *Taxodium* (see p. 245). A few tropical trees, notably figs, show 'prop roots'. These leave the trunk at about a metre above ground level and descend obliquely towards the substratum. They soon become secondarily thickened and the tree comes to be supported by a cone of such outgrowths around its base. In a number of dicotyledonous trees markedly asymmetrical thickening of the principal roots around the foot of the trunk leads to the production of remarkable buttresses.

Associations between roots and microorganisms are not infrequent. In forest trees the upper rootlets, usually found proliferating in litter, often possess mycorrhizal fungi. The roots of orchids are also mycorrhizal, but it has been discovered that these plants, which are all herbaceous, can be grown in pure culture in the absence of the fungus provided suitable nitrogenous substances are supplied in the medium. In the Leguminosae (the family containing the peas, beans and several important fodder plants) there is a regular association, involving anatomical modifications, between the roots

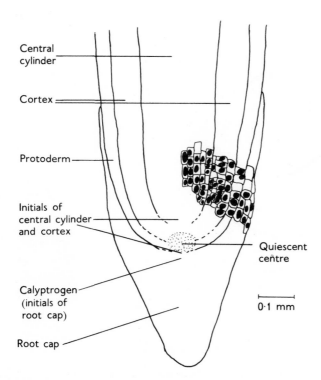

Fig. 8.18 *Trillium* sp. Longitudinal section of root apex.

and certain soil bacteria capable of fixing atmospheric nitrogen. Similar associations with other nitrogen-fixing microorganisms have been confirmed in *Alnus* (alder), *Hippophaë* (sea buckthorn) and *Myrica* (myrtle), and probably exist in a number of other plants.

The manner of growth of roots of angiosperms has been studied in some detail, since it is not complicated by the presence of leaf primordia. Apart from the root cap, which is maintained by a meristem adjacent to the surface of the apex proper, meristematic activity lies principally in the summit of the apex (Fig. 8.18). There is, however, convincing evidence, supported by experiments with radioactive isotopes, that a group of cells at the centre of the apex experiences few if any divisions. These cells, forming a so-called 'quiescent centre', can be stimulated into division by wounding and radiation damage, so they are not deficient in essentials. Their normal inactivity must therefore depend upon the physiological organization of the intact apex, but the precise factors involved are still unknown. Behind the meristematic area the cells elongate, and measurements have shown that there is considerable synthesis of ribonucleic acid and protein in the cells concerned at this time. In some cells, often those which produce root hairs, the nuclei undergo a curious form of mitosis without associated division (endomitosis), leading to a polyploid condition.

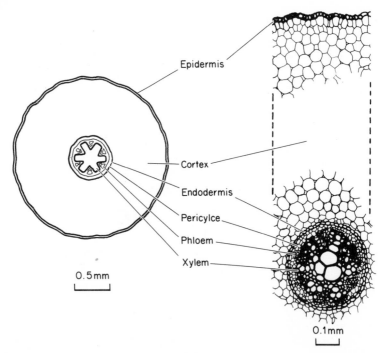

Epidermis

Cortex

Endodermis

Pericylce

Phloem

Xylem

0.5 mm

0.1mm

Fig. 8.19 *Ranunculus acris*. Transverse section of young root, showing the broad cortex and narrow stele. The protoxylem is exarch. Differentiation proceeds inwards yielding a central core of metaxylem. The phloem lies between the arms of xylem.

A transverse section of a young root of a dicotyledon shows a central core of xylem with radiating arms of protoxylem, between which lies the phloem. In dicotyledons the number of protoxylem groups is usually small (Fig. 8.19), but in the monocotyledons the number is much greater, usually in the region of 15 or 20. An endodermis, separated from the vascular tissue by a zone of parenchyma (*pericycle*), is usually present. The root is the only region in the plant body of an angiosperm where this layer of peculiar cells occurs with regularity. The Casparian strip, a band of fatty material embedded in the radial walls, provides a hydrophobic barrier to the lateral movement of dissolved ions in the apoplastic space. The passage of salts into and out of the vascular tissue is thus effectively under the control of the endodermal protoplasts, probably a feature of great importance in the regulation of mineral nutrition. The meristems of branch roots arise in the pericycle, an origin termed **endogenous**, and very different from the more superficial (**exogenous**) origin of branches of the stem. Where cork is produced, the cork cambium arises in the pericycle, thus causing the whole of the primary tissue external to it ultimately to die and slough off. In herbaceous plants the parenchymatous tissues of the roots sometimes become locally distended with food materials, occasionally forming distinct tubers (e.g. *Dahlia*, Fig. 8.20). These serve as organs of perennation.

Roots, especially those that are tuberous, are frequent sources of drugs and folk medicines. Aconitine, obtained from the roots of *Aconitum napellus*, frequent in alpine meadows and seen occasionally by shaded streams in Britain, is a well-known example. Derris, a powder made by grinding the roots of *Derris elliptica*, a climbing shrub of Asia, is a powerful insecticide harmless to man.

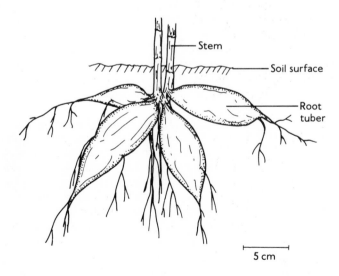

Fig. 8.20 *Dahlia* sp. The production of root tubers.

CORRELATION WITHIN THE PLANT BODY   The angiosperms, more than any other class of plants, have been investigated with the object of discovering the factors responsible for co-ordinating the growth of the plant as a whole. In the presence of an apical bud, for example, lateral buds commonly remain inactive, an instance of correlative inhibition. Correlations of this kind are clearly of great importance in determining the morphology of the mature plant.

Correlation is an example of cell interaction, but the interacting cells are often separated by others not directly concerned. These interactions at a distance are brought about by growth-regulating substances which are able to move through the plant in a manner not yet wholly understood. Research has now revealed a whole series of such substances, which for present purposes can be classified as auxins, kinins, abscissins, gibberellins and ethylene. Auxins (of which indole-3-acetic acid is the most familiar) are those substances which will cause curvature if placed asymmetrically on the tip of a decapitated oat coleoptile. They are usually produced by meristematic cells and are involved (amongst other effects) in maintaining apical dominance and initiating differentiation. In culture solutions indole-3-acetic acid (and its non-metabolizable analogue naphthalene acetic acid) are particularly effective in stimulating the production of roots, a property made use of in horticulture in the rooting of cuttings. Kinins (cytokinins, phytokinins) are substances which promote cell division. They are found in fruits and seeds, but may also be generated in damaged cells and cause proliferation around wounds and their subsequent repair. Abscissins, probably produced in leaves, are effective in causing leaf abscission, and dormancy in apical buds, common phenomena in trees of temperate regions. Gibberellins were first extracted from the fungus *Gibberella* which, when infecting rice, causes the plant to be excessively tall. They are now known to exist in higher plants, and early experiments showed that when administered to dwarf mutants as, for example, of maize (corn), these would grow to their normal stature. Additionally, they are probably involved in photoperiodic responses and the change from vegetative to reproductive growth. Ethylene has long been known to be produced by ripening fruits, and conversely the process of ripening in stored immature fruits can be controlled by regulation of the amount of ethylene in the atmosphere.

Although detailed investigation of the growth-regulating substances (which include others less well known in addition to those mentioned here) is the province of physiologists, they are also basic to the causal study of morphology since regulatory substances play an important role in directing the growth of a species in its characteristic and immediately recognizable way. It was inevitable that the flowering plants, because of their ubiquity, familiarity, and economic importance, should figure prominently in research into growth-regulating substances, but the study of simpler systems in algae and archegoniate plants might more readily reveal how they enter into the molecular biology of the cell and produce their effects.

Reproduction

In the angiosperms the onset of the reproductive phase is frequently (but not always) dependent on the length of day ('photoperiodism'), both short-day and long-day plants being clearly recognizable. The reproductive axes are ordinarily short and of limited growth. The axis itself and its attendant structures, which are often conspicuously coloured or shaped, is termed a flower. Flowers may be borne either singly (as in *Anemone nemorosa*) or severally, the branching system bearing the flowers then being called an ***inflorescence***. Inflorescences may have complex morphology.[48] Sometimes the inflorescence is contracted and superficially resembles a single flower (as in many Compositae), or even a strobilus (familiarly termed a catkin) or cone (as in many temperate trees). The parts of a plant giving rise to flowers or inflorescences are not normally very different from those that are vegetative, but distinct functional separation occurs in some tropical trees. In *Couroupita guianensis* (cannon ball tree), for example, the uppermost branches are densely leafy and solely vegetative. The reproductive branches, which are almost leafless and bear numerous flowers, arise directly from the trunk and hang down from the crown. In some other tropical trees, flowers are produced only from special regions towards the lower part of the trunk (as in the Jack Fruit (*Artocarpus*) of tropical gardens), a phenomenon known as *cauliflory*.

In annuals and certain longer-lived plants of warmer regions the production of flowers heralds the end of the life-span. Such plants are said to be ***monocarpic*** or ***hapaxanthic***. A spectacular example is the palm *Corypha* of

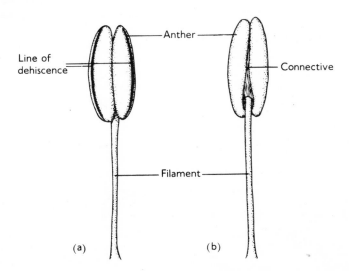

Fig. 8.21 Diagram showing the features of a typical angiosperm stamen. **(a)** Viewed from the front. **(b)** Viewed from the back. In this anther the pollen sacs face the floral axis (introrse), but the reverse position is also found (extrorse).

the Asiatic Tropics. Growth terminates with the production of an inflorescence reaching a height of 14 m and a breadth of 12 m, containing more than 100 000 flowers.

Reproduction in angiosperms may be either monoecious or dioecious. Monoecious species are further divisible into the monoclinous, where the male and female organs are together in the same flower, and the diclinous, where they are in separate flowers (as in the cucumber, *Cucumis sativus*). Functional separation of the sexes in bisexual flowers is often achieved by the organs of the two sexes maturing at different times, a phenomenon termed **protandry** if the male precede the female, and **protogyny** if the female precede the male. Some orchids show an extreme form of protandry in which the ovules are not formed unless the flower is pollinated. In some species self-pollination is discouraged or prevented by the flowers being in two or even three forms (see p. 309).

THE MALE REPRODUCTIVE STRUCTURES   The main sporangiophore of the angiosperms is termed a stamen (Fig. 8.21). Typically this consists of a stalk (filament) terminating in four pollen sacs, the sacs being in two pairs and these lying side by side and joined by the connective. The whole of this region is called the anther, and seen in transverse section (Fig. 8.22) resembles the male synangia of some pteridosperms, and of *Caytonia* (see p. 329). Apart from congenital fusion of stamens, discussed later, there is some variation in the form of the individual sporangiophore. In *Degeneria*, for example, and in some members of the Magnoliaceae, the 'filament' is in fact a broadly ovate scale, about 5.0 mm long and 2.0 mm wide, with four pollen sacs partially embedded on the abaxial surface (Fig. 8.23). At the other extreme the filament may be lacking, one or two anthers being attached directly to a modified

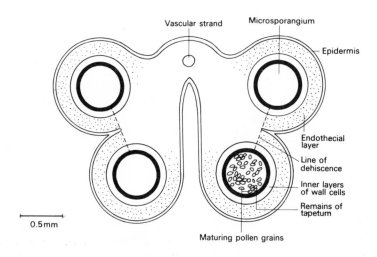

Fig. 8.22 *Lilium longiflorum*. Transverse section of almost mature anther.

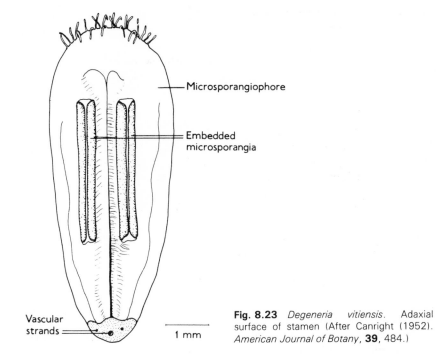

Vascular
strands

1 mm

Fig. 8.23 *Degeneria vitiensis*. Adaxial surface of stamen (After Canright (1952). *American Journal of Botany*, **39**, 484.)

floral axis, as in the complex flower of the orchids. Besides simple dehiscence along a longitudinal stomium, other methods of opening, such as the development of pores or the differentiation of distinct valves, are encountered in some species.

In contrast to microsporogenesis in *Pinus*, the development of angiospermous anthers is commonly synchronous. This is probably a consequence of broad cytoplasmic connections between the young spore mother cells, a phenomenon (*cytomixis*) not seen in *Pinus*. Subsequently the pollen mother cells become surrounded by thickened callose walls and all interconnections are obliterated. After meiosis the young microspores, while still in the tetrad, begin to be coated with sporopollenin. Concurrently cellular organization breaks down in the tapetum, metabolites are mobilized by hydrolase activity, and additional sporopollenin is synthesized. The young pollen grains are released into this medium and complete their development at its expense. Much of the sporopollenin is added to that already coating the grain, probably by simple accretion. The pattern of deposition and location of the pores (*colpi*) are established before the grain is liberated from the tetrad, but the manner of the inheritance of this pattern is still unknown (cf. *Sphaerocarpus*, p. 119). The structure and cytochemistry of the wall of a mature pollen grain are complex. The chambered exine often takes up hydrolases from the tapetum while the inner cellulose layer (intine) may contain secretions from the grain itself. While the former are sporophytic in origin,

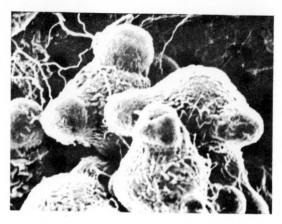

Fig. 8.24   *Oenothera organensis.* Pollen grains (each with three projecting colpi), enmeshed in threads of viscin (a polymer produced by the tapetum), lying on the stigma. (Photomicrograph by H. Dickinson.) (× 3000)

the latter are gametophytic. The patterns of pollen walls are often very distinctive (Fig. 8.24). Since sporopollenin, a highly polymerized mixture of esters of complex fatty acids, is extremely resistant to decay, pollen grains partially fossilized in peats and lake muds are often fully identifiable. Most pollen grains are spherical or ellipsoidal and land plants show only few examples of rod-like grains. Marine angiosperms ('sea grasses') however possess 'confervoid' pollen. These grains lack exine and may reach or exceed 5 mm in length. In some species (as in the Ericaceae (heaths) and the decorative plant *Salpiglossis variabilis*) the pollen remains stuck together in tetrads, a feature facilitating, as in *Sphaerocarpus* (p. 119), tetrad analysis.

Development of the male gametophyte normally begins while the pollen is still in the anther. Usually two nuclei are present, of which one (the vegetative nucleus) becomes large and diffusely staining and the other (the generative nucleus) dense and often transversely elongated (Fig. 8.31a). These nuclei lie on a radius of the original tetrad, and there is evidence that their conspicuously different behaviour depends upon gradients set up in the cytoplasms of the cleaving pollen mother cell and persisting in that of the grain. In about one third of all families (notably the grasses) the generative nucleus divides while still in the grain ('trinucleate pollen').

THE FEMALE REPRODUCTIVE STRUCTURES   The megasporangiophore is the distinguishing feature of the angiosperms for it is normally a closed body (termed a carpel), furnished with a distinct stigma (often elevated on a style) on which the pollen germinates (Fig. 8.25). In only a few genera (e.g. *Reseda*) are the carpels open at maturity. A carpel is dorsiventral in symmetry and the fertile region is adaxial. One or several ovules may be present, and if the latter they are commonly borne in two series along the so-called 'ventral suture' of the carpel (Fig. 8.26).

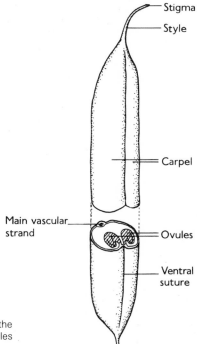

Fig. 8.25 The simple carpel characteristic of the family Leguminosae. This arrangement of the ovules is referred to as 'marginal placentation'.

The ovules are much smaller than those of most gymnosperms, but they are structurally similar. They may be upright, inverted, or occasionally more or less horizontal, these three orientations being termed **orthotropous**, **anatropous** and **campylotropous** respectively (Fig. 8.26). One integument may be present, but in many families there are two, and in some as many as four. The last formed integument, irrespective of the total number, sometimes becomes transformed into a fleshy aril in the fruit (as in the durian, *Durio*). The integuments normally develop uniformly, indicating the basic radial symmetry of the angiospermous ovule. A slender vascular trace enters tha stalk of the ovule, but in only a few families does this extend into an integumentary vascular system.

THE FEMALE GAMETOPHYTE   The development of the female gametophyte (Figs 8.27, 8.28), in the angiosperms termed the **embryo sac**, begins in a familiar fashion (Figs 8.27a, 8.28a). A cell in the upper part of the nucellus, immediately below the layer of cells at its surface, becomes conspicuously large and surrounded by a callosed wall. Meiosis then leads to a tetrad of megaspores (Figs 8.27b, 8.28b). Although the term 'megaspore', because of the evident homology of this cell with the megaspore of the gymnosperms, is retained, the megaspore in the angiosperms is frequently smaller than the microspore. In many instances only the inner of the tetrad of megaspores

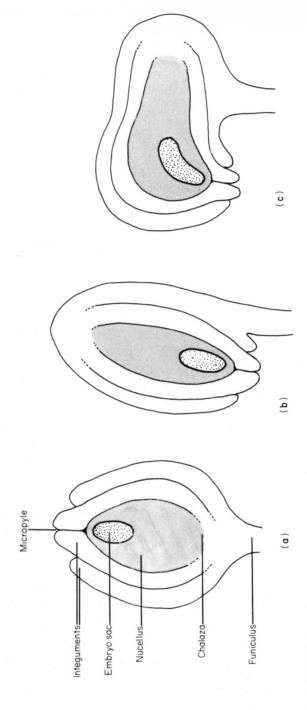

Fig. 8.26 Forms of ovule orientation (diagrammatic). **(a)** Orthotropous. **(b)** Anatropous. **(c)** Campylotropous.

Micropyle

Integuments

Embryo sac

Nucellus

Chalaza

Funiculus

(a)

(b)

(c)

undergoes further development, and the female gametophyte is consequently said to be *monosporic* in origin. The nucleus of the megaspore enters a succession of mitoses, and, in step with the expansion of the ovule as a whole, the multinucleate, acellular embryo sac is formed (Figs 8.27c, 8.28c–e). In a frequent type of embryo sac there are eight nuclei, produced by three successive mitoses, which associate themselves with cytoplasm and arrange themselves in a definite pattern (Fig. 8.28e). Eight cells are usually identifiable at this stage, each bounded by a delicate membrane. At the micropylar end of the sac is a group of three cells, of which one (frequently with a weakly staining nucleus) is the egg cell. The cells accompanying the egg are termed *synergid* cells. At the other end of the sac (the chalazal end) is another group of three cells, forming the so-called *antipodals*. The nuclei of the remaining two cells (termed polar nuclei) come together at the centre of the sac and eventually fuse, thus giving rise to a central diploid fusion nucleus (Figs 8.27d, 8.28f). During the development from the megaspore the boundary between the gametophyte and sporophyte remains distinct, and the mature embryo sac is commonly also surrounded by a callosed wall. This is known to be impermeable to large molecules such as nucleotides.

The foregoing is only one of the several kinds of development of the female gametophyte found in the angiosperms. In some families, notably the Onagraceae, the embryo sac develops from the outer and not from the inner megaspore, a feature of unknown cytological and physiological significance. Sometimes, as in *Allium* (onion), meiosis in the formation of the megaspores is incomplete, a diad of spores being produced instead of a tetrad. The embryo sac, which develops directly from the inner spore, is then said to be *bisporic* in origin. Finally, in another kind of development, termed *tetrasporic*, the nucleus of the megaspore mother cell divides reductionally, but the cell itself does not divide at all. Instead, mitotic divisions follow and the cell expands without interruption to form the sac. Bisporic and tetrasporic embryo sacs occur in a wide range of families (and are present in *Gnetum*), and there is no justification for regarding this kind of origin as abnormal. Another kind of variation lies in the number of nuclei in the mature sac. Although this is commonly eight, irrespective of the mode of origin of the sac, others are known containing four or sixteen. The cytology of development is also variable. Although where sexual reproduction is normal the egg nucleus remains haploid, the chalazal polar nucleus may contain more than one set of chromosomes so that the central fusion nucleus becomes triploid or even, as in *Fritillaria*, tetraploid.

THE ARRANGEMENT OF THE FERTILE REGIONS IN THE FLOWER   The arrangement of the reproductive organs in the flower is often of considerable complexity, and floral morphology is a specialized field with its own terminology.[20] Here we shall use only as much of this terminology as is necessary for general discussion, and for further details the reader must consult specialist works. To facilitate illustration of the principles involved in the structure of the flower, we shall first consider a bisexual flower in which the components

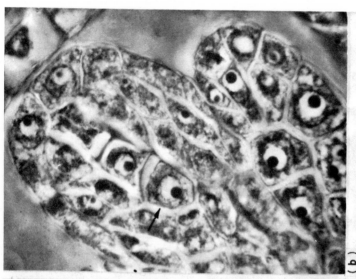

(b)

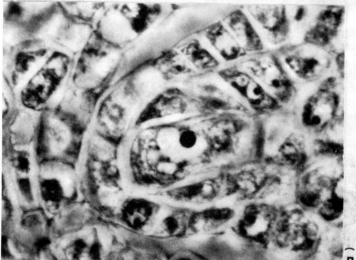

(a)

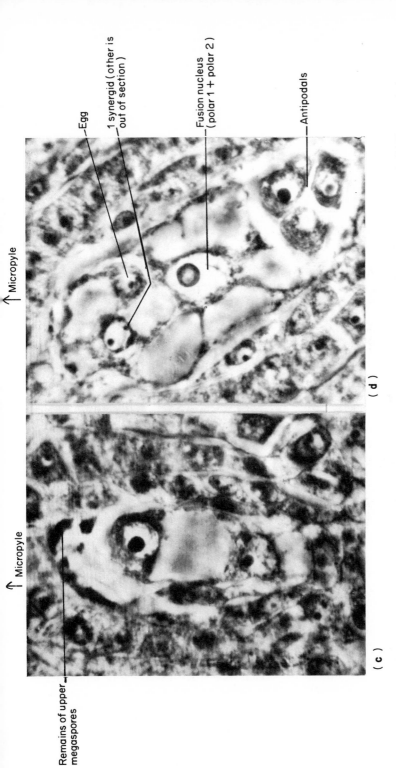

Fig. 8.27 *Myosurus minimus* (mousetail). Stages in the development of the female gametophyte. **(a)** Megaspore mother cell. **(b)** The tetrad of spores resulting from meiosis. The innermost (indicated by arrow) becomes the megaspore. **(c)** The two-nucleate embryo sac, with the remains of the three upper megaspores still visible above the sac. **(d)** The mature, eight-celled embryo sac. **((a)** and **(b)** × 1350; **(c)** × 1500; **(d)** × 1200)

Micropyle

Remains of upper megaspores

(c)

Micropyle

Egg

1 synergid (other is out of section)

Fusion nucleus (polar 1 + polar 2)

Antipodals

(d)

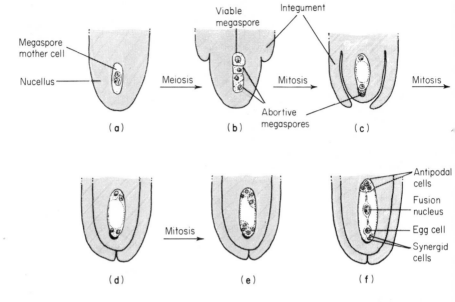

Fig. 8.28   The sequence of divisions leading up to the production of a monosporic, eight-nucleate embryo sac. From flowers of *Myosurus minimus*. Diagrammatic; not to scale.

remain separate (Fig. 8.29), a situation seen, for example, in *Ranunculus*. The flower terminates an axis, and the transition from the vegetative to the reproductive region is marked by the occurrence of one or more whorls of sterile structures resembling rudimentary leaves. These form collectively the **perianth**. Where two whorls are present the segments of the outer whorl are frequently green (and termed sepals), whereas those of the inner are brightly coloured (and termed petals), and the two parts of the perianth are termed **calyx** and **corolla** respectively. Following closely upon the perianth are the stamens, forming collectively the **androecium**. The stamens may be either indefinite in number and spirally arranged (as in *Ranunculus*), or definite and arranged in whorls, the positions of individual stamens often bearing a clear relation to each other and to those of the preceding perianth segments.

The termination of the reproductive axis, often termed the **receptacle**, bears the carpels, this female region being called the **gynaecium** or ovary. The carpels may again be spirally arranged, as in *Ranunculus*, or in a single whorl. In the latter event the carpels usually alternate in position with the stamens of the uppermost whorl.

In unisexual flowers the organs of one sex are absent or non-functional, although the bisexual condition can sometimes be brought about by treating the very young flower with growth substances such as gibberellic acid. Unisexual flowers may approach the limits of reduction, a feature well seen in catkin-bearing trees. In *Populus* (poplar), for example, the male flower consists of nothing more than a disk bearing up to 20 stamens, and in *Salix*

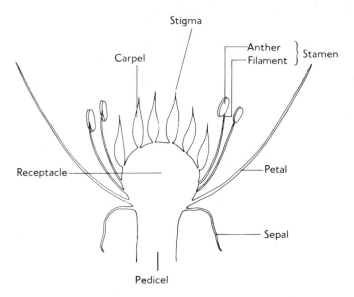

Fig. 8.29 Longitudinal section of a flower with separate components. Diagrammatic.

(willow), where the male flower is similar, the number of stamens is commonly no more than two. Congenital fusion of parts of similar nature is widespread in flowers. Both the calyx and corolla independently may become tubular, the composite nature of the tube being indicated by lobes or teeth at its mouth. Filaments of whorled stamens may also fuse for most or only part of their lengths, leading to a cylindrical androecium. Fusion of carpels leads to a *syncarpous* ovary, in which the individual carpels are often represented by compartments (or loculi), as in the Liliaceae. Externally the composite nature of the ovary is often indicated by lobing, or by the style, which rises from the centre of the ovary, breaking up into branches equivalent in number to the constituent carpels. The ovary may sometimes be sunken in the receptacle (as in *Rosa*) or otherwise surmounted by the perianth and androecium, leading to clearly *epigynous* flowers. Sterilization of one or more stamens may occur, giving rise to *staminodes*, which may either persist as rudimentary structures (as in *Scrophularia*) or occasionally become enlarged and brightly coloured, and play a conspicuous part in the organization of the flower (as in *Canna* and *Iris*). Both sterilization and congenital fusion of parts are involved in the complex flowers of the orchids.

THE ONTOGENY OF THE FLOWER   Although in the gymnosperms the apices giving rise to the reproductive and vegetative axes differ little in organization, in angiosperms they are usually easily distinguishable. In becoming reproductive an apex flattens considerably and may even become concave, regions previously quiescent now showing numerous mitoses. This has given rise to the concept of the vegetative apex containing a *meristème d'attente* active only in the reproductive phase. The perianth whorls and sporangiophores are often produced in acropetal sequence, the members of successive whorls alternating in position, but this is not always so. The direction of initiation may even become locally reversed, and primordia appear between the whorls of those already laid down.[16] This phenomenon affects particularly the androecium, with conspicuous results in the symmetry of the mature flower. In some Caryophyllaceae, for example, a whorl of five stamen primordia appears, each stamen alternating in position with the preceding petals. A second whorl of stamens is then initiated basipetally, the stamens alternating with those of the first whorl, and hence standing opposite the petals. In this way an *obdiplostemonous* flower is formed.

POLLINATION AND GROWTH OF THE MALE GAMETOPHYTE   Pollination in angiosperms is occasionally dependent upon aerial dispersal of the grains (as in catkin-bearing trees), but frequently involves the participation of insects as carriers. Various mechanisms have been evolved which ensure that insects visiting flowers in search of nectar (usually secreted by glands situated towards the interior of the flower) become dusted with pollen, which then becomes transported to stigmatic surfaces in other flowers. A familiar example of such a pollination mechanism is provided by *Salvia pratensis* (sage), a species pollinated by the bumble-bee. The stamens have short filaments, but a greatly elongated connective (Fig. 8.30). The longer upper portion of the connective terminates in an anther, but the shorter lower part in a plate blocking the approach to the nectary. A bee attempting to reach the nectar displaces the plate. Since the connective is hinged about the filament, the anther is in this way forced on to the insect's back, coating it with pollen (Fig. 8.30b). A bizarre form of pollination is found in some orchids (e.g. *Ophrys speculum* of the western Mediterranean) where the pattern and conformation of the lip recall the female of the insect concerned, and even an odour may be released resembling the pheromone of the species. The male insect, in attempting to copulate with the lip of the flower, detaches the pollen (which coheres in masses called *pollinia*) and carries it to another flower. A few tropical flowers are pollinated by birds and bats, and *Aspidistra*, once a favourite pot plant, by snails. The appropriate authorities must be consulted for more detailed accounts of pollination mechanisms.[31]

The pollen, having been brought into contact with the stigma (to which it may be firmly attached as a consequence of electrostatic forces[10]), germinates freely. A pollen tube breaks through a colpus in the exine of the grain (Fig. 8.31) and, provided there are no incompatibility barriers, penetrates the style. The growth of the tube is largely confined to the tip, behind which the wall becomes callosed. Its passage towards the ovules often follows a tract of specialized, thin-walled, transmitting tissue, the direction of growth probably being controlled by metabolic gradients of a quite simple kind. The pollen tube enters the ovule either by way of the micropyle, or by growth through the chalazal end, and ultimately reaches the embryo sac. Meanwhile the generative nucleus in each tube has divided into two gametic (sperm)

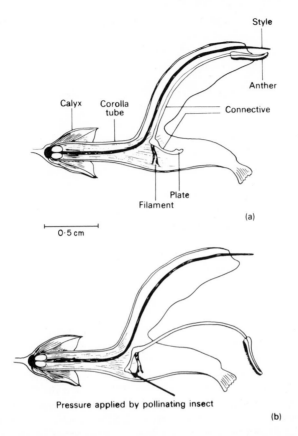

Fig. 8.30   *Salvia pratensis*. **(a)** Half flower showing the normal position of the stamen. **(b)** The same after the plate has been depressed. (After Kerner von Marilaun and Oliver (1902). *The Natural History of Plants*. Blackie, London.)

nuclei and these, each associated with a little cytoplasm, move towards the tip (Fig. 8.31c).

Various devices are found which discourage or prevent self-pollination, and its possibly deleterious genetic effects.[24] Conspicuous amongst these are those involving dimorphic flowers. In *Primula*, for example, the two kinds of flowers are termed 'thrum-eyed' with a short style and elevated stamens, and 'pin-eyed' with a long style and lower stamens (Fig. 8.41). An insect searching for nectar at the base of the corolla tube will tend to pass pollen from the stamens of the 'pins' to the styles of the 'thrums' and *vice versa*. This morphological device is accompanied by a physiological difference which causes the imperfect growth of 'pin' and 'thrum' pollen in their own styles. The mechanism, which is genetically controlled, depends upon the presence of antibodies in the style which precipitate essential enzymes in the pollen tubes of pollen from the same or like flowers. The pollen of the thrum flowers

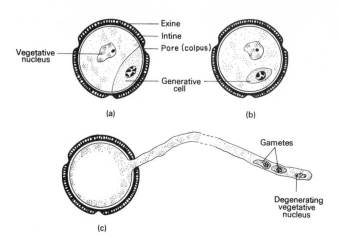

Fig. 8.31   Development of the pollen grain. **(a)** Condition in the anther during final stages of the growth of the wall. **(b)** Condition during liberation and pollination. **(c)** The final nuclear division during the growth of the pollen tube. Diagrammatic.

is some 50 per cent greater in diameter than that of the pin, and this is cor- related with a greater rate of extension of the pollen tube of the thrum pollen. *Lythrum salicaria* and some other species have a similar mechanism involving three kinds of flowers, the diameter of the pollen grains again being related to the position of the stamens producing them. Physiological systems of self-incompatibility unaccompanied by structural devices in the flower are wide-spread. In *Linum*, for example, successful growth of the pollen tube can take place only if the ratio of the osmotic pressure of the style to that of the pollen is the order of 1:4, a relationship never present between pollen and style of the same flower. Other mechanisms are seen in *Raphanus* (radish) and *Brassica* (cabbage). Here the pollen germinates following selfing, but the tube is prevented from penetrating the stigma by the rapid development of pads of callose in the stigmatic cells beneath its tip.[11] In *Oenothera* the incompatible tube penetrates, but in the style it becomes plugged and growth ceases.

In general two kinds of incompatibility can be distinguished, sporophytic and gametophytic. In the former the failure of pollination is a consequence of interaction between sporophytic gene products, those of the male being carried in the exine of the pollen grain and derived from the tapetum, and those of the female being present in stigma and style. In gametophytic incom-patibility the sporophytic gene products of stigma and style are again involved, but in this instance the interaction is with the gametophytic gene products in the pollen protoplast. The opportunity for interaction is delayed until the pollen tube, with its thin permeable wall, penetrates the pistil.

In some plants flowers produced late in the flowering season do not open, but are nevertheless fertile. Such flowers are termed *cleistogamous* (and the normal *chasmogamous*). A familiar example is provided by *Viola canina*. The normal flowers of spring produce little seed, but the bud-like cleistogamous flowers of summer are fully fertile. They contain only minute petals and all but the two abaxial stamens are abortive. The pollen of these, however, germinates in the anther and the pollen tubes grow through the wall into the stigma.

FERTILIZATION   A pollen tube usually enters the embryo sac at its micropylar end, and a region near the tip of the tube, possibly in consequence of enzymes produced by synergid cells, soon breaks open. The contents of the tube are discharged into the sac, and the 'double fertilization', characteristic of the angiosperms, then ensues. One male nucleus penetrates the egg; the other moves through the sac and unites with the central fusion nucleus, forming the polyploid primary endosperm nucleus. In cotton (*Gossypium*), and possibly in some other flowering plants, the sperm cells pass through one of the synergids, and in so doing lose their cytoplasm. Organelles from the male consequently do not contribute to the zygote. In *Pelargonium* however transmission of male organelles does occur. Following fertilization the contents of the sac are commonly resorbed, the zygote and the endosperm remaining the only sites of further growth.

THE FEATURES OF ENDOSPERM   The formation of the endosperm usually begins before the division of the zygote. The primary nucleus and its daughters undergo successive mitoses, giving rise to an endosperm that is either cellular from the first, or initially acellular (Fig. 8.32a) and only later partly or wholly cellular. Cytologically the endosperm is a remarkable tissue, with unique cytoplasmic and nuclear properties, significantly different from the haploid 'pseudo-endosperm' of the gymnosperms (see p. 239). It is a rich source of growth-regulating substances or their immediate precursors, many of them not yet chemically identified. The successful artificial culture of a number of plant tissues, for example, is still impossible in fully defined media, but is facilitated by the addition of coconut milk (the liquid endosperm of the coconut seed). Endosperm itself can sometimes be obtained in pure culture and, since the cell walls of young endosperm often have little thickening, such cultures have been used to study the details of mitosis *in vivo*. In some endosperms the nuclei reach high and irregular levels of endopolyploidy, and amitotic nuclear division has been reported in a number of instances.

In keeping with the function of providing for the nutrition of the embryo and young plant, the cells of the mature endosperm are often filled with food materials. Carbohydrates, fats and proteins are present in various labile forms. The occurrence of these materials is sometimes so abundant that the seeds concerned acquire vast economic importance. The cereal grains, where the endosperm yields starch and protein in a form readily palatable to humans, and the oil seeds, such as *Ricinus* (castor bean) and *Linum* (linseed), are familiar examples. Polymers of mannose (mannans) also occur as reserve products, a peculiar example being provided by *Phytelephas*, a palm. The large endosperm of this species becomes so heavily indurated with a mannan that it enters commerce under the name of 'vegetable ivory', once used for the

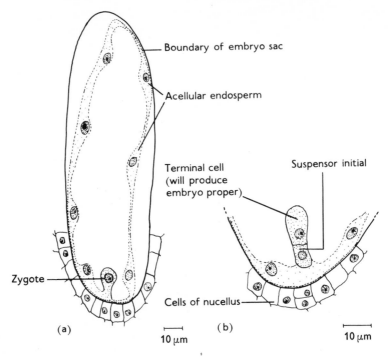

Boundary of embryo sac

Acellular endosperm

Terminal cell
(will produce
embryo proper)

Suspensor initial

Zygote

Cells of nucellus

(a)

(b)

10 μm

10 μm

Fig. 8.32   *Myosurus minimus.* Longitudinal section of embryo sac following fertilization. **(a)** After several divisions of the primary endosperm nucleus. **(b)** After the first division of the zygote. The zygote is at the micropylar end of the sac.

manufacture of buttons and billiard balls. In a few plants, notably the orchids, the endosperm undergoes only trifling development and in others it is remarkably aggressive. In *Pedicularis*, for example, the endosperm produces haustoria which invade the integumentary tissue, leading eventually to its complete resorption.

EMBRYOGENY[26]   As in other seed plants, the first dividing wall of the zygote lies transverse to the longitudinal axis of the ovule (Fig. 8.32), and the subsequent embryogeny is endoscopic. Free nuclear division is not, however, a characteristic of the development of the proembryo, as it is in many gymnosperms. Instead, embryogenesis begins with a series of precise cell divisions, showing little variation between species, and meristematic activity does not pause until the embryo differentiates and the seed is formed. The mature embryo (except in the orchids where it remains undifferentiated) possesses a stem apex (*plumule*), one or two cotyledons, and a root apex (*radicle*). Even though the mature embryo may be folded (as in *Capsella bursa-pastoris* (shepherd's purse), Fig. 8.33), the radicle is always directed towards the micropyle. The extent to which the fully formed embryo has drawn upon the food reserves of the endosperm varies widely with the species. In

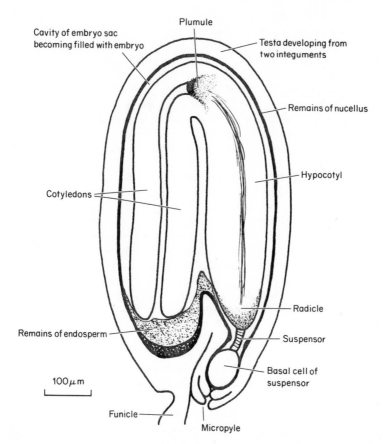

Fig. 8.33 An almost mature seed of *Capsella bursa-pastoris* sectioned longitudinally showing the orientation of the embryo.

'endospermous' seeds this reserve remains considerable, but in the 'non-endospermous' little remains, and much of the food materials in these seeds often becomes transferred to the cotyledons. Examples of such seeds are those of the legumes where the swollen cotyledons entirely replace the endosperm in both space and function (Fig. 8.34a).

THE MATURATION OF THE SEED AND FRUIT Associated with the maturation of the embryo are considerable changes in the nucellus and integuments. In some seeds food reserves are laid down in the nucellus, and in the mature seed this then forms a distinct tissue known as the *perisperm*. Frequently, however, only a little of the nucellus remains, and this together with the integument becomes transformed into the outer covering of the seed. This often involves the deposition of much cutin and lignin, and the resulting seed coat (*testa*) is consequently often remarkably impervious.

As the seeds mature the ovary becomes a fruit. There are numerous forms of fruits, but the wall is formed either from the carpel (*pericarp*), or by the fusion of the pericarp with surrounding tissues. The pericarp may become hard, as in nuts, or dry and brittle, as in the capsules of the Caryophyllaceae, or partly or wholly fleshy, as in common edible fruits. In berries (e.g. tomato, *Lycopersicon*) the pericarp is wholly fleshy, in drupes (e.g. cherry, *Prunus*) the outer part becomes fleshy while the inner hardens to form a stone, and in pomes (e.g. apple, *Malus*) a thin inner layer of pericarp hardens while the remainder fuses with the surrounding receptacle and forms the flesh. Where fruits are palatable to man or animals the seeds commonly have a testa able to survive passage through the digestive tract.

Seeds are often distributed while still in the fruit, and the fruit in this instance commonly bears hooks or spines (as in the burdock, *Arctium*), or wings (ash, *Fraxinus*), or parachute-like fringes of hairs (as in the pappus of many Compositae), which assist dispersal by attachment to animals or by wind. Other angiosperms have evolved mechanisms whereby tensions are set up in the carpel wall as it dries, causing eventually an explosive dehiscence which scatters the seeds far and wide. This brief summary of a wealth of morphological detail shows how the angiosperms have evolved a great variety of ways which ensure wide dissemination,[36] a feature in which they differ sharply from the gymnosperms, and which has undoubtedly contributed to their present success.

DORMANCY AND GERMINATION   The cause of the cessation of growth of the embryo as the seed matures is not altogether clear, but the partial dehydration of the interior of the seed at this time is probably an important factor. Certainly the imbibition of water is essential for renewed growth, but this by itself is often insufficient for the production of viable seedlings. A period of dormancy ('after ripening') must often ensue before successful germination will occur. The seeds of some species also need to be chilled, a treatment which apparently activates the enzymes which make available to the embryo the food reserves of the endosperm. This cold requirement also provides a biological advantage, since the seeds will not germinate until after the winter, and the seedlings thus avoid the rigours of this unfavourable season. Other seeds, for example those of lettuce, germinate only after illumination, and the phytochrome system (which depends upon a protein sensitive to red light) is here involved in the renewal of growth. In *Banksia* and some other Australian Proteaceae the fruits open only after bush fires (cf. 'closed cone' pines, p. 259), and the seeds germinate in the ash layer. Not only is this a more favourable substratum than loose humus, but the seeds of at least one species need heat treatment before they will germinate. The seeds of many tropical plants, on the other hand, germinate immediately in moist conditions, and soon lose their viability if stored. The tropical mangroves, plants of coasts and estuaries, are outstanding in that in some species the embryo continues to develop in the ovules, and a swollen radicle protrudes from the withered flower. The young plant eventually falls and, with the radicle correctly oriented for penetration of the mud beneath, it rapidly establishes itself.

As a germinating seed imbibes water there is usually considerable swelling and eventually rupture of the testa. The radicle is the first organ to emerge, followed closely by the plumule. The cotyledons may remain within the seed at or below soil level (*hypogeal* germination), as in *Vicia faba* (broad bean) (Fig. 8.34a) or become elevated (*epigeal* germination) and photosynthetic, as in *Sinapis* (mustard) (Fig. 8.34b). Germination is basically similar in monocotyledons and dicotyledons, but in the grasses the plumule and radicle are first enclosed in sheaths called the coleoptile and coleorhiza respectively. Coleoptiles show marked phototropic and geotropic responses and, being devoid of appendages, are particularly suitable for the experimental investigation of these phenomena.

Asexual reproduction

In addition to sexual reproduction, the angiosperms show many forms of asexual reproduction. One form of such reproduction, termed *apomictic*, superficially resembles sexual reproduction. Although its presence can be readily inferred from the genetics of the species concerned, the elucidation of the accompanying cytology has sometimes demanded very careful investigation. In one kind of apomixis, parthenogenesis, the egg develops without fertilization. Where this occurs regularly, the nuclei involved in the forma-

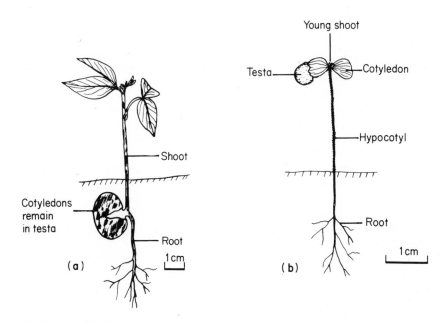

Fig. 8.34 **(a)** *Vicia faba*. Hypogeal germination. **(b)** *Sinapis* sp. Epigeal germination.

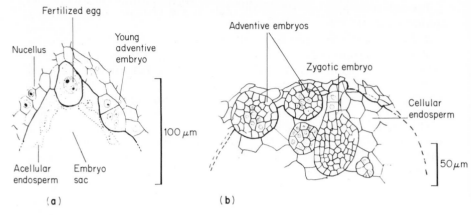

Fig. 8.35   *Citrus trifoliata.* Longitudinal section of micropylar end of embryo sac in which adventive embryony is occurring. **(a)** Immediately after the fertilization of the egg. **(b)** After the endosperm has become cellular. Only the true embryo has a suspensor. (Both after Osawa, from Maheshwari (1950). *An Introduction to the Embryology of Angiosperms.* McGraw-Hill, New York.)

tion of the embryo sac and the nucleus of the egg contain an unreduced number of chromosomes. Such embryo sacs may result from a modification of the meiosis which yields the megaspores (as in *Taraxacum*, dandelion), or from the spatial and functional replacement of the megaspore by an adjacent nucellar cell (as in *Hieracium*, hawkweed). Although the formation of the embryo occurs without fertilization, the stimulus of pollination is often needed before the process will begin (a phenomenon known as *pseudogamy*), and an endosperm may even be formed with the participation of one of the sperm nuclei. The pollen, although without genetic effect in the species producing it, is sometimes able to form hybrids with related species which reproduce sexually. This parallels the situation in apogamous ferns, such as *Dryopteris borreri*, where the male gametes, although normally functionless, are not impotent (see p. 220).

Another form of apomixis involves the formation of embryos, often in addition to the normal zygotic embryo, directly from the cells of the nucellus. A classical example of this phenomenon, known as *adventive embryony*, is provided by *Citrus* (Fig. 8.35), where the mature seeds may contain several viable embryos. Although in some species adventive embryony is independent of pollination, there are others (probably including *Citrus*) in which the embryos do not mature unless it occurs. In these instances the germinating pollen probably provides a chemical factor essential for continued growth. It is significant in this connection that in one orchid, *Zygopetalum machaii*, adventive embryony, although dependent upon pollination, is quite as effectively stimulated by pollen from another genus as by that from the species itself. Multiple seedlings from normally one-seeded fruits are a feature of many tropical trees, and this suggests that apomixis may be widespread in the Tropical Rain Forest.

The only remaining form of apomixis is more conspicuous since it involves the formation of bulbils or dwarf shoots in place of flowers. *Saxifraga cernua* and *Festuca vivipara* provide examples. Some species (e.g. *S. cernua* (Fig. 8.36)) producing these structures are also able to reproduce in a normal sexual manner, bulbils being borne in the lower part of the inflorescence and flowers in the upper. In some grasses, and probably in other plants displaying this phenomenon, the way the plant reproduces can be influenced by the length of day in which it is grown.

Other forms of asexual reproduction are purely vegetative. Stems may arch over and root themselves, as in *Rubus*, and leaves of a number of species produce plantlets either at the summit of the petiole, as in *Tolmiea* (a frequent house plant), or marginally, as in *Kalanchoë*. Species which grow from bulbs usually reproduce themselves by the production of axillary buds at the base of the axis which develop into daughter bulbs. Corms multiply

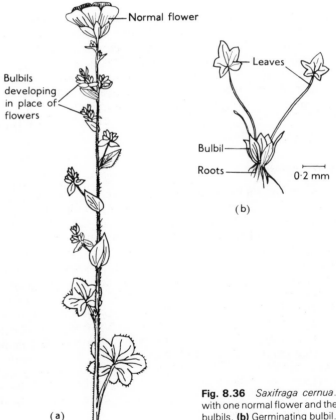

Normal flower

Bulbils developing in place of flowers

Leaves

Bulbil

Roots

0·2 mm

(b)

(a)

1 cm

Fig. 8.36 *Saxifraga cernua.* **(a)** Inflorescence with one normal flower and the rest developing as bulbils. **(b)** Germinating bulbil. (Both after Kerner von Marilaun and Oliver (1902). *The Natural History of Plants*. Blackie, London.)

themselves by a similar process. Fleshy roots which are able to give rise to buds largely account for the success with which plants such as *Convolvulus arvensis* (bindweed) and *Lepidium draba* (hoary cress) are able to multiply themselves. In addition to these various forms of vegetative reproduction occurring in Nature, layering, budding and grafting are frequently used in horticulture, and are the only means of propagating many valuable varieties whose sexual reproduction is defective.

Most remarkable of all, perhaps, is that in experimental conditions isolated cells and even pollen grains will give rise to whole plants of normal growth and form. Those raised from pollen grains are particularly useful in breeding. They initially contain the reduced number of chromosomes, but this can be doubled by colchicine so that a pure homozygote is obtained which shows normal sexual reproduction.

Evolution within the angiosperms

The evolution of the angiosperms has many different aspects. Of particular interest to systematists is the evolution of individual species, genera and families. Another approach, of more appeal to morphologists, is to consider the evolution of certain morphological or anatomical features – such as the form of leaves and the nature of vascular tissue – in the angiosperms as a whole. Unfortunately the fossil record of the angiosperms has little information to offer on any of these points. As soon as the angiosperms become well represented in the fossil floras of the Cretaceous, they are largely referable to modern families and even genera.

Recent evolution in angiosperms

Knowledge of recent evolution in the angiosperms has come principally, as with the ferns, from studies of chromosome pairing in hybrids, often produced under experimental conditions. Hybridization of diploid species, followed by allopolyploidy, appears to have been the origin of a number of well known plants, among them *Spartina anglica* (cord-grass) and *Galeopsis tetrahit* (hemp nettle). Observations of chromosome size may assist in assessing evolutionary relationships of a wider character. The family Commelinaceae, for example, includes some genera in which the chromosomes are small, others in which they are strikingly large, and yet others in which the nuclei contain chromosomes of both sizes. A few fossil series of recent age are known, mostly of angiospermous seeds, but they indicate little other than minor changes in structure and shape. Seeds of the water plant *Stratiotes*, for example, are found in successive strata throughout the Tertiary period. There was evidently a slight, but progressive, lengthening and narrowing of the seed during this period, together with minor changes in the relative development of the parts.

Another source of evidence relating to recent evolution is comparative biochemistry. In *Aesculus*, for example, an investigation of some curious and distinctive amino acids produced as metabolites has given a number of clues to interspecific relationships. Comparison of extractable proteins by serological methods is another promising technique for detecting affinities. Qualitative and quantitative comparisons of the occurrence of a number of different substances in extracts at one and the same time is made possible by chromatography. Two species which yield chromatograms which are closely similar are reasonably regarded as having diverged in relatively recent time.

The growth forms of the earliest angiosperms[2,46]

So far as we know the main trends in vascular plant evolution have occurred in equable humid environments, where conditions are particularly favourable for plant life. Today such conditions are found in the Tropics and warm temperate regions, to which about three-quarters of the families of living angiosperms are confined. The terrestrial vegetation in these regions consists predominantly of woody plants. This together with the apparent absence of herbaceous forms from the pteridosperms (the most likely ancestors of the flowering plants) has led to the view that the first angiosperms were also probably woody. It is significant that the bush forests of the Tropics and sub-Tropics contain several species in which the only tracheary elements in the xylem are tracheids similar to those of *Lyginopteris* (p. 225), and hence longer than most vessel segments. Caution however is necessary in assessing primitiveness from living species alone. Since woody plants generally reproduce less rapidly than herbaceous, there is less opportunity for genetic recombination and the selection of new forms. Hence, although primitive angiospermous features may be more frequently represented in woody species than in herbaceous, it does not necessarily follow that the original forms were woody. Nevertheless, it seems reasonable to conclude from present knowledge of the fossil record and from comparative anatomy that the earliest flowering plants were trees or shrubs, but this conclusion needs to be continually reviewed.

The evolution of xylem

Consideration of the evolution of morphological and anatomical features within the angiosperms as a whole is facilitated by the general structural relationship between the angiosperms and the gymnosperms. There are, of course, many features which are peculiar to the angiosperms, but these features are not present in all species. Since the fossil evidence, as will be discussed later (p. 326), indicates that the angiosperms were derived from some gymnospermous source, the more gymnosperm-like an anatomical or morphological character in an angiosperm, the more likely it is to portray the stage of that character in the earliest representatives of the Class.

This concept can be usefully employed in considering the evolution of xylem, especially of the woody plants. Examination of large numbers of dicotyledonous woods has shown that the lengths of the vessel segments, although varying little within a species, show wide variation between species.

There is also variation in the form of end-plate. Various combinations of length and form of end-plate are found, indicating that these two features can vary independently, but certain combinations occur more frequently than others. Long vessel segments (1.3–2.0 mm) tend to be associated with obliquely placed end-plates with scalariform perforations, and short segments with transverse end-plates with single large pores (see Fig. 8.10). Long vessel segments are also associated with a number of other features, such as the angularity of the outline in transverse section, small cross-sectional area and uniform thickening of the walls. These associations, which are based upon the examination of large numbers of woods, are statistically sound and consequently require a biological explanation.

The most satisfactory interpretation of these data makes use of the 'principle of correlation'. This maintains that in allied organisms undergoing evolution the primitive states of the evolving features, even though the features are evolving independently, will tend to remain associated in the group as a whole. Thus, in relation to dicotyledonous woods, we can identify long vessel segments and the features most frequently associated with them, which are all suggestive of tracheids, as indicating the most primitive state of xylem vessels. Conversely, the features associated together at the other end of the range, namely shortness of vessel segments (<0.3 mm), circularity of outline in transverse section, and transversely placed and simply porous end-plates, are those of the most recent kind of vessel. Using this principle of correlation it is possible to identify the primitive and advanced states of other features of the xylem, such as the arrangement of the vessels in relation to each other and the nature of the parenchymatous rays.

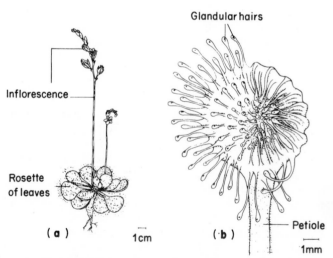

Glandular hairs

Inflorescence

Rosette
of leaves

(a)   1cm

(b)

Petiole

1mm

Fig. 8.37   *Drosera rotundifolia.* **(a)** Habit. **(b)** Single leaf in which some of the glandular hairs have closed on an entrapped insect. (After Kerner von Marilaun and Oliver (1902). *The Natural History of Plants.* Blackie, London.)

The evolution of leaves

Although little can be said with certainty concerning the evolution of leaves, there are many indications of stem-like properties in the leaves of angiosperms, indicating their affinity with megaphylls (p. 222) rather than microphylls (p. 150). Apart from the stem-like branching of some leaves (p. 285) and the readiness of others to yield shoot buds (p. 317), many pinnate leaves are hardly distinguishable from lateral shoots of limited growth, bearing leaves in a single plane. In *Chisocheton*, a tree of south-east Asia with a large pinnate leaf, the apex of the main rachis even remains active for several seasons, producing a pair of pinnae annually. Inbreeding may also reveal the developmental similarity of stems and leaves. In some improved varieties of tomato, for example, buds arise freely along the rachis of the leaf, and the distal part often continues growth as a shoot.

Comparatively recent evolution in the form of leaves has evidently been concerned in some instances with modification for a particular function. The insect traps of *Drosera* (sundew) (Fig. 8.37) and of *Nepenthes* (pitcher plant) (Fig. 8.38), are examples of this kind of development (see also Fig. 8.39). Sometimes evolution has been in relation to, and perhaps coupled with, that of a particular insect, usually a species of ant, which lives in and is nourished by the plant concerned. In several tropical species of *Acacia*, for example, the long, thorn-like stipules are hollow and provide shelter, while the leaflets terminate in small globules of parenchyma (Belt's corpuscles) which are devoured by the ants (Fig. 8.40). Succulence of leaves (the anatomical and

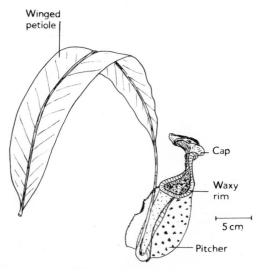

Fig. 8.38 *Nepenthes*. Leaf modified to form a pitcher-like insect-trap. (After *Botanical Magazine*, from Wettstein (1935). *Handbuch der systematischen Botanik*. Deuticke, Leipzig.). The inner surface of the pitcher is covered with tile-like plates of wax making it impossible for the insect to climb out.

Fig. 8.39   *Dionaea muscipula* (Venus fly trap): an insectivorous plant in which the leaf is highly modified. The terminal lamina of the leaf is in two halves which rapidly close together when the sensitive hairs on the surface are touched, the marginal teeth interlocking. Entrapped insects are digested by proteolytic enzymes coming from glandular hairs on the inner (adaxial) surface of the lamina. (× 7.5)

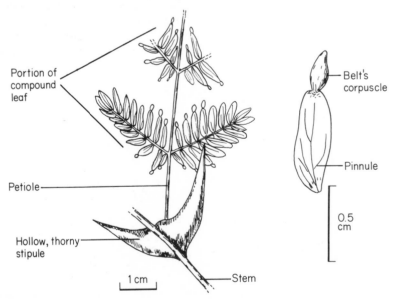

Fig. 8.40   *Acacia sphaerocephala*. Portion of leaf showing the hollow stipules and Belt's corpuscles.

physiological modification of the mesophyll so that it is able to retain large quantities of water) also reaches its highest development in the angiosperms and is a feature of many desert and maritime species.

The evolution of flowers

The evolution of flowers presents numerous complex problems, in many instances evidently bound up with the evolution of particular insects or other organisms upon which the species have become dependent for pollination, or with an outbreeding mechanism (Fig. 8.41) (see also p. 309). The first problem, however, is to identify the most primitive kind of flower. Here we can again make use of the principle of correlation (p. 320), and the significant fact emerges that in those woody species in which the xylem consists wholly of tracheids the members of each region of the flower are separate from each other. Such *chorichlamydeous* flowers are not of course confined to the vessel-less dicotyledons, they are also characteristic of the herbaceous Ranunculaceae and the monocotyledonous Alismataceae. Nevertheless congenital fusion of parts (which involves no actual fusion, but is a consequence of the delayed appearance of separated meristems in development), both of the perianth and reproductive region, can be looked upon with fair certainty as a later feature in the evolution of the flower.

Quite apart from fusion of parts within the flower, another tendency, clearly evident in some alliances, has been towards the evolution of compact inflorescences. The ultimate form of such an inflorescence is the dense capitulum, occurring in a number of families, but characteristic of the Compositae. In many instances, owing to the differentiation of disc and ray florets, capitula have come superficially to resemble simple flowers, a feature well shown by the common daisy, *Bellis perennis*. In *Syncephalantha*, a Central American composite, we even find a racemose aggregate of capitula

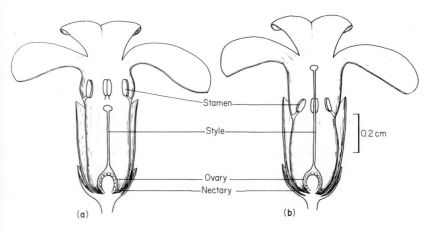

Fig. 8.41 *Primula* sp. Half flowers showing **(a)** thrum arrangement; **(b)** pin arrangement.

which, taken as a whole, also has a flower-like form owing to the asymmetry of the marginal capitula. The sporadic occurrence of these flower-like inflorescences in a number of families indicates that the radiate form of the simple flower has biological significance. Natural selection would account for the repeated emergence of this pattern despite the increasing morphological complexity of the reproductive region.

The evolution of spurred and otherwise zygomorphic petals, distinctive scents, nectaries and peculiar stamens (see p. 308), and of striking pigmentation of the perianth are all usually associated with insect (and rarely bird) pollination. Sometimes the life cycle of both plant and insect are so closely interwoven that one could not exist without the other. Reproduction in the figs provides a particularly striking example. The female flowers, the first to open in the urn-shaped receptacle, are pollinated by particular wasps which force their way through the narrow entrance and bring in pollen from elsewhere. Eggs are laid in some of the ovaries and the emergence of the new generation of wasp coincides with the opening of the male flowers. In escaping from the fruit they take pollen with them and the cycle is repeated in another developing fruit. In some species the cycle is even more complicated, two species of wasp being involved. These situations raise profound problems of evolution, since at some stage flower and insect must have begun to evolve together, leading ultimately to complete mutual dependence.

Physiological evolution

To conclude this outline of evolution within the angiosperms, mention should also be made of the striking physiological evolution that in some species and families has evidently accompanied the morphological. Occasionally this has extended to the abandonment of autotrophy, the species concerned becoming in consequence obligate heterotrophs, a transition already seen in other Divisions of the Plant Kingdom. Examples amongst the angiosperms are provided by *Epipogium*, an orchid which lacks chlorophyll and derives its nutrition from a highly developed mycorrhiza, and such parasites as *Lathraea* (toothwort), *Orobanche* (broomrape) and *Cuscuta* (dodder) (Fig. 8.42). A number of angiosperms, such as *Viscum* (mistletoe) and *Euphrasia* (eyebright), have become parasitic without losing the ability to photosynthesize, so they are not entirely dependent upon their hosts.

Physiological evolution has also extended to the appearance of some species or races tolerant to relatively high concentrations of normally toxic metals. The lead- and copper-tolerant forms of the grasses *Agrostis tenuis* and *Festuca ovina*, found on spoil heaps of old mines, are striking examples of recent evolution of this kind. These are generally considered to have arisen by selection of resistant mutants appearing spontaneously in natural populations. Remarkable ecological adaptation is also shown by those species which have evolved means of eliminating ions taken up in excess with the soil water. The salt glands of *Limonium*, a salt marsh plant, provide a notable example, and offer a model system for the study of membrane transport.

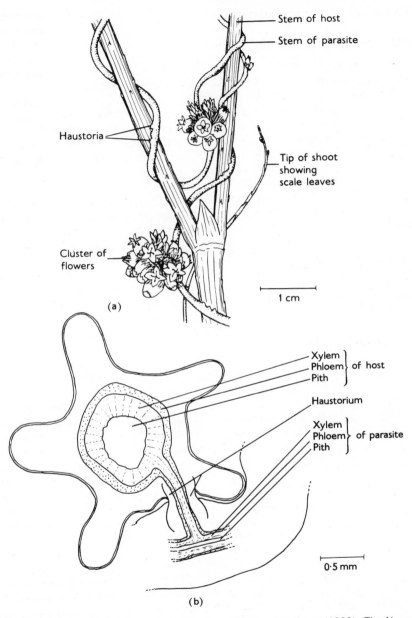

Stem of host

Stem of parasite

Haustoria

Tip of shoot showing scale leaves

Cluster of flowers

1 cm

(a)

Xylem ⎫
Phloem ⎬ of host
Pith ⎭

Haustorium

Xylem ⎫
Phloem ⎬ of parasite
Pith ⎭

0·5 mm

(b)

Fig. 8.42 *Cuscuta.* **(a)** Habit. (After Kerner von Marilaun and Oliver (1902). *The Natural History of Plants*. Blackie, London.) **(b)** Transverse section of host stem showing the penetration of the vascular tissue of the parasite into that of the host.

The origin of the angiosperms

The age of the angiosperms

There remains the problem of the origin of the angiosperms, and this inevitably raises that of the time of their first appearance in the Earth's vegetation. Unfortunately, upon this cardinal point there is no general agreement. Although some have envisaged the angiosperms appearing in upland regions (and hence away from sites of fossilization) as early as the end of the Palaeozoic, this view owes more to inference than direct evidence, and has been disputed by many. There are admittedly several reports of vegetative remains of angiosperms from early Mesozoic rocks, but these should for the present be accepted with caution since the existence of angiosperms at this time has not yet received support from the study of fossil pollen.[19] Investigation of the fine details of the organization of the pollen wall has now reached a high level of precision, and it has become clear that pollen with columnar exine structure is unknown outside the angiosperms. So far as sampled, rocks older than the Cretaceous have yielded no well-established records of pollen of this kind, and there are consequently few grounds for believing that the angiosperms were in existence before the Cretaceous period. In those few instances where remains whose angiospermous nature seems beyond question have been described as pre-Cretaceous, there is a variety of geological reasons why the age of the beds containing them cannot be taken as certain.

If the angiosperms arose at the end of the Jurassic or at the beginning of the Cretaceous period, this would imply that in about 25 million years (i.e. in about half the Cretaceous period) they had become the dominant component of the earth's vegetation. Provided, once the transition had been made, evolution within the angiosperms was comparatively rapid this time scale is in no way unreasonable. The angiosperms may in fact have enjoyed from the time of their origin the advantage of relatively rapid sexual reproduction. In the angiosperms the interval between the production of the flower and the setting of the seed is usually a matter of weeks, sometimes only of days (e.g. in the so-called 'desert ephemerals'), whereas in most gymnosperms the equivalent process takes at least a year. The angiosperms thus have the possibility of more frequent sexual reproduction and consequently the opportunity for more rapid evolutionary advance (see p. 319).

Once initiated, if better suited to the environment than the existing gymnosperms, the angiosperms no doubt soon became dominant. The rapidity with which better adapted immigrant species will replace native is very striking. In parts of central Portugal, for example, the arboreal vegetation has been largely replaced in less than a century by introduced Australian species.

The nature of the immediate precursors of the angiosperms

The immediate ancestors of the angiosperms should therefore probably be sought in late Jurassic and early Cretaceous rocks. No fossils of this age have

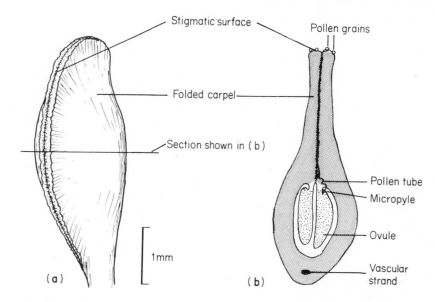

Fig. 8.43 *Drimys piperita.* (a) Carpel with stigmatic surface along the ventral suture. (b) Transverse section of carpel. (Both after Bailey and Swamy, from Foster and Gifford (1949). *Comparative Morphology of Vascular Plants.* Freeman, San Francisco. Copyright © 1949.)

yet been generally accepted as intermediate between gymnosperms and angiosperms. Despite this absence of direct evidence, however, examination of the carpels of living angiosperms, particularly those possessing primitive wood (p. 319), gives strong indications of how angiospermy developed. In several of these genera the carpels are simple folded structures, with stigmatic surfaces lying along part or whole of the ventral suture. This is well shown in *Drimys piperita* (Fig. 8.43), a vessel-less dicotyledon. In *Degeneria* (Fig. 8.44), the carpel is similar, but is even more striking. The ventral margins of the carpel do not fuse, but become interlocked by papillae, between which the pollen tubes force their way. The principle of correlation thus indicates that the earliest angiospermous carpel was a conduplicate structure with a lateral stigma along the approximated margins. The carpel with a capitate stigma elevated on a style would then be regarded as a later development.

Fossils of possibly great significance in the evolution of the angiosperms have come from the Permian of the Southern hemisphere. Curious reproductive organs are found associated with a reticulately-veined leaf and conifer-like wood and all are now thought to be parts of one plant, *Glossopteris*. The megasporangiophores were leaf-like, but were inserted in the axils of foliage leaves (Fig. 8.45). They bore several ovules on the abaxial side, each about 1.5 mm long, and the fertile area was largely covered by the recurved margins of the sporangiophore. The male sporangiophores were also leaf-like with lateral tassels of pollen sacs. The pollen was winged in the manner of that of

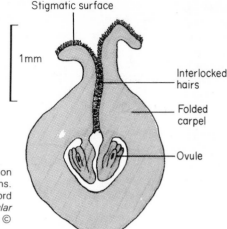

Fig. 8.44   *Degeneria* sp. Transverse section of conduplicate carpel with infused margins. (After Swamy, from Foster and Gifford (1949). *Comparative Morphology of Vascular Plants*. Freeman, San Francisco. Copyright © 1949).

Pinus. Glossopteris was probably a bush or small tree. The importance of *Glossopteris* is that it reveals that already by the end of the Palaeozoic era there was a clear tendency for ovules to become enclosed in carpel-like structures. This tendency is seen again in the Caytoniales, frequent in the Yorkshire Jurassic, and in Greenland. Although the growth form of these plants remains obscure the reproductive organs and leaves are well known. The female organ consisted of an axis, 5 cm or more in length, and possibly dorsiventral in symmetry. This axis bore several short pinnae, arranged in more or less opposite pairs, each pinna terminating in a hollow,

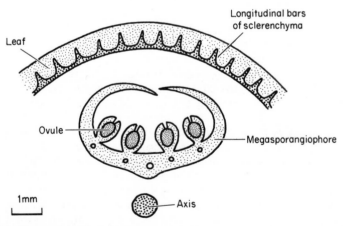

Fig. 8.45   *Glossopteris*. Diagrammatic representation of the female reproductive region. The leaf-like megasporangiophore was borne in the axil of a foliage leaf. (From information in Gould and Delevoryas (1977). *Alcheringa*, **1**, 387.)

spherical body about 0.5 cm in diameter (Fig. 8.46). This sphere contained the ovules, and, because of its analogies with the structures partially enclosing the ovules of some Carboniferous pteridosperms, it is termed a cupule. At the base of the cupule, on the upper side, and concealed by a small flap of tissue, was a pore communicating with the interior. Up to 32 upright ovules, each about 2.5 mm in length, were arranged in lines on the inner surface of the cupule, more or less opposite the basal pore (Fig. 8.46b). The symmetry of the male reproductive organ was similar to that of the female, but the pinnae terminated in several short irregular branches, each of which bore one or more synangia (Fig. 8.46c). Each synangium was divided longitudinally into four pollen sacs (and consequently, although symmetrical, bore a striking resemblance to the anther of a modern flowering plant), the loculi containing winged pollen grains. The leaves consisted of a rachis ending in 3–6 leaflets, the leaflets palmately arranged and reticulately veined.

It was presumably from gymnosperms similar to *Glossopteris* and the Caytoniales that the angiosperms were evolved. If the flap at the entrance to the *Caytonia* cupule had acquired stigmatic properties the cupule would have become a carpel not entirely dissimilar to those of angiosperms with lateral stigmas (e.g. *Drimys*) and basally inserted styles (e.g. *Comarum*).

Some have suggested that the angiosperms had more than one origin, i.e. that they, like the gymnosperms, are polyphyletic. This, however, seems unlikely in view of the general occurrence of the characteristic 'double fertilization' in the embryo sac. Nevertheless, the relationship between some monocotyledons, especially the palms, and dicotyledons certainly appears

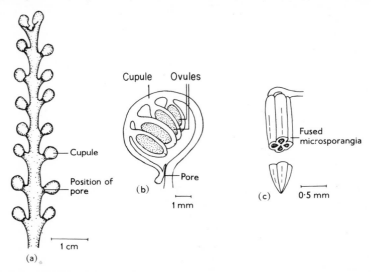

Fig. 8.46 *Caytonia nathorsti.* **(a)** Reconstruction of fertile shoot. **(b)** Longitudinal section of cupule. (**(a)** and **(b)** after Thomas, from Andrews (1961). *Studies in Paleobotany.* Wiley, New York.) **(c)** *Caytonanthus kochi.* Reconstruction. (After Harris, from Andrews (1961). *Studies in Paleobotany.* Wiley, New York.)

remote, and we must assume that dicotyledons and monocotyledons have followed more or less parallel lines of evolution for a considerable period.

The main trends of angiosperm evolution

To summarize, we can envisage the main evolutionary trends in the angiosperms to have been as follows. They emerged from the gymnosperms, probably, in view of the radial symmetry of the ovules and certain anatomical similarities, from a group evolving from the pteridosperms of the late Palaeozoic. The angiosperms thus inherited a seed-habit of long standing, and a serviceable form of axial structure in which lateral branch systems had long ago become megaphylls, and in which megaphylls in turn were becoming differentiated into petioles and laminae with reticulate venation. They improved upon this structure, replacing coarse and angular ramification by elegant branching. Anatomically, strengthening tissues tended to become reduced to those required by mechanical laws, and stems and branches consequently acquired a springiness that is a protection against storms. Vessels were also developed from tracheids, and a more highly organized phloem containing sieve tubes and companion cells from simple sieve cells and parenchyma, both developments probably having advantageous physiological consequences. The reproductive regions became highly specialized, their evolution often being correlated with that of pollinating organisms.

We have considered the primitive angiosperms to be woody (p. 319), and the evolution of herbaceous angiosperms consequently a later phenomenon, certainly in the grasses correlated with that of the grazing animals. In tropical regions many herbaceous angiosperms are in fact epiphytes upon angiospermous trees and shrubs, and it seems clear that these followed rather than preceded the arboreal vegetation. Because of the rapidity with which the relatively succulent leaves of angiosperms decay in tropical conditions, humus is soon formed wherever they accumulate. Consequently, fertile substrata are found in abundance on horizontal surfaces and in the crotches of branches. This leads to an ecological situation different from that in gymnospermous forests, where the leaves mostly reach the ground and the humus, owing to the slow decay of the leaves, is covered by harsh and unconsolidated material unfavourable for the establishment of delicate herbaceous forms. It thus seems reasonable to suppose that the epiphytic angiosperms (and ferns) emerged in response to and are in the process of exploiting the new ecological habitat created by the rise of the angiosperm forests. Their evolution would then have been, in geological time, a relatively recent event.

Today the angiosperms in some way resemble successful individuals reaching affluent middle age, determined to let no human experience escape them. They experiment freely with metabolic novelties, such as fragrant oils, peculiar alkaloids and unusual carbohydrates (such as inulin). In sexual reproduction, both as regards pollination and the cytology of the gametophytes, they show almost every conceivable variation. We probably witness the angiosperms at the height of their success. What they will be like in their senescence we can safely leave to our successors.

Glossary

Terms relating to special features defined in the text are not repeated here. These may be traced through the Index.

Abaxial, of lateral organs. The side away from the main axis (the same as DORSAL).

Abscission. The separation of structures from the stem of a plant after the formation of a layer of specialized cells (abscission layer).

Achene, of flowering plants. A dry, indehiscent, one-seeded FRUIT.

Acropetal. Developing from the base towards the apex, i.e. the youngest structures are adjacent to the apex, or, with regard to the movement of materials, towards the apex.

Actinomorphic. Radially symmetrical.

Adaxial, of lateral organs. Said of the side towards the main axis (the same as VENTRAL).

Adventitious. Said of plant organs which appear in unusual positions, e.g. roots from stems, or buds not in the axils of leaves.

Adventive embryony. The development of EMBRYOS from cells other than EGG cells, and without FERTILIZATION.

Aerenchyma. A plant tissue composed of unthickened, often irregularly shaped, cells surrounding large air spaces.

Aerophore. A negatively GEOTROPIC root produced by certain plants growing in waterlogged conditions and believed to assist the aeration of the root system.

Akinete, of algae. A single-celled, non-motile, resting SPORE in which the original cell wall forms part of the spore wall.

Allopolyploidy. The type of POLYPLOIDY found in organisms of hybrid origin, at least two complete sets of CHROMOSOMES from each of the original parents being present in the SPOROPHYTIC phase.

Alternation of generations. An outmoded term referring to the manner in which in a life cycle a SPOROPHYTE is succeeded by a GAMETOPHYTE and this in turn by a sporophyte.

Amphiphloic. Having PHLOEM tissue on each side of the XYLEM.

Amyloplast. A PLASTID without CHLOROPHYLL, involved in the synthesis and storage of starch.

Anemophily. Pollination by wind.

Aneuploid. Possessing a chromosome number not an exact multiple of the HAPLOID number.

Anisogamy. The production or fusion of GAMETES which are visibly different.

Annual. A plant which completes its life cycle in one growing season and then dies.

Anthesis, of flowering plants. The time of opening of the FLOWER and exposure of the mature reproductive organs.

Anticlinal. Of cell divisions in which the new cell wall is perpendicular to the outer surface of the region in question.

Antithetic (relating to life cycles). The view that the SPOROPHYTE has evolved by elaboration of a transient ZYGOTIC phase interpolated between successive GAMETOPHYTES.

Aplanospore, of algae. A single-celled, non-motile SPORE which is liberated from the parent cell.

Apogamy. The ASEXUAL production of a SPOROPHYTE from a GAMETOPHYTE.

Apomixis, of seed plants. Reproduction by SEEDS which have not developed as a consequence of a sexual process.

Apoplast. The space formed by the hydrated framework of cell walls outside the protoplasts in which soluble substances and ions may freely move.

Apospory. The production of a GAMETOPHYTE directly from the SPOROPHYTE without the intervention of MEIOSIS.

Archesporium. The group of cells giving rise to the SPORE mother cells (MEIOCYTES).

Areola (Areole). Area surrounded by VEINS in a RETICULATELY veined leaf.

_____, of some cacti. The tuft of spines indicating a lateral BUD.

Aril. A fleshy outer covering of the OVULE of certain SEED plants.

Asexual. Reproduction not involving the fusion of GAMETES.

Autogamous, of FLOWERS. Normally self-pollinated.

Autopolyploidy. The occurrence of three of more identical sets of CHROMOSOMES in the nucleus.

Autosome. A CHROMOSOME that is not a sex chromosome.

Auxin. A general term for organic substances (naturally occurring or synthetic) which promote or regulate plant growth.

Axil. The upper angle which a leaf makes with a stem.

Basipetal. Developing from the apex towards the base, i.e. the youngest structures are the most distant from the apex, or, with regard to the movement of materials, towards the base.

Berry, of flowering plants. A many-seeded FRUIT with a fleshy PERICARP, produced from a single flower.

Biennial. A plant in which the life cycle requires two growing seasons for completion, reproduction being followed by death.

Bifid. Divided deeply into two parts.

Bifurcate. To divide into two branches.

Bilateral symmetry. Symmetry with respect to a single plane; when divided by the plane, the two halves are mirror-images.

Blade. The broad, flattened portion of a leaf (the lamina).

Bordered pit. A PIT in which the thin area of primary wall is overhung by lignified secondary wall. The centre of the pit membrane is often thickened, forming the torus (see Fig. 7.21).

Bract, of seed plants. A reduced leaf with a reproductive structure in its AXIL.

Bract scale, of gymnosperms. The scale (in some species becoming woody) in the female CONE in the AXIL of which arises the ovuliferous scale.

Bracteole, of flowering plants. A reduced leaf produced on a FLOWER stalk.

Bud. The apical MERISTEM of a shoot, surrounded by leaf primordia and sometimes enclosed in scale leaves.

Bulb, of flowering plants. An underground PERENNATING and storage organ, consisting of a reduced stem surrounded by fleshy leaves.

Callus. A PARENCHYMATOUS tissue formed over wounds or in culture, often capable of being sub-cultured indefinitely in nutrient media.

Cambium. A secondary MERISTEMATIC tissue: the VASCULAR CAMBIUM produces secondary XYLEM and PHLOEM; the CORK CAMBIUM produces CORK and secondary CORTEX.

Capitulum, of flowering plants. An INFLORESCENCE in which the FLOWERS are closely grouped on a disc.

Carotenoids. A group of yellow pigments occurring in chloroplasts, CHROMOPLASTS and elsewhere.

Carpel, of flowering plants. The OVULE-bearing structure, usually regarded as a MEGASPOROPHYLL.

Casparian band (strip). A girdle of SUBERIZED thickening occurring on the radial and upper and lower walls of the cells of the ENDODERMIS.

Cauliflory, of flowering plants. The continued lateral production of FLOWERS on old woody stems.

Cauline. Appertaining to the stem.

Cell sap. The watery contents of the large VACUOLE usually present in plant cells.

Centrifugal. Developing from the centre towards the outside, i.e. the oldest structures are at the centre, and the youngest at the outside.

Centripetal. Developing towards the centre from the outside, i.e. the oldest structures are at the outside, and the youngest at the centre.

Chemotaxis. The movement of a motile organism, SPORE, or GAMETE in response to a chemical stimulus.

Chimaera. An organism with tissues of more than one GENOTYPE. It may be produced by mutation or by grafting.

Chlorophyll. A group of structurally similar green or blue-green PHOTOSYNTHETIC pigments common to all photoautotrophic plants.

Chromatin. A complex of deoxyribonucleic acid and basic proteins present in CHROMOSOMES and defined by its staining properties.

Chromatophore. An intracellular body containing pigments; hence a general class containing chloroplasts, but not often used for the latter.

Chromoplast. A PLASTID deficient in or lacking CHLOROPHYLL, but accumulating other pigments, principally CAROTENOIDS.

Chromosomes. The organized structures in the NUCLEUS, carrying the Mendelian genes.

Cilium. A short, thread-like, locomotory ORGANELLE which beats in a regular manner. Usually numerous.

Circumnutation. The sweeping movements made by the tip of a stem, particularly marked in twining plants, caused by unequal rates of growth around the axis.

Cladode. A short, flattened shoot of limited growth, replacing the leaves as the site of photosynthesis.

Coccoid, of algae. A unicellular, non-motile, spherical condition of the adult organism.

Coenobium. A COLONY of cells firmly attached to each other in which there is some degree of co-ordination.

Coenocyte. A multinucleate plant body or cell.

Coleoptile, of flowering plants. The sheath around the PLUMULE of certain MONOCOTYLEDONOUS plants.

Coleorhiza, of flowering plants. The sheath around the RADICLE of certain MONOCOTYLEDONOUS plants.

Collateral bundle. A VASCULAR bundle in which the XYLEM and PHLOEM are CORADIAL, the xylem lying towards the centre of the axis and the phloem away from it.

Collenchyma. A strengthening tissue composed of columnar cells with heavy deposits of cellulose at the angles of the primary walls.

Colloid. A system of minute, charged particles suspended in a liquid. The colloid itself may be in the form of a liquid (sol) or solid (gel).

Colony, of algae. A group of adhering unicellular organisms, all of which are equivalent and independent of one another.

Colpus. Each of the thin areas in the EXINE of a POLLEN grain, through one of which the pollen tube usually emerges.

Columella. A central, sterile tissue found in certain SPORANGIA.

Companion cells, of flowering plants. The relatively small, nucleate cells with dense PROTOPLASM associated with SIEVE TUBES in the PHLOEM tissue.

Complanate. Arranged in a single plane.

Conceptacle, of brown algae. The flask-shaped cavity containing the sex organs in certain genera.

Concrescent. Grown together, coalesced.

Cone. An axis bearing SPOROPHYLLS (usually closely packed).

Congenital (as used in *Congenital fusion*). Said of an inherited characteristic which becomes apparent during growth.

Coplanar. In the same plane.

Coralloid. Coral-like as a consequence of repeated branching.

Cordate. Heart-shaped.

Cork. The outer tissue formed by the activity of a special CAMBIUM and consisting of cells which become SUBERIZED and impervious to water.

Corm, of flowering plants. A subterranean storage and PERENNATING organ consisting of a short, swollen, stem base.

Corolla, of flowering plants. A collective term for the PETALS of a FLOWER.

Corpus. The central part of the apex of a stem in which cell divisions occur in a variety of planes.

Cortex. The outer tissue of a multicellular axis, often little differentiated.

Cotyledons, of seed plants. The first leaf or leaves of the EMBRYO.

Crenulate. Toothed, but the teeth small and rounded.

Cupule, of seed plants. An extra cup-shaped organ surrounding the ovule of certain fossil plants, or a cup formed by concrescent bracts at the base of the fruit of certain living genera (e.g. acorn).

Cuticle. The non-cellular waxy coating of the EPIDERMIS of all land plants.

Cymose branching. The type of branching in which the apices abort successively, growth being continued by laterals in each instance.

Cytoplasm. The PROTOPLAST of a cell, apart from the NUCLEUS.

Deciduous plant. One which sheds its leaves at intervals, in temperate regions usually at the end of the growing season, by means of an ABSCISSION layer.

Decumbent, of a shoot. Either Initially, or soon becoming, prostrate.

Decussate. Arranged in opposite pairs, alternate pairs being at right angles to each other.

Dehiscence. Opening at maturity, by splitting, especially of SPORANGIA and FRUITS.

Diarch xylem. XYLEM tissue consisting of two strands, or having two PROTOXYLEM groups.

Dichogamy. The maturing of the male and female organs of a FLOWER at different times.

Dichotomous branching. Branching into two equal parts.

Diclinous, of flowering plants. Having STAMENS and CARPELS in separate FLOWERS, but the flowers borne by the same individual. Also used correspondingly of ARCHEGONIATE plants.

Dicotyledonous. Having two COTYLEDONS.

Differentiation. The development of cells, tissues and organs during which differences arise in structure and function.

Dimorphic. Existing in two separate forms.

Dioecious. Bearing male and female sex organs on different individuals.

Diploid (2n). Having twice the basic (HAPLOID) number of CHROMOSOMES.

Distal. Distant from a point of reference or symmetry (ant. PROXIMAL).

Distichous. Arranged in two ranks, diametrically opposed.

Dorsal, of lateral organs. The same as ABAXIAL.

———, of prostrate plants. The side away from the substratum.

Dorsiventral. Having distinct upper and lower sides.

Drupe, of flowering plants. A one-SEEDED FRUIT, formed from a single flower, in which the PERICARP is part fleshy and part stony.

Egg. A non-motile, female GAMETE.

Elaters. Sterile cells or parts of cells interspersed with spores and assisting dispersal by hygroscopic movements.

Emarginate, of leaves. Having a small depression at the apex.

Embryo. The young, partly developed SPOROPHYTE.

Embryo sac, chiefly of angiosperms. The mature female GAMETOPHYTE.

Embryogeny. The development of the EMBRYO from the ZYGOTE.

Endarch, of xylem. XYLEM tissue which differentiates CENTRIFUGALLY, leaving the PROTOXYLEM nearest the centre.

Endodermis. The inner layer of cells of the CORTEX of a stem or root, distinguished by the presence of CASPARIAN BANDS, and often by starch grains.

Endogenous. Developing or emanating from the inside.

Endolithic. Living within pores or narrow channels in rocks.

Endomitosis. Duplication of the CHROMOSOMES without nuclear division, resulting in POLYPLOIDY.

Endophyte. A plant which lives inside another organism, but is not PARASITIC upon it.

Endopolyploidy. POLYPLOIDY which is the result of ENDOMITOSIS.

Endoscopic embryogeny. EMBRYOGENY in which the first division of the ZYGOTE is such that the apex of the EMBRYO develops from the inner cell.

Endosperm, of flowering plants. The nutritive tissue in SEEDS derived from the primary endosperm cell of the EMBRYO SAC. Notable for the frequent presence of TRIPLOID nuclei and as a source of AUXINS.

Endosporic gametophyte. A GAMETOPHYTE which develops within the SPORE wall.

Endosymbiosis. SYMBIOSIS between two organisms, one of which lives inside the other.

Entomophily. Pollination by insects.

Enzyme. A soluble protein, or protein complex, which catalyses a biochemical reaction.

Epicotyl. That part of the axis of a seedling immediately above the COTYLEDONS.

Epidermis. The primary outer layer of cells of all plant organs.

Epiphyte. A plant which grows attached to another, but is not PARASITIC upon it.

Etiolation. The abnormal type of growth occurring in heavy shade or darkness, involving particularly reduction in the amount of CHLOROPHYLL, and an exaggerated elongation of the INTERNODES.

Eustele. A STELE consisting of a ring of COLLATERAL bundles.

Exarch. XYLEM tissue which differentiates CENTRIPETALLY, leaving the PROTOXYLEM furthest from the centre.

Exine. The outer layer of the wall of a SPORE or POLLEN grain.

Exogenous. Developing or emanating from the outside.

Exoscopic embryogeny. EMBRYOGENY in which the first division of the ZYGOTE is transverse, the apex of the EMBRYO developing from the outer cell.

Exoskeleton, of algae. A firm outer layer, especially calcium carbonate, secreted by certain species.

Facultative. Of a response which depends upon conditions; implying that the organism in other conditions is able to respond in another way.

Falcate. Sickle-shaped.

Fascicular. Relating to a cluster or bundle, usually implying association with VASCULAR bundles. Hence FASCICULAR CAMBIUM.

Fertilization. The fusion of GAMETES.

Fibre cell. An elongated cell with a thick lignified wall, tapering at each end.

Filiform. Thread-like.

Fission. The ASEXUAL division of a unicellular organism into two similar organisms.

Flagellum. A whip-like organ of locomotion, single, paired, or many, found in motile algae and GAMETES.

Flower. A reproductive region of a non-archegoniate SEED-bearing plant, usually consisting of a PERIANTH and one or more STAMENS or PISTILS, or both, and terminating the PEDUNCLE or PEDICEL.

Foot. The portion of the EMBRYO in the bryophytes and lower tracheophytes which anchors the embryo in the GAMETOPHYTE and may absorb some nutrients from it.

Frond. The leaf of a fern, or of a cycad.

Fruit. The SEED-containing structure derived from the OVARY, sometimes associated with other parts.

Funicle. The delicate stalk of an OVULE of a flowering plant, sometimes (as with inverted ovules) wholly or partly concrescent with the outer INTEGUMENT.

Fusion, in relation to organs. Used figuratively, implying CONCRESCENCE either during development or in the course of evolution.

Gametangium. An organ producing GAMETES.

Gamete. A mature sex cell, capable of fusing with another to form a ZYGOTE.

Gametophyte. The (normally) HAPLOID phase producing GAMETES which fuse to form the DIPLOID SPOROPHYTE.

Gemma, chiefly of bryophytes. A multicellular, ASEXUAL, reproductive structure which becomes detached from the parent and is capable of growing into a new plant.

Generative cell, of seed plants. The cell of the male GAMETOPHYTE which by division gives rise to the two male GAMETES.

Genotype. The genetic constitution of an organism.

Geotropism (Gravitropism). A growth movement of a plant dependent upon the direction of gravitation.

Guard cell. One of a pair of cells, containing chloroplasts, which occur in the EPIDERMIS and form a STOMA.

Halophyte. A plant living in a habitat of high sodium chloride concentration, e:g. a saltmarsh.

Haploid (n). Possessing a single set of CHROMOSOMES.

Hapteron. Same as HOLDFAST.

_____, in *Equisetum*. A synonym for ELATER.

Haustorium. An absorptive organ of PARASITES which penetrates the HOST tissue.

Heterogamy. The condition in which GAMETES are morphologically distinct.

Heteromorphic. Having more than one growth form.

Heterophyllous. Having more than one leaf form.

Heterospory. The production of spores of two sizes (MEGASPORES and MICROSPORES).

Heterothallic, of algae. The condition in which GAMETES from the same individual or strain of individuals either cannot fuse or, if they fuse, yield an inviable ZYGOTE.

Heterotrichous, of algae. Having two kinds of FILAMENTOUS growth, one prostrate and the other erect, in the same individual. Also used in relation to the PROTONEMATA of bryophytes.

Hilum. Region of attachment of FUNICLE to OVULE.

Histone. A highly basic protein often associated with deoxyribonucleic acid in the CHROMOSOMES.

Holdfast, of algae. The organ or cell which anchors the plant to the substratum.

Homologous. Said of organs believed to be identical in nature (but not necessarily in form), often (but not always) implying an evolutionary relationship.

Homospory. The production of only one type of SPORE.

Homothallic, of algae. The opposite of HETEROTHALLIC – the condition in which GAMETES from the same individual commonly fuse and yield viable ZYGOTES.

Host. The organism from which a PARASITE obtains its food.

Hyaline. Clear, transparent.

Hydathode. A water-secreting gland or pore found in the leaves of many plants.

Hygroscopic. Water absorbing.

Hypocotyl. The stem of a seedling between the ROOT and the COTYLEDONS.

Hypogynous, of flowering plants. A FLOWER in which the perianth and androecium are inserted below the gynaecium.

Incompatibility. The inability of conspecific GAMETES to fuse and form a ZYGOTE, resulting from a number of different mechanisms in different kinds of plants.

Inflorescence. A group of FLOWERS with a common stalk.

Integuments, of seed plants. The envelopes surrounding the NUCELLUS.

Intercalary. Of MERISTEMS that lie between two apices.

Internode. The portion of a stem occurring between two NODES.

Intine. The inner layer of the wall of a SPORE or POLLEN grain.

Involucre. A whorl of sterile structures beneath a terminal reproductive region.

Isodiametric. Of equal height, length and breadth.

Isogamy. The condition in which GAMETES are morphologically identical.

Isomorphic. Of the same form.

Lacuna. A cavity.

Lamina. See BLADE.

Lanceolate. Spear-shaped; flattened, and tapering to a point, with the widest part near the centre.

Leaf gap. A gap in the VASCULAR cylinder of a stem immediately above the departure of a LEAF TRACE.

Leaf trace. The VASCULAR supply to a leaf, consisting of one or more strands.

Lenticel. A channel filled with loosely packed CORK cells permitting the diffusion of gases in stems (and occasionally in roots) formed by a characteristically shaped CORK CAMBIUM.

Leucoplast. A PLASTID lacking CHLOROPHYLL and other pigments, but often containing starch.

Lignin. A complex macromolecule formed by dehydrogenative polymerization of p-hydroxycinnamyl alcohols and deposited in the walls of certain cells.

Ligule, of grasses. A collar-like outgrowth at the inner junction of the leaf-sheath and BLADE.

_____, of lycopods. A scale inserted into the upper surface of the leaf or SPOROPHYLL.

Littoral zone. The zone on the sea shore between the high- and low-tide marks.

Loculus, of flowering plants. The cavities (loculi) in an OVARY, each containing one or more OVULES.

Lumen. The space bounded by a cell wall, more often used in relation to dead cells than living.

Lysigenous (of spaces). Formed by the dissolution of a cell or cells (cf. SCHIZOGENOUS).

Medulla. The central region of an axis or organ.

Megasporangiophore. A structure bearing MEGASPORANGIA.

Megasporangium. A sporangium producing only MEGASPORES.

Megaspore. A SPORE which on germination produces a female GAMETOPHYTE.

Megasporocyte. The cell whose nucleus undergoes MEIOSIS and which yields MEGASPORES.

Megasporophyll. A structure bearing MEGASPORANGIA, and believed to be equivalent to a leaf.

Meiocyte. A cell in which MEIOSIS occurs.

Meiosis. The two consecutive nuclear divisions, in the first of which the CHROMOSOME number is halved, so that each NUCLEUS ultimately contains only one set of chromosomes. The pairing of chromosomes in the first division offers the opportunity for genetic interchange.

Meristele. The unit of a DICTYOSTELE, composed of a central strand of XYLEM completely surrounded by PHLOEM and delimited by an ENDODERMIS.

Meristem. A group of undifferentiated cells retaining the capacity to divide indefinitely.

Mesarch. XYLEM tissue which differentiates both CENTRIFUGALLY and CENTRIPETALLY, the PROTOXYLEM consequently lying within the METAXYLEM.

Mesophyll. The PARENCHYMATOUS tissue of leaves lying between the upper and lower epidermis; the site of most of the chloroplasts.

Metaxylem. The portion of XYLEM tissue which is derived from the PROCAMBIUM (and is hence PRIMARY) and which DIFFERENTIATES after the PROTOXYLEM.

Micropyle, of seed plants. The minute pore in the distal end of the INTEGUMENTS of the OVULE.

Microsporangiophore. A structure bearing MICROSPORANGIA.

Microsporangium. A sporangium producing only MICROSPORES.

Microspore. A SPORE which on germination produces a male GAMETOPHYTE.

Microsporophyll. A structure bearing MICROSPORANGIA, and believed to be equivalent to a leaf.

Midrib. The central VEIN of a leaf.

Mitochondrion. An ORGANELLE of variable shape bounded by a distinct membrane, and occurring in all eukaryotic cells. The site of the major part of respiration.

Mitosis. Nuclear division in which two identical NUCLEI are formed.

Monoclinous, of flowering plants. Having functional STAMENS and CARPELS in the same FLOWER. Also used correspondingly of ARCHEGONIATE plants.

Monocolpate. Said of a SPORE or POLLEN grain which possesses only one COLPUS.

Monocotyledonous, of flowering plants. An EMBRYO possessing a single COTYLEDON.

Monoecious. The condition in which both male and female sex organs are produced on one plant.

Monolete spore. The kind of SPORE which, having been formed in a tetrad without tetrahedral symmetry, bears a linear scar on its PROXIMAL face.

Monopodial branching. The type of branching in which the main apex remains active, the shoots produced from AXILLARY BUDS remaining clearly lateral.

Morphology. The study of the form and related anatomy of living organisms.

Mycorrhiza. A SYMEIOSIS beween the roots of a plant and a fungus.

Myrmecophily. The association between certain plants and ants, possibly of advantage to both organisms.

Nectar, of flowering plants. A sugary fluid produced by many species, often in FLOWERS, and collected by insects.

Nectary, of flowering plants. A gland which secretes nectar.

Nerve, of mosses. A bundle of elongated cells normally at the centre of the leaf.

Node. The position on a stem at which leaves and branches are attached.

Nucellus, of seed plants. The tissue in which the MEGASPOROCYTE arises.

Nucleolus. A body lying within the NUCLEUS, rich in ribonucleic acid.

Nucleus. A conspicuous body in the protoplasm containing most of the deoxyribonucleic acid of the cell, and the site of the Mendelian genes.

Organelle. A part of a cell with certain definite functions and characteristic morphological features.

Organism. An individual animal or plant.

Ovary, of flowering plants. The female region of the FLOWER.

Ovule, of seed plants. The NUCELLUS (containing the EMBRYO SAC) and INTEGUMENTS yielding the SEED after FERTILIZATION.

Palisade tissue. The portion of the leaf MESOPHYLL (usually the uppermost) which consists of regular, columnar cells, containing most of the CHLOROPLASTS.

Palmate. Shaped like a hand with finger-like lobes.

Palmelloid state, of algae. The aggregation of normally separate individuals into a gelatinous, more or less PALMATE COLONY.

Paraphyses, of algae and bryophytes. Sterile hairs found among the reproductive organs in certain genera.

Parasite. An organism which lives at the expense of another, the HOST, upon which it confers no benefits.

Parenchyma. A tissue composed of VACUOLATED, thin-walled, more or less ISODIAMETRIC cells, functioning chiefly as a ground and storage tissue, the cells often retaining the ability to divide.

Parthenogenesis. The development of a female GAMETE into a new individual without

FERTILIZATION.

Parthenospore, of algae. A female GAMETE which develops into a resting SPORE without FERTILIZATION.

Pedicel, of flowering plants. The stalk of a single FLOWER in an INFLORESCENCE.

Peduncle, of flowering plants. The main axis of the INFLORESCENCE, or the stalk of a single FLOWER if it is solitary.

Pellicle, of algae. An elastic membrane surrounding a unicellular organism, found chiefly in the Euglenophyta.

Peltate. Umbrella-shaped.

Perennating organ. A specialized part of a plant ensuring its survival from one season to the next.

Perennial. A plant with an indefinite life span.

Perianth, of flowering plants. Collective term for the PETALS and SEPALS of a FLOWER.

Pericycle. The tissue lying between the vascular tissue and the ENDODERMIS.

Periderm. The collective term for the PHELLEM, PHELLOGEN and PHELLODERM.

Perigynous, of flowering plants. A FLOWER in which the PERIANTH and androecium are borne on the rim of a concave receptacle, separate from the central gynaecium.

Perispore, of ferns. An irregular accretion around the spore, derived from the TAPETUM.

Petals, of flowering plants. The upper PERIANTH segments, especially when these are distinct by form or pigmentation from the lower.

Petiole. A leaf stalk.

Phellem. See CORK.

Phelloderm. The layer of cells produced towards the inside by the PHELLOGEN (secondary CORTEX).

Phellogen. The CAMBIUM producing the CORK and PHELLODERM.

Phenotype. The visible, or chemically or biologically detectable, manifestation of the GENOTYPE produced as a consequence of growth and development.

Phloem. VASCULAR tissue, composed of SIEVE CELLS or SIEVE TUBES, and associated PARENCHYMATOUS and FIBROUS cells.

Phototaxis. The movement of a whole organism in response to light.

Phototropism. The change in the direction of a plant's growth in response to unequal illumination.

Phyllode. A PETIOLE which is broad and leaf-like, the LAMINA often being little developed.

Phyllotaxis. The arrangement of the leaves on a stem.

Phylogeny. The evolutionary history of an organism, or group of organisms.

Physiology. The study of chemical and physical processes which occur in living organisms.

Pinna. The primary division of a PINNATE leaf.

Pinnate. Of leaves which have a series of leaflets on each side of a common RACHIS.

Pinnule. The ultimate division of a PINNATE leaf.

Pistil, of flowering plants. The ovary, STYLE and stigma.

Pit. A thin area of cell wall traversed by PLASMODESMATA through which substances may pass from cell to cell.

Pith. The core of PARENCHYMA which may occur at the centre of a STELE.

Placenta, of flowering plants. The region of attachment of the OVULE to the OVARY wall.

Plagiogeotropism. The orientation of growth at a definite angle to the gravitational field.

Plagiotropic. Tending to take up a position at a definite angle to an orientating influence (e.g. PLAGIOGEOTROPISM).

Plankton. Collective term for the microscopic organisms, both plant (phytoplankton) and animal (zooplankton), which occur at the surface of fresh and salt water.

Plasma membrane. Same as PLASMALEMMA.

Plasmalemma. The membrane which bounds the PROTOPLASM of a cell and lies next to the cell wall.

Plasmodesma. A tubular connection between PROTOPLASTS traversing the cell wall.

Plastid. Collective term for chloroplasts, LEUCOPLASTS, AMYLOPLASTS and CHROMOPLASTS.

Platyspermic, of seed plants. Having flattened, BILATERALLY SYMMETRICAL SEEDS.

Plumule. The embryonic shoot.

Polarity. DIFFERENTIATION in structure or function between two ends of an axis.

Pollen, of seed plants. The MICROSPORES, often showing ENDOSPORIC germination.

Pollination, of seed plants. The process by which POLLEN is transferred from the ANTHER to the CARPELS or OVARY.

Pollinium, of certain genera of flowering plants. A mass of POLLEN grains held together by a sticky secretion.

Polycyclic. Arranged in concentric circles.

Polyembryony, of seed plants. The development of more than one EMBRYO in an OVULE.

Polyploidy. The possession of three or more sets of CHROMOSOMES per NUCLEUS.

Polystelic. Having several STELES.

Primary tissues. Tissues produced as a direct consequence of the activity of apical MERISTEMS.

Procambium. A strand of relatively UNDIFFERENTIATED cells which later gives rise to the PRIMARY XYLEM and PRIMARY PHLOEM tissues.

Proembryo, of gymnosperms. The structure developing immediately from the ZYGOTE.

Prothallus. A little DIFFERENTIATED, free-living, GAMETOPHYTE.

Protocorm, chiefly of lycopods and orchids. A tuberous structure developing from a portion of the young EMBRYO.

Protonema, chiefly of mosses. The FILAMENT or small plate of cells produced by the germinating SPORE.

Protoplasm. The contents of a living cell.

Protoplast. That part of a living cell remaining after removal of the wall.

Protoxylem. The portion of the PRIMARY XYLEM which DIFFERENTIATES first.

Proximal. Adjacent to a point of reference or symmetry (ant. DISTAL).

Pseudo-endosperm, of gymnosperms. The nutritive tissue in SEEDS derived from the female GAMETOPHYTE (and hence HAPLOID).

Pseudogamy. The development of a female GAMETE into a new individual after stimulation by a male GAMETE, but without actual FERTILIZATION.

Pseudoparenchyma, in algae and fungi. A tissue consisting of closely adpressed, often inter-woven FILAMENTS, giving the impression of PARENCHYMA.

Raceme, of flowering plants. An INFLORESCENCE in which FLOWERS are borne on branches of the main axis, those at the base maturing first.

Rachis. The main axis of a pinnate leaf.

Radial symmetry. Symmetry about a central axis; when divided longitudinally along any diameter, the two halves are mirror-images.

Radicle. The embryonic root.

Radiospermic, of seed plants. Bearing RADIALLY SYMMETRICAL SEEDS.

Raphe, of diatoms. The longitudinal fissure in the wall of pennate diatoms.

_____, of flowering plants. A line marking the position of the FUNICLE in a seed developed from an anatropous OVULE.

Ray. A vertical sheet of PARENCHYMATOUS cells traversing the STELE radially.

Reniform. Kidney-shaped.

Reticulate. In the form of a network.

Rhizoids. Thread-like anchoring and absorbing organs produced by plants lacking ROOTS.

Rhizome. A PLAGIOTROPIC underground stem.

Root. That part of a VASCULAR plant which in branching produces only simple axes like itself, usually subterranean.

Root cap. A cap of tissue over the ROOT apex.

Root hair. Hair-like outgrowths of the EPIDERMAL cells of the young ROOT which absorb water and minerals.

Rootstock. A short, vertical, underground stem, bearing ROOTS; in horticulture the plant providing the root system on the stem of which another kind of plant is grafted.

Saprophyte. An organism obtaining its food in the form of complex molecules from dead organic matter.

Scalariform. In the form of a ladder.

Scandent. Climbing.

Scarious. Thin, dry and membranous.

Schizogenous (of spaces). Formed by the separation of cells (cf. LYSIGENOUS).

Sclerenchyma. A strengthening tissue composed of FIBRES or STONE CELLS.

Secondary tissues. Produced from MERISTEMS which arise after the DIFFERENTIATION of the PRIMARY TISSUES.

Seed, of seed plants. The product of the OVULE after fertilization, comprising the EMBRYO, with its surrounding food reserves and protective coverings.

Sepals, of flowering plants. The lowermost PERIANTH segments, especially when green, and thus distinguished from PETALS.

Septum. A partition.

Sessile. Without a stalk.

Shrub. A woody plant with no distinct trunk, the main branches arising near ground level.

Sieve cell. A cell with a PROTOPLAST but lacking a distinct NUCLEUS, concerned with the transport of organic materials. A principal component of PHLOEM.

Sieve plate. The perforate area of the cell wall between two SIEVE CELLS or SIEVE TUBE elements.

Sieve tube, of flowering plants. A column of SIEVE CELLS (here usually called SIEVE TUBE elements) with more or less transverse end walls bearing SIEVE PLATES.

Sinuose. Undulating.

Siphon(ac)eous. Tubular.

Spike, chiefly of flowering plants. A RACEMOSE INFLORESCENCE in which the FLOWERS are SESSILE.

Spongy tissue. The lower portion of the MESOPHYLL of a leaf, consisting of irregular cells with large air spaces.

Sporangiophore. A structure bearing one or more SPORANGIA.

Spore. A unicellular, ASEXUAL reproductive cell, usually uninucleate.

Sporophyll. A structure bearing SPORANGIA, and believed to be equivalent to a leaf.

Sporophyte. The (normally) DIPLOID generation, producing HAPLOID SPORES which germinate to give the GAMETOPHYTE generation.

Stamen, of flowering plants. The MICROSPORANGIA (anther) together with their common stalk (filament), usually regarded as a MICROSPOROPHYLL.

Staminode, of flowering plants. An infertile STAMEN often highly modified or reduced.

Stele. The VASCULAR tissue of a root or stem, extending to the ENDODERMIS if present.

Stipule, of flowering plants. Outgrowths (usually two) at the base of the PETIOLE.

Stolon. A short, PLAGIOTROPIC shoot which develops a new plant at the tip, eventually severing connection with the parent.

Stoma. A pore in the EPIDERMIS bounded by a pair of GUARD CELLS.

Stone cells. More or less isodiametric, heavily LIGNIFIED cells.

Strain. A mating group within a species; or a variety within a species with distinctive PHYSIOLOGICAL or MORPHOLOGICAL features.

Strobilus. See CONE.

Style, of flowering plants. An elongation of the distal end of the CARPEL, bearing the stigma.

Suberin. An impervious waxy substance with which CORK and other cells are impregnated.

Subsidiary cells. Cells occurring adjacent to the GUARD CELLS of STOMATA, which are usually of constant shape and position, and distinguishable from other EPIDERMAL cells.

Suspensor, chiefly of seed plants. The cell or group of cells arising with the EMBRYO proper from the ZYGOTE, and pushing the young EMBRYO into the nutritive material of the SEED.

Swarmer. See ZOOSPORE.

Symbiosis. An association between two organisms in which there is usually mutual benefit.

Symplast. The continuum formed by PROTOPLASTS linked together by the PLASMODESMATA.

Sympodial branching. The type of branching in which the terminal bud ceases to grow, growth being continued in each instance by the uppermost lateral branch. A form of CYMOSE BRANCHING.

Synangium. A composite structure formed by the CONCRESCENCE of SPORANGIA.

Syngamy. The fusion of GAMETES.

Tangential. Perpendicular to a radius. Used principally to describe the orientation of longitudinal sections of axes, or of cell division.

Tapetum. A layer of nutritive cells surrounding the SPORE mother cells in the SPORANGIUM. Conspicuous only in plants with VASCULAR TISSUE.

Taproot. A persistent primary root, often swollen with food reserves.

Tendril, of flowering plants. A modified leaf, leaflet or stem, often sensitive to contact and able to coil round objects it touches, thus supporting and anchoring the plant.

Testa, of flowering plants. The covering of the SEED, derived from the INTEGUMENTS.

Tetrad. The group of four cells produced by a MEIOTIC division, frequently genetically dissimilar.

Thallus. A relatively undifferentiated plant body.

Tonoplast. The membrane surrounding the VACUOLE, often also called the vacuolar membrane.

Trabeculum. A bar, or elongated cell, traversing a cavity.

Tracheid. An element of the XYLEM tissue consisting of an elongated, elaborately pitted, lignified cell with oblique end walls.

Tree. A woody plant with a definite trunk; the main branching well above ground level.

Trichothallic growth, of algae. The type of growth in which cell divisions are confined to a region at the base of the FILAMENT.

Trilete spore. The kind of SPORE which, having been formed in a tetrahedral TETRAD, bears a tri-radiate scar on its PROXIMAL face.

Triploid. Having three times the basic (HAPLOID) number of chromosomes.

Trisomy. The presence of one chromosome in triplicate in the diploid nucleus. A form of ANEUPLOIDY.

Tuber. A rounded storage organ formed from a ROOT (root tuber) or a stem (stem tuber).

Tunica. The outer layer or layers of cells at the apex of a stem in which the divisions are principally ANTICLINAL.

Vacuole. The portion of the cell which contains the CELL SAP, bounded by the TONOPLAST.

Vascular tissue. A collective term for the XYLEM and PHLOEM.

Veins. The VASCULAR strands of leaves.

Venation. The pattern of VEINS in a leaf.

Ventral, of lateral organs. The same as ADAXIAL.

_____, of prostrate plants. The side towards the substratum.

Vessels, of flowering plants. Tubes occurring in the XYLEM and concerned with the conduction of water. Formed from columns of broad, short cells (vessel segments), the end walls of which break down during DIFFERENTIATION.

Xanthophyll. A class of yellow, CAROTENOID pigments associated with CHLOROPHYLL in the CHLOROPLAST.

Xeromorph. A plant possessing the features often found in XEROPHYTES, but not necessarily confined to dry places.

Xerophyte. A plant growing characteristically in dry situations, and often possessing anatomical and physiological features enabling it to withstand prolonged drought.

Xylem. VASCULAR tissue which may comprise VESSELS, TRACHEIDS, FIBRES, together with some PARENCHYMA.

Zoosporangium. An organ producing ZOOSPORES.

Zoospore. A motile SPORE.

Zygomorphic. BILATERALLY SYMMETRICAL.

Zygote. The cell formed by the fusion of two GAMETES.

References

1. Ashton, N.W., Grimsley, N.H. and Cove, D.J. (1979). Analysis of gametophytic development in the moss, *Physcomitrella patens*, using auxin and cytokinin resistant mutants. *Planta*, **144**, 427–35.
2. Bews, J.W. (1925). *Plant Forms*. Longmans, Green, London.
3. Bower, F.O. (1923–28). *The Ferns*. 3 vols. Cambridge University Press.
4. Brook, A.J. (1981). *The Biology of the Desmids*. Blackwell Scientific Publications, Oxford.
5. Buetow, D.E. (Ed.) (1968, 1982). *The Biology of Euglena*, 1, 2 and 3. Academic Press, New York.
6. Carr, N.G. and Whitton, B.A. (Eds) (1982). *The Biology of Cyanobacteria*. Blackwell Scientific Publications, Oxford.
7. Chaloner, W.G., Hill, A.J. and Lacey, W.S. (1979). First Devonian platyspermic seed and its implications in gymnosperm evolution. *Nature, London*, **265**, 233–5.
8. Church, A.H. (1981). *Revolutionary Botany, 'Thalassiophyta' and Other Essays*. Oxford University Press, London.
9. Clowes, F.A.L. (1961). *Apical Meristems*. Oxford University Press, London.
10. Corbet, S.A., Beament, J. and Eisikowitch, M. (1982). Are electrostatic forces involved in pollen transfer? *Plant Cell and Environment*, **5**, 125–9.
11. Dickinson, H.G. and Lewis, D. (1973). Cytochemical and ultrastructural differences between intraspecific compatible and incompatible pollinations in *Raphanus*. *Proceedings of the Royal Society, London*, **183B**, 21–38.
12. Dixon, P.S. (1972). *Biology of the Rhodophyta*. Oliver and Boyd, Edinburgh.
13. Dougherty, E.C. (1957). Neologisms needed for structures of primitive organisms. 1. Types of nuclei. *Journal of Protozoology*, **4** (Suppl.), 14.
14. Duckett, J.G. and Bell, P.R. (1977). An ultrastructural study of the mature spermatozoid of *Equisetum*. *Philosophical Transactions of the Royal Society, London*, **277**, 131–58.
15. Dyer, A.F. (Ed.) (1979). *The Experimental Biology of Ferns*. Academic Press, London.
16. Eichler, A.W. (1954). *Blüthendiagramme*, 2nd edition. Koeltz, Eppenheim.
17. Fritsch, F.E. (1935, 1945). *Structure and Reproduction of the Algae*, 1 and 2. Cambridge University Press.
18. Gillespie, W.H., Rothwell, G.W. and Scheckler, S.E. (1981). The earliest seeds. *Nature, London*, **293**, 462–4.
19. Hughes, N. (1976). *Palaeobiology of Angiosperm Origins*. Cambridge University Press.
20. Knuth, P. (1898–1905). *Handbuch der Blütenbiologie*. Engelmann, Leipzig.
21. Konar, R.N. and Moitra, A. (1980). Ultrastructure, cyto- and histochemistry of

female gametophyte of gymnosperms. *Gamete Research,* **3**, 67–97.
22. Lewin, R.A. (Ed.) (1976). *The Genetics of Algae.* Blackwell Scientific Publications, Oxford.
23. Lewin, R.A. (1977). *Prochloron,* type genus of the Prochlorophyta. *Phycologia,* **16**, 217.
24. Lewis, D. (1979). *Sexual Incompatibility in Plants.* Studies in Biology, no. 110. Edward Arnold, London.
25. Lobban, C.S. and Wynne, M.J. (Eds) (1981). *The Biology of Seaweeds.* Blackwell Scientific Publications, Oxford.
26. Maheshwari, P. (1950). *An Introduction to the Embryology of Angiosperms.* McGraw-Hill, New York.
27. Menon, M.K.C. and Lal, M. (1977). Regulation of a sub-sexual life cycle in a moss: evidence for the occurrence of a factor for apogamy in *Physcomitrium. Annals of Botany,* **41**, 1179–89.
28. Mereschkowsky, C. (1905). Über Natur und Ursprung der Chromatophoren im Pflanzenreiche. *Biologisches Zentralblatt,* **25**, 593–604.
29. Moitra, A. and Bhatnagar (1982). Ultrastructure, cytochemical and histochemical studies on pollen and male gamete development in gymnosperms. *Gamete Research,* **5**, 71–112.
30. Morris, I. (1980). *Physiological Ecology of Phytoplankton.* Blackwell Scientific Publications, Oxford.
31. Proctor, M. and Yeo, P. (1973). *The Pollination of Flowers.* Collins, London.
32. Raunkiaer, C. (1934). *The Life Forms of Plants.* Oxford University Press, London.
33. Reid, G.K. (1961). *Ecology of Inland Waters and Estuaries.* Reinhold, New York.
34. Remy, W. (1982). Lower Devonian gametophytes: relation to the phylogeny of land plants. *Science,* **215**, 1625–7.
35. Richards, P.W. (1979). *The Tropical Rain Forest,* new edition. Cambridge University Press.
36. Ridley, H.N. (1930). *The Dispersal of Plants.* Reeve, Ashford.
37. Round, F.E. (1973). *The Biology of Algae,* 2nd edition. Edward Arnold, London.
38. Round, F.E. (1981). *The Ecology of Algae.* Cambridge University Press.
39. Schultes, K.E. and Hoffman, A. (1980). *The Botany and Chemistry of Hallucinogens,* 2nd edition. Thomas, Springfield.
40. Singh, H. (1978). *Embryology of Gymnosperms.* Borntraeger, Berlin.
41. Sporne, K.R. (1976). *The Morphology of Pteridophytes.* Hutchinson, London.
42. Sporne, K.R. (1974). *The Morphology of Gymnosperms.* Hutchinson, London.
43. Sporne, K.R. (1975). *The Morphology of Angiosperms.* Hutchinson, London.
44. Stainier, R.Y. and van Niel, C.B. (1962). The concept of a bacterium. *Archiv für Mikrobiologie,* **42**, 17–75.
45. Stewart, W.D.P. (1974). *Algal Physiology and Biochemistry.* Blackwell Scientific Publications, Oxford.
46. Takhtajan, A. (1969). *Flowering Plants.* Oliver and Boyd, Edinburgh.
47. Taylor, F.J.R. (1983). *The Biology of the Dinoflagellates.* Blackwell Scientific Publications, Oxford.
48. Troll, W. (1964). *Die Infloreszensen.* Fischer, Stuttgart.
49. Watson, E.V. (1978). *The Structure and Life of Bryophytes.* Hutchinson, London.
50. Werner, D. (1977). *The Biology of Diatoms.* Blackwell Scientific Publications, Oxford.

Index

Bold page numbers indicate definitions.